Handbook of NON-INVASIVE DRUG DELIVERY SYSTEMS

Non-Invasive and Minimally-Invasive Drug Delivery Systems for Pharmaceutical and Personal Care Products

Edited by

DR VITTHAL S. KULKARNI

ELSEVIER

AMSTERDAM • BOSTON • HEIDELBERG • LONDON NEW YORK • OXFORD • PARIS
SAN DIEGO • SAN FRANCISCO • SINGAPORE • SYDNEY • TOKYO
William Andrew is an imprint of Elsevier

William Andrew is an imprint of Elsevier
The Boulevard, Langford Lane, Kidlington, Oxford, OX5 1GB, UK
30 Corporate Drive, Suite 400, Burlington, MA 01803, USA

First edition 2010

Notices
Knowledge and best practice in this field are constantly changing. As new research and experience broaden our understanding, changes in research methods, professional practices, or medical treatment may become necessary.

Practitioners and researchers must always rely on their own experience and knowledge in evaluating and using any information, methods, compounds, or experiments described herein. In using such information or methods they should be mindful of their own safety and the safety of others, including parties for whom they have a professional responsibility.

To the fullest extent of the law, neither the Publisher nor the authors, contributors, or editors, assume any liability for any injury and/or damage to persons or property as a matter of products liability, negligence or otherwise, or from any use or operation of any methods, products, instructions, or ideas contained in the material herein.

British Library Cataloguing in Publication Data
Kulkarni, Vitthal S.
Handbook of non-invasive drug delivery systems.
1. Drug delivery systems. 2. Transdermal medication.
3. Oral medication.
I. Title
615.6-dc22

Library of Congress Control Number: 2009935943

ISBN: 978-0-8155-2025-2

For information on all William Andrew publications visit our website at elsevierdirect.com

Printed and bound in the United States of America

09 10 11 12 11 10 9 8 7 6 5 4 3 2 1

CONTENTS

Preface	*xi*
Editor Biography	*xiii*
Contributors	*xv*
Acknowledgement	*xxiii*

Chapter 1 Topical and Transdermal Drug Delivery **1**

H.N. Shivakumar and S. Narasimha Murthy

1.1 Structure of the skin	1
1.1.1 Dermis and hypodermis	3
1.2 Topical vs. transdermal drug delivery	3
1.2.1 Topical drug delivery	3
1.2.2 Transdermal patch systems	4
1.3 Percutaneous absorption pathways	5
1.4 Approaches to enhance passive cutaneous drug absorption	6
1.4.1 Supersaturation of drug	6
1.4.2 Eutectic systems	7
1.4.3 Prodrug approach	7
1.4.4 Ion pair formation	8
1.4.5 Complexation	8
1.4.6 Liposomes	9
1.4.7 Microemulsions	13
1.4.8 Organogels	14
1.4.9 Solid lipid nanoparticles	15
1.4.10 Liquid crystalline systems	16
1.4.11 Chemical permeation enhancers	16
1.5 Conclusions	26
References	27

Chapter 2 Micellar Nanoparticles: Applications for Topical and Passive Transdermal Drug Delivery **37**

Robert W. Lee, Dinesh B. Shenoy and Rajiv Sheel

2.1 Introduction	37
2.1.1 MNP composition and structures	39
2.1.2 Physicochemical characterization	42
2.1.3 Antimicrobial properties	44

2.2 Transdermal drug delivery applications of MNP technology 45
2.2.1 Estrasorb™ – commercial validation of MNP technology 45
2.2.2 Raloxifene MNP product 47
2.2.3 Nicotine MNP product 51
2.3 Topical drug delivery applications of MNP technology 53
2.4 Conclusion 54
References 56
Further reading 58

Chapter 3 Emulsions and Microemulsions for Topical and Transdermal Drug Delivery 59

Guang Wei Lu and Ping Gao

3.1 Overview of basic aspects of emulsions and microemulsions 59
3.1.1 Introduction 59
3.1.2 Methods of characterization 63
3.1.3 Stability evaluation 69
3.2 Application of emulsions for topical and transdermal drug delivery 71
3.2.1 Drug transport from an emulsion to the skin 71
3.2.2 Investigation of topical and transdermal delivery in human subjects 78
3.2.3 Formulation strategies for emulsion-based products 80
3.3 Advances in using microemulsions as topical and transdermal drug delivery systems 86
References 91

Chapter 4 Iontophoretic Transdermal Drug Delivery 95

Longsheng Hu, Priya Batheja, Victor Meidan and Bozena B. Michniak-Kohn

4.1 Introduction 95
4.2 History and development of iontophoresis 97
4.3 Advantages and limitations of iontophoresis 97
4.4 Iontophoretic delivery principles and system components 98
4.5 Drug transport in iontophoresis 100
4.6 Pathways for iontophoretic transport and factors affecting the same 102
4.7 Properties of the drug/ion: concentration, competing ions, and molecular size 103
4.8 Properties of the iontophoretic system: pH, current 103
4.9 Combination strategies to improve transdermal iontophoretic drug delivery 105
4.9.1 Iontophoresis with electroporation 106
4.9.2 Iontophoresis with chemical enhancers 106
4.9.3 Iontophoresis with microneedles 107
4.9.4 Iontophoresis with sonophoresis 108
4.9.5 Ion-exchange materials 109
4.9.6 Other combinations 110

4.10 Reverse iontophoresis 110
4.11 Applications of iontophoresis 111
4.12 Current and future development 113
References 114

Chapter 5 Ultrasound-based Technology for Skin Barrier Permeabilization 119

Victor Meidan and Bozena B. Michniak-Kohn

5.1 Introduction 119
5.2 What is ultrasound? 120
5.3 Ultrasound bioeffects 121
5.3.1 Cavitation 121
5.3.2 Heating 122
5.3.3 Microstreaming 122
5.4 Historical background 122
5.5 Sonophoresis for enhanced drug delivery 123
5.5.1 Mechanisms 125
5.6 Ongoing developments 127
5.6.1 Effective delivery of macromolecules 127
5.6.2 "Reverse sonophoresis" for transdermal diagnostics 128
5.6.3 Advances in transducer fabrication 129
5.7 Current challenges and trends 130
References 131

Chapter 6 Microneedles – Minimally Invasive Transdermal Delivery Technology 135

Diksha Kaushik, Brian Kilfoyle, Rashmi Thakur and Bozena B. Michniak-Kohn

6.1 Introduction 135
6.2 Microneedle geometry 136
6.2.1 Solid microneedles 137
6.2.2 Hollow microneedles 139
6.2.3 Biodegradable microneedles 140
6.3 Types of approach 140
6.3.1 Poke with patch technique 142
6.3.2 Coat and poke technique 142
6.3.3 Dip and scrape technique 142
6.4 Factors influencing transdermal permeation 143
6.4.1 Depth of permeation 143
6.4.2 Coating strategies 143
6.4.3 Microneedle geometry and force of penetration 148
6.5 Microneedle technology in drug delivery 151
6.6 Commercialization of microneedle technology 157
6.7 Conclusion 160
References 161

Chapter 7 Recent Advances in Ophthalmic Drug Delivery **165**

James N. Chang

7.1 Introduction 165
7.2 The anatomy of the eye 167
7.3 Delivery systems for the anterior segment 169
7.3.1 Viscous solutions 170
7.3.2 Gels 171
7.3.3 Bioadhesive polymers 173
7.3.4 Nanosuspensions 175
7.3.5 Inserts 176
7.3.6 Liposomes 177
7.3.7 Others 178
7.4 Retinal drug delivery systems 179
7.4.1 Non-erodible implants 180
7.4.2 Erodible implants 181
7.4.3 Bolus injections 182
7.4.4 Periocular drug delivery 182
7.4.5 Others 186
7.5 Conclusions 186
References 186

Chapter 8 Nasal Delivery of Micro- and Nano-encapsulated Drugs **193**

Joe McDonough, Hong Dixon and Michael Ladika

8.1 Introduction 193
8.2 Physiology of the nose 198
8.3 Drug delivery formulation 199
8.4 Micro- and nano-encapsulation 201
8.4.1 PLGA 201
8.4.2 Chitosan 203
8.4.3 Other polymers 205
8.5 Summary 206
References 206

Chapter 9 Pulmonary Drug Delivery **209**

Mohammed Ali

9.1 General introduction 209
9.1.1 Non-invasive drug delivery 209
9.1.2 The human respiratory system 211
9.1.3 Physical properties of drug aerosol particles 212
9.2 Pulmonary drug deposition mechanisms 214
9.3 Pulmonary drug delivery methods 218
9.3.1 Metered-dose inhalers 219
9.3.2 Dry powder inhalers 225
9.3.3 Nebulizers 227

9.3.4 Comparable and contrasting features of MDIs, DPIs, and nebulizers 229
9.4 Analytical methods 231
9.4.1 Non-invasive scintigraphic techniques 232
9.4.2 Inertial impaction 233
9.4.3 Optical microscopy 237
9.4.4 Laser-induced tools 237
9.5 Challenges and advantages 239
9.6 Conclusions 241
References 242

Chapter 10 Regulatory Aspects of Nasal and Pulmonary Spray Drug Products 247

Dino J. Farina

10.1 Introduction 247
10.2 Setting specifications and performing product characterization studies 253
10.2.1 Tests related to product labeling 253
10.2.2 Tests related to patient use/misuse 259
10.2.3 Tests related to general properties 267
10.3 Conclusion 287
References 287

Index 291

PREFACE

Topical application of pharmaceuticals has a long tradition in drug delivery dosage forms; however, it is typically a passive means of skin penetration and not efficient for systemic drug delivery. Therefore the efficacy of topical drug delivery is not comparable to that of oral or injectable routes of drug administration. The unique advantages of topical drug delivery include local application, ease of use, and self-administration. Conversely, concerns for oral dosage forms include nausea and, gastrointestinal upset/stomach irritation, and repeated doses are often needed whereas concerns for injectable (invasive) dosage forms include pain and fear of injection, and requiring professional assistance for drug administration. These concerns exert limitations on patient compliance for treatment by oral or invasive medication forms.

To reap the potential of topical drug delivery for systemic circulation, it is necessary to improve the skin penetration of topically applied medicines. This can be achieved to a limited degree by improving the formulation and to a greater degree by applying external electromechanical energy for enhancing skin penetration. Alternatively to the topical/transdermal route of drug administration, other routes including nasal, pulmonary, and ophthalmic are quite effective for entering the systemic circulation without the pain and concerns of invasive or oral drug delivery routes. Therefore, a medication that is delivered systemically by routes other than the oral or injectable route are considered non-invasive or minimally invasive modes of drug administration. For successful systemic delivery of drugs, a non-invasive route, the requirements are efficient electromechanical devices that are safe, easy to use, and compatible with drug formulation, and affordability for regular use by anyone in need of treatment. In the past 10 years, significant improvements in drug delivery device technologies have taken place, making them safe, compatible, and affordable; consequently, the systemic delivery of drugs by a minimally invasive or non-invasive route has become a reality. More and more medications are formulated or reformulated for systemic delivery via topical/transdermal, nasal, pulmonary, and ophthalmic routes. This book aims to review the current status of a variety of minimally invasive drug delivery systems.

Transdermal drug delivery is a non-invasive system; however, more efficient drug delivery systems that enhance skin penetration by employing external devices have shown strong commercial potential. Drug delivery systems that employ external devices to enhance skin penetration include ultrasonication, iontophoresis and microneedles, and these devices are expected to be in common use in the near future. Other systems included in this book are ophthalmic, nasal, and pulmonary drug delivery systems. The nasal route has shown potential for bypassing the blood–brain barrier to deliver drugs to the brain and spinal cord. In addition to the pharmacological advantages of nasal or pulmonary routes, self-administration, ease of administration, convenience, and accessibility make these minimally invasive drug delivery systems effective and patients readily comply with the treatment. Consequently, this route of drug administration has become popular and is expected to grow its market share in the future.

This book is comprised of 10 chapters that provide readers with thorough and focused reviews on topical and transdermal emulsions, microemulsions and nanoparticles, device-aided enhanced skin penetration, and nasal, pulmonary, and ophthalmic drug delivery systems. Regulatory requirements for nasal and pulmonary drug delivery are complex, and include investigating the compatibility of the formulation with a container-closure system (CCS) and establishing the stability of the formulation in a complex CCS, such as pressurized metered-dose inhalers or dry powder inhaler systems. The last chapter of this book is exclusively focused on the regulatory aspects of nasal and inhaled drug products.

The contributors to this book come from both academia and industry, and provide a wealth of current and thorough information on the topic of minimally invasive drug delivery systems. I sincerely believe that minimally invasive drug delivery systems will be in great demand in the near future; therefore, this topic warrants a thorough review and realization of the potential for this new form of medication. For the first time, an attempt has been made to review the various minimally invasive drug delivery routes in this book, which is intended to be read by industrial and academic professionals in the area of novel drug delivery. I hope that readers will be enriched with knowledge of the current and emerging trends of minimally invasive drug delivery systems and regulatory aspects.

Dr Vitthal S. Kulkarni

EDITOR BIOGRAPHY

Dr Vitthal S. Kulkarni
Scientific Advisor,
DPT Laboratories, San Antonio, TX.

Dr Vitthal S. Kulkarni has been based at DPT Laboratories since 2003 in the Research and Development Department and leads the preformulation/research group.

Dr Kulkarni is highly experienced in developing and characterizing various non-invasive drug delivery systems including topical drug delivery systems comprised of liposomes, and nano-particles, emulsions and nasal spray products. After earning a PhD in Chemistry at National Chemical Laboratory, Pune, India in the area of surface/colloid chemistry, Dr Kulkarni studied lipid monolayers, bilayers (liposomes), lipid nano-structures, and protein electron crystallography at various institutions including Université de Provence, Marseille, France, JEOL Ltd., Tokyo, Japan, Vanderbilt University, Nashville, TN, University of Minnesota, Austin, MN and Yale University, New Haven, CT before moving into industry.

Dr Kulkarni has been an invited speaker at Drug Delivery Conferences and he has published several research papers in peer-reviewed journals, industrial periodicals, and book chapters in the area of lipid bilayers and monolayers, molecular self-assemblies, liposome technology and topical drug delivery systems. Dr Kulkarni is also an adjunct faculty member at San Antonio College and he is an active member of the American Association of Pharmaceutical Scientists and the American Chemical Society.

CONTRIBUTORS

Dr Mohammed Ali

Mohammed Ali is currently an Assistant Professor in the Department of Technology at Jackson State University, Jackson, Mississippi, where he is the Director of the Laboratory for Aerosol and Respiratory Drug Delivery. His technical articles have been published in 10 peer reviewed journals and 18 national and international conference proceedings in the areas of aerosol medicine, pulmonary drug delivery and pollution in the workplace environment. Dr Ali has received several awards, such as: Research Achievement Award (2008) from the American Association of Bangladeshi Pharmaceutical Scientists; NIEHS Intramural Research Training Award (2007); Best Student Research Award (2007) from the Arkansas Society for Public Health Education; and Alpha Epsilon Lambda (2006). He is the editorial board member of the AAPS PharmSciTech Journal, Journal of Recent Patents on Inflammation and Allergy Drug Discovery, and the Journal of Management and Engineering Integration. Dr Ali is an active member of the American Association of Aerosol Research (AAAR), American Association of Pharmaceutical Scientists (AAPS), and the Association of Technology, Management and Applied Engineering (ATMAE). Other recognitions and awards bestowed on Dr Ali include Fullbright Scholarship from the Government of France for Graduate Study at the Asian Institute of Technology, Bangkok, Thailand (1993); United States National Collegiate Award (2001) by the US Achievement Academy; Upsilon Pi Epsilon (2001); and Biltmore Who's Who in the World (2009).

Priya Batheja

Priya Batheja is a research assistant at the Drug Delivery Laboratory, NJ and is currently pursuing her PhD in pharmaceutical sciences at Rutgers University, NJ. Post her Bachelors degree in Pharmacy and Masters in Business Management, she has pursued a career in pharmaceutical sciences with a focus on drug delivery and formulation development. Priya is an author of eight publications and two book chapters and is also a co-inventor on a patent. She is a recipient of a number of awards including the "Innovative controlled delivery concept" award by the CRS NJ student chapter (2008),

"Outstanding Poster" at the GRASP (Graduate Research Association of Students in Pharmacy) 2007, best poster (third place) at the "New Approaches and Technologies in Oral Drug Delivery" conference organized by the New Jersey Chapter of CRS (2009), and the AstraZeneca Travelship award for the 2008 AAPS Annual Meeting.

Dr James N. Chang
James N. Chang is currently the Chief Executive Officer of TaiMed Biologics, a biotech company focused on the development of antiviral drugs. Dr Chang received his doctorate in Chemical Engineering from the University of Washington in 1986. Prior to his current appointment, he had a long career in the pharmaceutical/biotech industry working at Procter & Gamble and Allergan, where he was in charge of formulation development and drug delivery for a number of years. Dr Chang is an expert in mathematic modeling and polymeric drug delivery systems, and he has direct experience in taking drug delivery systems from conception, through development and regulatory approval, to the market place. Dr Chang is the author of over 30 patents and peer-reviewed papers in ophthalmic drug delivery. He has also been invited to speak at international conferences.

Dr Hong Dixon
Hong Dixon is the Senior Research Scientist in the Microencapsulation and Nanomaterials Department at Southwest Research Institute, San Antonio, TX. Dr Dixon is focusing on developing novel drug delivery technology, including targeted and triggered drug delivery, micro- and nano-encapsulation formulation, and novel polymer and nanocomposite material synthesis. She holds 12 US and foreign patents.

Mr Dino J. Farina
Dino J. Farina is the Founder and CEO of Proveris Scientific Corporation. Previously, Mr Farina developed quantitative imaging systems for temperature and fluid flow measurement as a NASA Graduate Student Fellow at Stanford University and the NASA–Ames Research Center. Mr Farina is a named inventor on 18 US and international patents related to pharmaceutical spray product testing and analysis. Mr Farina is considered an expert in the field of spray drug product characterization having authored and co-authored numerous international journal articles and given technical presentations to the FDA in this field. Farina holds an MS and Master of Engineering from Stanford University and a BS from Northeastern University in mechanical engineering.

Dr Ping Gao

Ping Gao currently holds a Director position, Global Pharmaceutical and Analytical Sciences, Abbott Laboratories. He has over 18 years' working experience in the pharmaceutical industry. He developed his scientific and technical expertise in areas including early discovery support and preformulation, material characterization, formulation sciences and drug delivery technologies, and commercial scale manufacturing. His long-standing interest and scientific investigation have focused on physicochemical/biopharmaceutical properties of drugs, design and development of controlled/extended release dosage forms, enabling drug delivery technologies for poorly soluble drugs, and *in vitro* dissolution methods and the *in vitro–in vivo* relationship. He has 47 publications including research articles in peer-reviewed journals and book chapters, and has made 48 podium presentations at national and international conferences, workshops and short courses, universities, and the US-FDA. He has 15 US patents and 17 patent applications pending. He is a member of both ACS and AAPS and was elected as AAPS Fellow in 2009.

Mr Longsheng Hu

Longsheng Hu is working as a product development scientist at the Global Topical HealthCare Research and Development Department of Johnson & Johnson, Morris Plains, NJ. Currently Mr Hu is also part time PhD student working on iontophoretic transdermal drug delivery systems under the guidance of Dr Bozena "Bo" B. Michniak-Kohn, at the College of Pharmacy, Rutgers University, Piscataway, NJ. He has Masters degrees in both Chemical Engineering and Pharmaceutical Science from the University of Cincinnati. He has one patent publication. He is a member of AAPS.

Diksha Kaushik

Diksha Kaushik is currently a PhD student in the Department of Pharmaceutics in the Ernest Mario School of Pharmacy at Rutgers – the State University of NJ. She holds a Master's degree in Pharmaceutics as well as a Bachelor's degree in Pharmacy from Hamdard College of Pharmacy, New Delhi, India. She has contributed to several review articles and book chapters. Her research interests are in the fields of transdermal/topical drug delivery and the pharmacokinetics of the actives.

Mr Brian Kilfoyle

Brian Kilfoyle is currently a graduate student working towards his PhD in the Department of Pharmaceutics at the Ernest Mario School of Pharmacy at Rutgers, the State University of New Jersey. His current research areas include the optimization of nanosphere-drug formulations as delivery vehicles into and across biological membranes, permeation analysis across biological and synthetic membranes, and the development of a synthetic membrane for transdermal testing. Prior to his graduate studies, Brian worked in the drug delivery field specializing in pulmonary delivery systems at MicroDose Technologies and Alkermes Inc. and transdermal delivery systems at MacroChem Corporation.

Mr Michael Ladika

Michael Ladika is the Senior Manager for Institute Business Development at Southwest Research Institute, San Antonio, TX and is also the Chairman of the Southwest Research Consortium. Mr Ladika has served on corporate innovation councils to address technology and business issues requiring multi-disciplinary, multi-organizations resources. He has led programs in the research, development, engineering, and production of chemical and mechanical systems. He holds two US Patents and has 22 classified invention disclosures.

Dr Robert W. Lee

Dr Lee is Vice President of Pharmaceutical Development at Particle Sciences Inc. He is responsible for product development at Particle Sciences as well as providing support to clinical manufacturing operations and business development. His responsibilities include oversight of formulation development, drug delivery, analytical sciences, quality control, and quality assurance.

Before joining Particle Sciences, Rob held senior management positions at Novavax, Inc., Lyotropic Therapeutics, Inc., and Imcor Pharmaceutical Co. He holds BSs in Biology and Chemistry from the University of Washington and a PhD in Physical Bioorganic Chemistry from the University of California, Santa Barbara.

Rob has published articles in numerous peer-reviewed journals and three book chapters plus holding 11 issued patents and 14 provisional or PCT patent applications.

He has more than 18 years of experience in pharmaceutical research and development of both therapeutic drugs and diagnostic imaging agents. Rob maintains strong academic ties, including an appointment as Adjunct Associate Professor of Pharmaceutical Chemistry at the University of Kansas in 1992, and serving as a reviewer for both the International Journal of Pharmaceutics and Journal of Pharmaceutical Sciences.

Dr Guang Wei Lu

Dr Guang Wei Lu is currently a Research Fellow in the Parenteral Center of Emphasis, Pfizer in Groton, CT. His recent interests focus on the development of liquid and semisolid formulations, and drug delivery through topical, transdermal, and ocular routes of administration. Some of his past positions include Associate Professor at the Shanghai University of Traditional Chinese Medicine (China); Visiting Scholar at the Medical School of Osaka City University, Department of Pharmacology (Japan); Senior/Staff Scientist in Johnson & Johnson Consumer products (USA); and principal research scientist and associate research fellow at Pharmacia (USA). Dr Lu has written and published more than 30 research articles, book chapters, and patent applications. He has also been invited to lecture at numerous national and international conferences as well as universities. In 2006, he received the Pfizer Global Research and Development Achievement Award.

Dr Joe McDonough

Joe McDonough is the director for the Microencapsulation and Nanomaterials Department at Southwest Research Institute, San Antonio, TX. Dr McDonough is currently developing novel drug delivery technology, including cellular delivery and nano-stabilization. He oversees programs in ocular, nasal, transdermal, oral and intramuscular delivery, primarily focusing on micron-and nanoparticle formulations of new and existing pharmaceuticals. He has 20 years of experience managing research efforts in pharmaceutical development, including bulk and dose-form manufacturing, regulatory strategy and compliance for filings and licensing of new chemical entities.

Dr Victor Meidan

Victor Meidan is a lecturer at the Strathclyde Institute of Pharmacy and BioMedical Sciences (SIPBS) located at Strathclyde University in Glasgow, Scotland. His main research interest is in the transfer of drugs across biological barriers, particularly skin and mucosal surfaces. His current publications include seven book chapters and 30 full-length papers in internationally recognized, peer-reviewed journals.

Dr Bozena B. Michniak-Kohn

Bozena B. Michniak-Kohn obtained her PhD in Pharmacology from Leicester Polytechnic, Leicester, UK now renamed the DeMontfort University. Dr Michniak-Kohn is a tenured Full Professor in Pharmaceutics at the Ernest Mario School of Pharmacy, Rutgers – the State University of New Jersey, Piscataway, NJ and is the Director of the Laboratory for Drug Delivery of the New Jersey Center for Biomaterials (NJCBM). Dr Michniak-Kohn has directed over 26 PhD students, six Masters students, and 200+ undergraduates and the work resulted in over 300 abstracts, three patents, two books, and more than 100 papers and book chapters. The work is funded from both federal (NIH/NSF/DoD/US Army Natick/DARPA) and industrial funds.

Dr Michniak-Kohn is also the US Editor of the Controlled Release Society Newsletter. She is a Fellow of AAPS and is a reviewer for over a dozen pharmaceutical and drug delivery journals.

Dr S. Narasimha Murthy

Dr S. Narasimha Murthy is currently an Assistant Professor of Pharmaceutics at the University of Mississippi, USA. Dr Murthy obtained his PhD from Bangalore University, India and completed his post doctoral research training at Roswell Park Cancer Institute, Buffalo, NY. "Murthy research group" focuses on developing noninvasive drug delivery techniques including transdermal drug delivery systems. Dr Murthy has about 50 publications in peer reviewed international journals and has presented several papers in scientific meetings. His research programs are funded by National Institutes of Health and Pharmaceutical industries.

Mr Rajiv Sheel

Mr Sheel is Director of US Corporate Development & Strategy for Orgenus Pharma (a wholly-owned subsidiary of Orchid Pharmaceuticals). He is responsible for growing Orchid's footprint in the US via mergers/acquisitions, joint ventures and product/technology licensing. His responsibilities include the scientific and commercial evaluation of new products and drug delivery technologies.

Before joining Orgenus Pharma, Rajiv held key business development and strategic marketing with Novavax, Inc., ImmunoGen and Smith and Nephew Plc. He holds a BS in Biological Sciences from the University of Massachusetts-Lowell and an MBA from Babson College, and received graduate training in Molecular Neurobiology from MIT.

He has more than 12 years of experience in commercial evaluation, planning and launch of new products and technologies in the pharmaceutical, biotech & medical device industries and has significant background in the selection and pursuit of new applications for novel technologies.

Dr Dinesh B. Shenoy

Dr Shenoy is General Manager – Formulation Development, at Cadila Pharmaceuticals Ltd (CPL), India. He is responsible for product development (for both pharmaceuticals and biologicals) at CPL – with a focus on innovative and novel delivery systems.

Before joining CPL, Dr Shenoy worked as Director (Formulation Development) at Novavax, Inc., where he worked on development and expansion of micellar nanoparticle platform technology. He has also served at the Department of Pharmaceutical Sciences, Northeastern University, Boston and did his post-doctoral studies at Max Planck Institute of Colloids and Interfaces, Potsdam/Golm, Germany. He holds Masters and PhD degrees in Pharmaceutical Sciences from Manipal College of Pharmaceutical Sciences, Manipal University, India.

Dr Shenoy has published 29 research and review articles in peer-reviewed journals and six book chapters. He is an inventor in 10 provisional or PCT patent applications.

He has more than nine years of experience in pharmaceutical and biotechnology formulation research and development. Dr Shenoy maintains strong academic ties, including an appointment as Adjunct Faculty of Pharmaceutical Sciences at Manipal College of Pharmaceutical Sciences.

Dr H. N. Shivakumar

Dr H. N. Shivakumar is currently working as a research scientist and is an active member of Dr S. N. Murthy's research group at the Department of Pharmaceutics, the University of Mississippi, MS, USA. Dr Shivakumar obtained his PhD from Rajiv Gandhi University of Health Sciences, Bangalore, Karnataka, India. Before joining the research group he worked as a Professor of Pharmaceutics at KLES's College of Pharmacy, Bangalore, India. He has guided nine Master level projects and two PhD projects to date. He has 20 research publications and over 20 conference presentations to his credit. His research area include topical drug delivery, chronotherapeutic drug delivery, multiparticulate drug delivery and formulation development by statistical optimization techniques.

Dr Rashmi Thakur

Rashmi Thakur currently works as a Senior Scientist in R&D at McNeil Consumer Healthcare in Pennsylvania. She has contributed to six technical articles and five book chapters in the general area of drug delivery and has two US patent applications. Dr Thakur has received a number of research awards from the American Association of Indian Pharmaceutical Scientists and Graduate Research Association of Students in Pharmacy. She received her PhD in Pharmaceutical Sciences from Rutgers, the State University of New Jersey and a Masters in Management studies from the University of Mumbai. She is a member of the American Association of Pharmaceutical Scientists.

ACKNOWLEDGEMENT

Many accomplished professionals in the field of pharmaceutical research and development provided me with helpful discussions and valuable input during my editing of this book. I take this opportunity to sincerely acknowledge these individuals for their great help; they are Dr Frank Sorgi, Vice President R&D, Dr Charles Shaw, and Kay Mary Harrell from DPT Laboratories, Dr Duncan Aust, Sr Vice President, HealthPoint, Inc., and Dr Majumdar, Department of Pharmacy, University of Mississippi.

I thank all contributors to this book for their time and effort and for their cooperation during the review and re-submission process. Without the efforts of the contributors, this book would not have come into existence.

And last but not least I would like to thank my wife, kids and parents for their love, support, and encouragement during all stages of this book. I dedicate this book to my late father, Shriniwas K. Kulkarni, a dedicated and accomplished teacher, who valued and encouraged hard work, perseverance and a pursuit for knowledge.

CHAPTER 1

Topical and Transdermal Drug Delivery

S. Narasimha Murthy *and* **H. N. Shivakumar***
Department of Pharmaceutics, the University of Mississippi, MS 38677, USA

Contents

1.1 Structure of the Skin 1
  1.1.1 Dermis and hypodermis 3
1.2 Topical vs. transdermal drug delivery 3
  1.2.1 Topical drug delivery 3
  1.2.2 Transdermal patch systems 4
1.3 Percutaneous absorption pathways 5
1.4 Approaches to enhance passive cutaneous drug absorption 6
  1.4.1 Supersaturation of drug 6
  1.4.2 Eutectic systems 7
  1.4.3 Prodrug approach 7
  1.4.4 Ion pair formation 8
  1.4.5 Complexation 8
  1.4.6 Liposomes 9
  1.4.7 Microemulsions 13
  1.4.8 Organogels 14
  1.4.9 Solid lipid nanoparticles 15
  1.4.10 Liquid crystalline systems 16
  1.4.11 Chemical permeation enhancers 16
1.5 Conclusions 26
References 27

1.1 STRUCTURE OF THE SKIN

Skin is the largest organ of our body, which acts as a protective barrier against the entry of foreign material and possible invasion of pathogens. The skin also prevents the loss of excessive endogenous material such as water (Brown et al., 2006). In addition, the skin serves to reduce the damaging impact of solar UV radiation (Hadgraft, 2004).

The structure of human skin is portrayed in Figure 1.1. The skin is about 0.5 mm thick and is made up of two distinct layers, the inner dermis and the

* Author to whom correspondence should be addressed.

Handbook of Non-Invasive Drug Delivery Systems ISBN 9780815520252

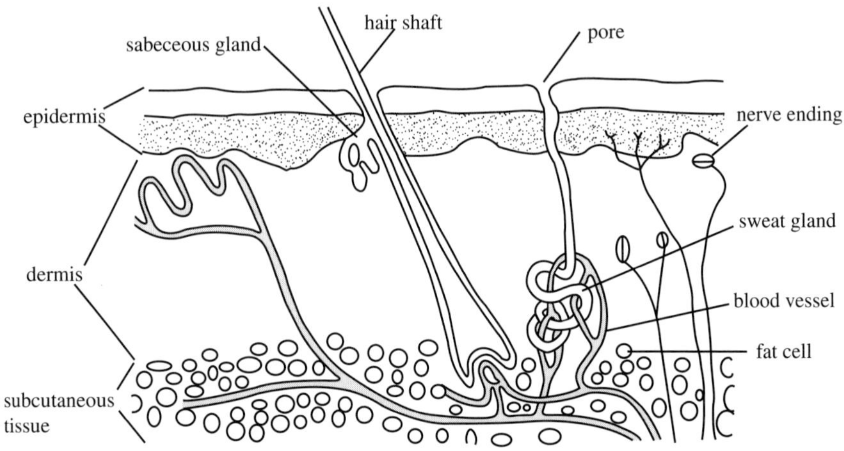

Figure 1.1 Anatomy of skin. Transverse section of skin showing different strata and the associated regions in the skin.

overlaying epidermis. The dermis that forms the bulk of the skin (1–2 mm thick) is made up of connective tissue elements. Dermis is highly vascular and filled with pilosebaceous units, sweat glands, adipose cells, mast cells, and infiltrating leukocytes (Menon, 2002). The epidermis is avascular in nature, consisting of several types of cell (corneocytes, melanocytes, Langerhans cells, and Merkel cells) and a variety of catabolic enzymes (esterases, phosphatases, proteases, nucleotidases, and lipases) (Jansen and Hopsu-Havu, 1969; Mier and van den Hurk, 1975). The stratified epidermis is about 100–150 μm thick and comprises four distinct layers, namely the stratum basale, stratum spinosum, stratum granulosum, and stratum corneum.

The stratum corneum is the outermost layer of skin that forms the main barrier for diffusion of the permeants through the skin (Wertz and Downing, 1989). Stratum corneum consists of 18–21 layers of flat, roughly hexagonal cells called corneocytes that are constantly shed and renewed (Menon, 2002). These keratin-rich dead cells, measuring 20–40 μm in diameter, are interspersed within crystalline lamellar lipid matrix to assume a “bricks and mortar” arrangement (Elias, 1983). The extracellular lipid contributes 10% of the dry weight of this layer, while 90% is the intracellular keratin. The barrier function of the skin can be attributed to the lamellar lipids that are synthesized in the granular layer and subsequently organized into the extracelluar lipid bilayer domains of the stratum corneum (Landmann, 1986). The barrier function of the skin depends on specific ratios of various lipids present and studies reveal that relatively polar lipids play a critical role in maintaining the barrier integrity of the stratum corneum (Elias et al., 1984; Menon et al., 1986; Elias and Feingold, 1988).

The viable epidermis is made up of keratinocytes at various stages of differentiation. Lipid catabolic enzymes, namely acid lipase, phospholipase, sphingomyelinase, and steroid sulfatase, are distributed throughout the epidermis, though mainly found in the stratum granulosum and stratum corneum (Menon et al., 1986). The phospholipid content decreases while the sphingolipid and cholesterol content gradually increases as the cells differentiate during their migration to the surface.

1.1.1 Dermis and hypodermis

The dermis is rich in blood vessels, lymphatic vessels, and nerve endings. An extensive capillary network connects to the systemic circulation with substantial horizontal branching from the arterioles and venules in the papillary dermis. These in turn form plexus and supply capillaries to the hair follicles and the glands. The lymphatic vessels serve to drain the excess extracellular fluid and clear the antigenic materials. The dermis is filled with scattered fibroblasts, macrophages, leukocytes, and mast cells, in addition to the hair follicles, sebaceous glands, and sweat glands. On average, about 10 hair follicles, 15 sebaceous glands, 12 nerves, 100 sweat glands, 360 cm of nerves, and three blood vessels are present in one square centimeter of skin (Barry, 1983). The hypodermis constitutes the deepest layer of the skin, and consists of the subcutaneous tissue filled with fat cells, fibroblasts, and macrophages.

1.2 TOPICAL vs. TRANSDERMAL DRUG DELIVERY

1.2.1 Topical drug delivery

Topical drug delivery is the term used for localized treatment of dermatological condition where the medication is not targeted for systemic delivery (Osborne, 2008); examples include treatment of dermatological conditions like eczema or psoriasis by topical application. Examples of drugs delivered topically include corticosteroids, antifungals, antivirals, antibiotics, antiseptics, local anesthetics, and antineoplastics. Topical agents that act by physical action would include protectives, adsorbents, emollients, and cleansing agents, whereas the astringents, irritants, rubefacients, and keratolytic agents are the ones which act by chemical means.

Conventional topical drug delivery systems include semisolid dosage forms and liquid dosage forms. The semisolid dosage forms include ointments, creams, gels, or pastes, while the liquid dosage forms include lotions that may be an emulsion, suspension, or a solution (Buhse et al., 2005). Ointments usually contain <20% of water and >50% hydrocarbons, waxes, or polyols as vehicles. Ointments are used as vehicles for topical application of the actives and basically function as skin protective and emollient.

Creams are emulsion semisolid dosage forms usually containing more than 20% water or volatile components and typically less than 50% hydrocarbons, waxes, or polyols as vehicles (Osborne, 2008). A gel is a semisolid dosage form that contains a gelling agent to provide stiffness to the dispersion. Gels can be water based (hydrogels) or organic solvent based (organogels) (Gupta and Garg, 2002). A paste can be defined as a semisolid dosage form, containing a large proportion of solids (20–50%) finely dispersed into a suitable vehicle. A lotion may be in the form of a solution or a suspension or an emulsion. Typically these formulations are intended to be applied to the intact skin, generally without any friction.

1.2.2 Transdermal patch systems

Transdermal delivery is the term that is confined to a situation in which the drug diffuses through different layers of the skin into systemic circulation to elicit the therapeutic response (Brown et al., 2006). An example would be management of hypertension using a transdermal clonidine patch. In a broader sense transdermal delivery also includes local anesthetic patches in which the drug is intended to diffuse regionally in the skin to elicit the pharmacological action only in the treated area of the skin. Often, delivery of local anesthesia has been classified under topical drug delivery. An overview of cutaneous drug delivery system is shown in Figure 1.2.

Transdermal drug delivery systems also termed as "patches" are self-contained discrete dosage forms designed to deliver a therapeutically effective amount of drug through intact skin (Wokovich et al., 2006). Most commercially available transdermal drug delivery systems are of three different types, namely reservoir systems, matrix systems with rate-controlling membrane, and matrix systems without rate-controlling membrane. The reservoir system is made up of three major components, namely the drug

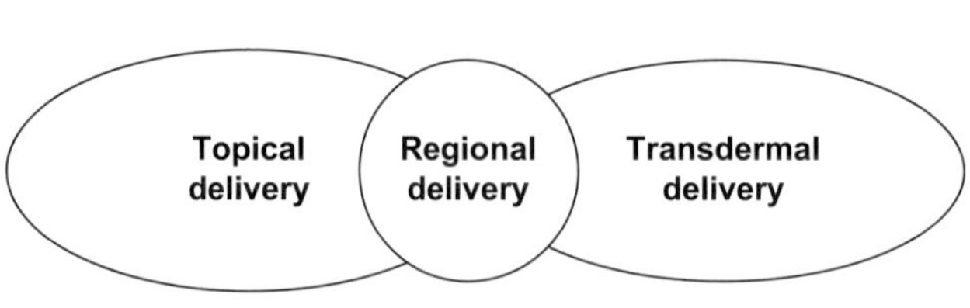

Figure 1.2 Transdermal and topical drug delivery.

reservoir, the rate-controlling membrane, and the adhesive. The drug present in the reservoir, along with the other excipients, has to permeate through the rate-controlling membrane before reaching the skin. The adhesive that holds the system in place on the skin can completely cover the drug release area or only the perimeter around the non-adhering drug release surface. In the matrix type, the drug may be embedded in the adhesive matrix. A rate-controlling membrane may be present between the drug-loaded matrix and the adhesive or sometimes the matrix itself can control the rate of release of the actives from the system.

The drugs that have made it into the transdermal market include scopolamine, nitroglycerine, nicotine, clonidine, fantanyl, estradiol, testosterone, lidocaine, and oxybutinin (Langer, 2004). Recent additions to this list include lidocaine-tetracaine, selegiline, methyl phenidate, and rotigotine. However, the future focus is production of transdermal systems capable of delivering peptides and proteins including insulin, growth hormone, and vaccine across the skin.

1.3 PERCUTANEOUS ABSORPTION PATHWAYS

The lipid-rich and structurally complex intercellular region of stratum corneum is said to play an important role in percutaneous absorption (Elias and Friend, 1975). The stratum corneum is known to be selectively permeable and allows relatively lipophilic molecules to diffuse to the lower skin layers (Brown et al., 2006). The transport of such molecules across the stratum corneum barrier is mainly by passive diffusion (Scheuplein and Blank, 1971). The permeation rate through the stratum corneum has been represented by a simple equation (equation (1.1)) (Barry, 1983):

$$\frac{\mathrm{d}m}{\mathrm{d}t} = \frac{DC_0K}{h} \tag{1.1}$$

where dm is the amount of the diffusant passed through the membrane in time dt, C_0 is the drug concentration in the donor solution, K is the partition coefficient of the diffusant between the membrane and the solution, D is the diffusion coefficient of the diffusant in the membrane, and h is the membrane thickness. Considering the tortuous intercellular pathway between the corneocytes, the diffusional path length for the permeants is much longer than the thickness of the stratum corneum and is estimated to be ~500 μm (Hadgraft, 2004).

The other potential routes of entry for the permeants from the skin surface to the subepidermal tissues are through the hair follicles with their associated sebaceous glands and via the sweat ducts or through the stratum corneum between these appendages (Barry, 2001). These follicles passing from the skin surface through the epidermis and reaching the dermis or even the underlying subcutaneous region are the most important appendages of human skin.

1.4 APPROACHES TO ENHANCE PASSIVE CUTANEOUS DRUG ABSORPTION

For a substance to be well absorbed from the skin it should have a molecular weight of less than 0.6 kDa (Schafer-Korting et al., 2007), and a balanced solubility in both oil and water with a log partition coefficient value between 1 and 3 (Hadgraft, 2004). Drug absorption across the skin could be enhanced by adapting one or more of the several strategies discussed subsequently in this chapter.

1.4.1 Supersaturation of drug

Skin absorption can be enhanced using supersaturated solutions that have greater thermodynamic activity or chemical potential than the saturated solutions (Hadgraft, 1999). This has been demonstrated based on equation (1.2) proposed by Higuchi (1960):

$$\frac{\mathrm{d}m}{\mathrm{d}t} = \frac{aD}{\gamma h} \tag{1.2}$$

where $\mathrm{d}m/\mathrm{d}t$ is the permeation rate, a represents the thermodynamic activity of the permeant in its vehicle, D is the diffusion coefficient of the drug, h is the membrane thickness, and γ is the effective activity in the skin. The degree of saturation can be increased by increasing the drug concentration in the vehicle or reducing the drug solubility in the vehicle. Thermodynamic activity of saturated solution is unity and a transient increase of the degree of saturation to greater than 1 can be achieved by supersaturation (Moser et al., 2001a). Supersaturation can take place due to water uptake by the skin (Kemken et al., 1992), or by evaporation of a volatile formulation component following application (Coldman et al., 1969; Kondo and Sugimoto, 1987; Kondo et al., 1987a, b; Chiang et al., 1989). Different methods have been used to produce supersaturated states such as solvent evaporation, mixing of two solvents wherein the drug solubility in one is much more than in the other, or mixing of two solutions of different pH where the solubility is pH dependent. An 18-fold increase in stratum corneum uptake and a 13-fold increase in the flux have been observed with a supersaturated solution of estradiol when compared to a saturated solution that was 18 times less concentrated (Megrab et al., 1995a). The transport of drugs such as hydrocortisone acetate (Davis and Hadgraft, 1991) and piroxicam (Pellett et al., 1994, 1997) has been reported to increase linearly with supersaturation.

The technique has been used recently to increase the permeation of a model lipophilic compound (Moser et al., 2001b).

Supersaturated solutions are thermodynamically unstable for long-term storage due to drug crystallization. They have been stabilized by addition of polymers, which act as anti-nucleants (Pellett et al. 1997; Schwarb et al., 1999; Ierovolino et al., 2000; Raghavan et al., 2000).

1.4.2 Eutectic systems

The melting point of a drug influences the solubility and hence the skin penetration. A simple empirical equation (equation (1.3)) that relates the permeability coefficient with the octanol water partition coefficient and the molecular weight of the permeant has been proposed (Potts and Guy, 1992):

$$\text{Log } K_p = -6.3 + 0.71 \log K_{o/w} - 0.0061 \times \text{MW} \qquad (1.3)$$

where K_p represents the permeability coefficient, $K_{o/w}$ is the octanol–water partition coefficient, and MW is the molecular weight. The permeant melting point was found to be inversely related to the lipophilicity and therefore the transdermal flux for a series of anti-emetics (Calpena et al., 1994). It has been postulated that the lower the melting point of the permeant, the greater is the solubility in a given solvent, including the skin lipids (Benson, 2005). One of the methods by which the melting point of the drug can be reduced is by the formation of a eutectic mixture. A eutectic mixture is a physical mixture of two components that do not interact to form a new chemical substance but at certain ratios inhibit each other's crystallization, resulting in a substance with a lower melting point than that of either of the components (Stott et al., 1998). A eutectic mixture is formed only when the two components are miscible in the liquid state but remain completely immiscible in the solid state. EMLA cream, a preparation based on a eutectic mixture of lignocaine and prilocaine, is known to provide effective anesthesia for pain-free venepuncture when applied under occlusive conditions (Benson, 2005). Eutectic systems of ibuprofen with terpenes (Stott et al., 1998), ibuprofen with methyl nicotinate (Woolfson et al., 2000), propranolol with fatty acids (Stott et al., 2001), lidocaine with menthol (Kang et al., 2000), and cannabidiol with phosphatidyl choline (Lodzki et al., 2003) have been reported to increase the transdermal permeation. The depression in the melting point to around or below the skin temperature was considered responsible for enhancing the solubility in the skin lipids and thereby the permeability in most of the cases.

1.4.3 Prodrug approach

Prodrugs have been utilized to enhance the dermal and transdermal permeation of drugs with unfavorable partition coefficients (Sloan, 1989). The prodrug approach involves addition of a promoiety to increase the partition coefficient and hence the permeation of the parent molecule across the skin. On reaching the viable epidermis, esterases release the parent molecule by hydrolysis of the prodrug. The permeability coefficient of the parent compound indomethacin, has been reported to increase by 100-fold by conversion to an ester prodrug (Jona et al., 1995). The rate of transdermal permeation of the esters through cadaver skin was found to be more than 10-fold higher when compared to the parent drug indomethacin. The effect of prodrug structure on the permeability of levonorgestrel through rat skin has been studied (Friend et al., 1988). The glycidol and hexanediol esters of levonorgestrel were found to partition better into the viable epidermis than did the parent compound. The *in vitro* flux values were found to be 20–40 times higher than that of the free drug. Attempts have been made to enhance the intrinsic permeability of 5-fluorouracil by conversion to 1-alkyloxycarbonyl-5-fluorouracil prodrugs (Beall and Sloan, 1996). The ethyl derivative was found to be the most effective, with a 25-fold increase in the flux across hairless mouse skin when compared to 5-fluorouracil (Beall et al., 1994). One of the marketed topical preparations containing the prodrug betamethasone-17-valerate is known to provide a better anti-inflammatory activity than that of the parent steroid (Benson, 2005).

1.4.4 Ion pair formation

Charged molecules are poorly permeable across the stratum corneum due to their low lipid partitioning ability. One approach to enhance transdermal permeation is to add an oppositely charged moiety to form a lipophilic ion pair that can partition into the stratum corneum lipids as the charges temporarily neutralize (Barry, 2001). On reaching the aqueous viable epidermis the ion pair dissociates, releasing the parent charged drug, which can partition into the epidermis or dermis and diffuse further (Megwa et al., 2000a, b; Valenta et al., 2000; Stott et al., 2001). Formation of ion pairs with organic acids has been reported to increase the transdermal permeation of indapamide across rat abdominal skin, with lactic acid showing the greatest effect (Ren et al., 2008). The permeation of lidocaine hydrochloride from transdermal films through rabbit ear skin was improved by formation of ion pairs with lauric acid (Padula et al., 2007). However, a mere twofold enhancement was observed in the permeation of propranolol by ion pairing with fatty acids (Stott et al., 2001).

1.4.5 Complexation

Cyclodextrins are cyclic dextrose polymers that are known to enhance the aqueous solubility and stability of drugs on complex formation. The ring has a hydrophilic surface with a lipophilic core in which organic molecules of appropriate size can be held to form noncovalent inclusion complexes, resulting in increased aqueous solubility and chemical stability (Loftsson and Brewster, 1996). Cyclodextrins are macromolecules with a molecular weight of greater than 1000 Da, and therefore not likely to readily permeate the skin. Hydroxypropyl β-cyclodextrin has been reported to increase the *in vivo* percutaneous absorption of liarozole by threefold, while skin pretreatment with an aqueous solution of dimethyl β-cyclodextrin provided a 9.4-fold enhancement (Vollmer et al., 1994). Randomly methylated β-cyclodextrin was found to enhance the permeation of piribedil through hairless rat skin (Legendre et al., 1995). Increased flux across the skin has also been attributed to extraction of stratum corneum lipids by the cyclodextrins (Bentley et al., 1997).

In contrast, a few studies report a decrease in the skin permeability of the drug after forming a complex with cyclodextrin (Loftsson and Brewster 1996; Williams et al., 1998; Simeoni et al., 2004). The drug permeation is reported to depend on the proportion of cyclodextrin in the complex. A reduced flux was generally seen at relatively higher cyclodextrin concentrations, while higher flux was observed at lower cyclodextrin concentrations (Loftsson and Masson, 2001). It is clear from recent findings that cyclodextrins can only enhance the transdermal delivery when the resistance of the unstirred water layer on the donor side is equal to or greater than the skin resistance (Loftsson et al., 2007). The greatest enhancement on complexation with cyclodextrins was only with smaller lipophilic molecules.

1.4.6 Liposomes

Liposomes are colloidal systems comprised of bilayered vesicles made of phospholipids. They may consist of a single (unilamellar) or few (oligolamellar) or many (multilamellar) concentric phospholipid bilayer(s) (Gregoriadis and Florence, 1993). Depending on the size of the vesicles they are further categorized as small unilamellar (SUV) or large unilamellar (LUV) vesicles. Liposomes are capable of encapsulating both hydrophilic and lipophilic molecules in their concentric bilayers, as represented in Figure 1.3. The hydrophilic drugs are usually entrapped in the inner aqueous compartment, while the lipophilic or amphiphilic and sometimes charged molecules are associated with the phospholipid bilayers (Honeywell-Nguyen and Bouwstra, 2005). When the vesicular bilayers are made up of non-ionic surfactants they are termed niosomes. The physicochemical properties like size, charge, lamellarity, and elasticity are governed by the composition of the vesicles. These properties are known to have a significant influence on the effectiveness of liposomes as drug delivery systems.

Liposomes are used as carriers to deliver the entrapped drugs *into* the skin. They act as permeation enhancers by virtue of the phospholipids that

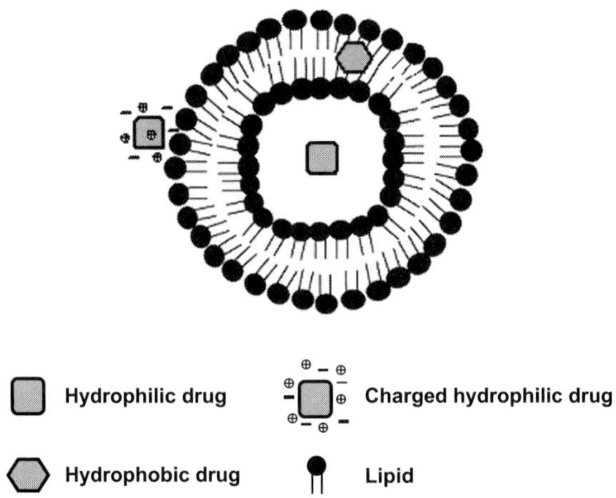

Figure 1.3 Schematic representation of unilamellar vesicle displaying a single lipid bilayer with encapsulated molecules.

penetrate into the stratum corneum and subsequently alter the skin lipid bilayers. They are known to act as a depot for sustained release of actives into the skin, and also modulate the rate and extent of systemic drug absorption (Honeywell-Nguyen and Bouwstra, 2005). Liposomal formulations are known to favor drug deposition in the skin, reduce irritation potential of drugs, and improve drug stability.

A four- to fivefold higher concentration of triamcinolone acetonide was observed in the epidermis and dermis in addition to lower systemic concentrations with liposomal lotions compared to conventional formulations (Mezei and Gulasekharam, 1980). Similar higher drug concentrations in epidermis and dermis have been observed in comparison to the systemic concentrations with liposomal gel formulations of triamcinolone acetonide (Mezei and Gulasekharam, 1982), progesterone, and econazole (Mezei, 1985). The topical delivery of interferon from liposomes constituted with stratum corneum lipids has been reported to be better when compared to water-in-oil emulsion or aqueous solution in a cutaneous herpes simplex guinea-pig model (Weiner et al., 1989).

The lipid composition, method of preparation, and thermodynamic state of the vesicular bilayers are shown to govern the extent of skin deposition of liposomes (Bouwstra and Honeywell-Nguyen, 2002; El Maghraby et al., 2006). Liposomes constituted with skin lipids were found to be more effective than phospholipid vesicles in delivering drugs to the skin (Egbaria et al., 1990, 1991; Fresta and Puglisi, 1996, 1997; Yu and Liao, 1996; Liu et al., 2004). The reduction in the liposomal cholesterol content increased

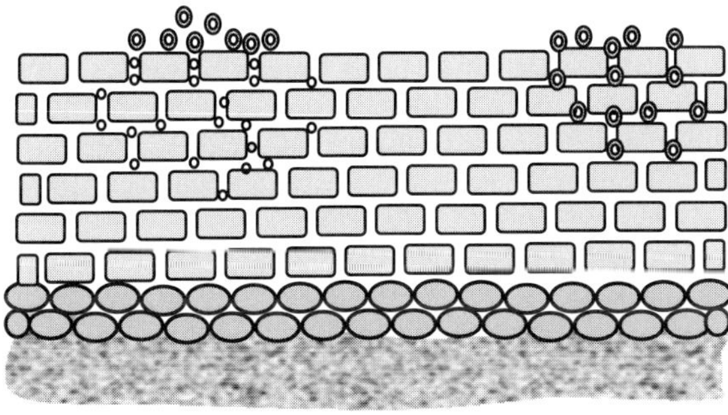

Figure 1.4 Illustration of the proposed mechanism to explain liposomal permeation across the skin. A: Drug permeates independently after leaving the vesicles. B: Penetration of drug loaded vesicles.

the fluidity of the bilayers, resulting in an increase in drug transport across the stratum corneum. The other physicochemical properties that may influence the liposomal delivery would be the size, lamellarity, and charge of the lipids (Yu and Liao, 1996; Katahira et al., 1999; Ogiso et al., 2001; Liu et al., 2004; Manosroi et al., 2004; Choi and Maibach, 2005; Sinico et al., 2005).

Application of the liposomes to target the skin appendages, i.e. the hair follicles with their associated sebaceous glands, has been extensively reviewed (El Maghraby et al., 2006). Studies revealed that vesicles made of unsaturated phospholipids produced higher percutaneous absorption and tissue distribution rather than skin accumulation of the radiolabeled marker [^{14}C]inulin (Fresta and Puglisi, 1996). However, it is evident that in most of the cases conventional liposomes have less potential as carriers for transdermal delivery as they fail to deeply penetrate the skin, but on the contrary remain confined to the upper layers of stratum corneum (Elsayed et al., 2007a).

Different mechanisms have been proposed by which the liposomes penetrate across the skin and thereby improve the skin delivery of drugs, as illustrated in Figure 1.4. According to the free drug mechanism, the drug is known to permeate the skin independently after leaving the vesicles (Ganesan et al., 1984). The second mechanism suggests changes in the ultrastuctures of the intercellular lipids on topical application of liposomes, thus leading to penetration of the vesicles themselves. The lipid components of the vesicles are said to fuse and mix with the skin lipids to loosen their structure and thereby enhance the drug permeation (Kirjavainen et al., 1996). It has been reported that particles larger than 10 μm tend to remain on the surface of the stratum corneum while those in the size range of

3–10 μm concentrate within the hair follicle, but particles below 3 μm in size can penetrate the stratum corneum (Schaefer and Redelmeier, 1996).

A new class of highly deformable elastic or ultraflexible liposomes termed transfersomes are known to penetrate into the skin layers effectively to reach the systemic circulation when applied in non-occlusive conditions (Cevc and Blume, 1992). Transfersomes are the first-generation elastic liposomes made up of phospholipids and an edge activator consisting of a single-chain surfactant. The edge activator is known to destabilize the liposomal lipid bilayers and increase their deformability (Cevc, 1996; Cevc et al., 1996). Sodium cholate, sodium deoxycholate, span 60, span 65, span 80, Tween 20, Tween 60, Tween 80, and dipotassium glycyrrhizinate could be used as edge activators (Cevc et al., 1998; El Maghraby et al., 1999, 2000a, b; Trotta et al., 2004; Garg et al., 2006; Oh et al., 2006).

A 14- to 17-fold increase in maximum estradiol flux and 9.2- to 11-fold increase in the skin deposition was observed with deformable liposomes through human cadaver epidermis (El Maghraby et al., 1999). An increase in the skin permeation and deposition of cyclosporin A (Guo et al., 2000), methotrexate (Trotta et al., 2004), and melatonin (Dubey et al., 2006) has been seen with deformable liposomes. A greater improvement in the skin deposition of ketotifen was observed in comparison to the amount permeated, indicating that the deformable liposomes would be more suitable for intradermal delivery rather than for the transdermal delivery (Elsayed et al., 2007b).

Deformable lipid vesicles termed "vesosomes" have been used as carriers for topical immunization (Gupta et al., 2005a). These vesicles were found to penetrate the skin to deliver the antigens to the Langerhan cells (Gupta et al., 2005b). Vesosomes have been used as carrier for topical delivery of hepatitis B surface antigen (Mishra et al., 2006). The immune response elicited was compared with intramuscularly administered and topically applied antigen solutions.

Ultradeformable vesicles are said to pass through the intact skin and move deep enough to be absorbed by the systemic circulation (El Maghraby et al., 2008). Transfersomes in the size range of 200–300 nm are reported to easily penetrate through the skin (Cevc et al., 1998). The transdermal hydration gradient has been considered to be the main force responsible to drive the ultradeformable vesicles through the intact stratum corneum and epidermis. Phospholipids, being hydrophilic in nature, try to follow the local hydration gradient and move to the deeper skin strata (Cevc and Blume, 1992; Cevc et al., 1995). The skin penetrating ability of these vesicles was ascribed to the high deformability resulting from the edge activators getting accumulated at the high stress sites due to their greater affinity for curvatures. This arrangement has been claimed to reduce the energy required for deformation. The presence of the edge activators helps the vesicles to deform and penetrate through the dense stratum corneum containing very small pores relative to the vesicle diameter (Cevc et al., 2002; Cevc and Gebauer, 2003). However, occlusive application and full skin hydration limit transdermal delivery from ultradeformable vesicles mainly due to inhibition of transdermal hydration gradient (Cevc et al., 1995).

Ethosomes are novel lipid vesicles composed of phospholipids, ethanol, and water (Touitou et al., 2000a, b). They are known to contain high concentrations of ethanol (20–45%) and display enhanced skin delivery. Several studies demonstrated the influence of ethanol concentration on the physicochemical attributes of these systems (Dayan and Touitou, 2000; Lopez-Pinto et al., 2005; Elsayed et al., 2007b). The size of the ethosomes was found to decrease with increase in the ethanol concentration in the range of 20–45% (Touitou et al., 2000a). The phospholipid concentration was found to have an influence on the ethosomal size (Touitou et al., 2000a; Elsayed et al., 2007b). Ethosomes were found to exhibit high encapsulation efficiency mainly by virtue of the multilamellarity of the vesicular structures (Touitou et al., 2000a).

Ethanol is known to initially disturb the stratum corneum organization and increases the lipid fluidity to facilitate the ethosome penetration of the disturbed stratum corneum lipid bilayers. The drug release in the deeper skin layers followed by transdermal absorption was due to fusion of ethosomes with the skin lipids (Touitou et al., 2000a). The high ethanol concentration is known to fluidize the ethosomal membranes to produce highly deformable vesicles that squeeze between the cells in the stratum corneum and carry large amounts of actives across intact skin (El Maghraby et al., 2001). Unlike deformable liposomes, ethosomes are reported to enhance skin delivery of drugs under occlusive (Dayan and Touitou, 2000; Ainbinder and Touitou, 2005; Lopez-Pinto et al., 2005) as well as non-occlusive conditions (Dayan and Touitou, 2000; Elsayed et al., 2007b).

1.4.7 Microemulsions

Microemulsions are clear, thermodynamically stable, isotropic liquid mixtures of oil, water, and surfactant, usually in combination with a cosurfactant (Kogan and Garti, 2006). The droplet size of the dispersed phase in a microemulsion is less than 100 nm. Microemulsions are known to increase the drug absorption on topical application. This could be due to the penetration enhancement effect of the carrier, usually composed of saturated or unsaturated fatty acids serving as the oil phase. The *in vitro* percutaneous permeability of diclofenac sodium across human skin was studied using ternary mixtures and microemulsions (Escribano et al., 2003). The soyabean lecithin microemulsion gel has been used as potential transdermal matrix for diclofenac and indomethacin (Dreher et al., 1997). Infrared spectroscopy, differential scanning calorimetry, and low-temperature scanning electron microscopy revealed that isopropyl palmitate altered the stratum corneum lipid organization. Studies demonstrated no significant irritancy of the microemulsion gels on application to human skin.

Microemulsions increased the transdermal flux of lidocaine fourfold compared to the conventional oil-in-water emulsions and the flux of prilocaine hydrochloride 10-fold compared to the hydrogels (Kreilgaard et al., 2000). The solubilities of the drugs in the microemulsion were found to be higher in comparison to the solubilities in the individual microemulsion components. Lidocaine displayed a 28–62% increase in the solubility while prilocaine showed a 24–40% increase. It was found that the higher solubilizing property of microemulsion resulted in high concentration gradients and higher transdermal flux. A microemulsion comprised of glyceryl oleate–polyethylene glycol (PEG)400 stearate produced a higher transdermal flux of lidocaine across rat skin in comparison to one consisting of glyceryl oleate–PEG400 hydroxylated castor oil (Sintov and Shapiro, 2004).

Triptolide is an immunosuppressive, antifertility, and anticancerous drug that undergoes hydrolysis on long-term storage in aqueous environment. Microemulsions containing oleic acid, Tween 80, and propylene glycol were found to have globule sizes in the range of 12–83 nm and a pH of 4–6. Permeation studies of microemulsion using mouse skin showed a steady increase in drug flux for 24 h, which indicated that the microemulsions were able to provide a prolonged release of triptolide. A microemulsion comprised of oleic acid (6%), Tween 80 (20%), propylene glycol (10%), and menthol (1%) in water (62.97%) was found to be the most efficient for the delivery of triptolide. The microemulsion was non-irritant, ensured adequate protection from drug degradation, provided prolonged release, and was found to be suitable for long-term use (Chen et al., 2004).

1.4.8 Organogels

Organogels are semisolid systems in which the organic liquid phase is immobilized by a three-dimensional network consisting of self-organized, entangled gelator fibers (Vintiloiu and Leroux, 2008). Typical organogel components include well-known permeation enhancers such as fatty acids, surfactants, alcohols, azone, *N*-methylpyrrolidone, urea, sulfoxides, essential oils, terpenes, terpenoids, and glycols (Williams and Barry, 2004). The most common organogels employed for topical and transdermal delivery of actives are lecithin organogels. The components of these organogels are known to efficiently partition with the skin and enhance the permeation of many drugs (Willimann et al., 1992; Bhatnagar and Vyas, 1994; Aboofazeli et al., 2002; Shaikh et al., 2006). The skin penetration enhancement effect observed was due to the lecithin as well as certain solvents like isopropyl myristate and ethyl oleate (Dreher et al., 1997; Agrawal et al., 2004). Owing to their amphiphilic nature, the lecithin organogels are able to solubilize and entrap permeants with diverse physicochemical nature (Scartazzini and Luisi, 1988). A leveling in subcutaneous tumor growth was reported on transdermal application of lecithin organogels containing tetrabenzamidine in mice (Nastruzzi and Gambari, 1994).

Organogels composed of non-ionic surfactants have been tested for their safety as topical applications. The transdermal delivery of piroxicam from organogels composed of glyceryl fatty acid esters in pharmaceutical oils has been investigated. The organogels were found to be superior in inhibiting edema when compared to conventional topical formulations (Penzes et al., 2005). The polymeric organogels that have been widely tested for drug delivery applications are those based on polyethylene. Polyethylene formulations have been used for the delivery of 5-iodo-2-deoxyuridine for the treatment of oral herpes simplex (Najjar et al., 1969).

1.4.9 Solid lipid nanoparticles

Solid lipid nanoparticles are particulate colloidal drug delivery systems comprised of solid lipids and ranging in size between 50 and 1000 nm. Solid lipid nanoparticles (SLNs) have been extensively investigated for topical application of actives for various skin diseases. SLNs meant for topical applications comprise lipids like glyceryl behenate, glyceryl palmitostearate or the wax cetyl palmitate (Schafer-Korting et al., 2007). By virtue of their smaller particle size and larger surface area they are known to possess strong adhesive properties (Muller-Goymann, 2004). This may result in film formation on the skin that would restore the previously damaged protective stratum corneum and enhance the moisturizing effect by occlusion (Jenning et al., 2000a; Wissing et al., 2001). The penetration of coenzyme Q_{10} across stratum corneum was found to be more effective when applied as SLNs compared to solutions in isopropyl alcohol and liquid paraffin (Müller et al., 2002). Increased levels of retinol and prednicarbate were seen in the epidermis rather than the dermis when applied as SLN formulations in comparison to conventional cream (Jenning et al., 2000b; Santos Maia et al., 2002). This is said to be of particular importance in the treatment of eczema, where the inflammatory process is more pronounced within the epidermis (Schafer-Korting et al., 2007). An increased concentration in the first 100-μm layer of pig skin is suspected due to the drug released from SLNs (Jenning et al., 2000a). SLNs increased the skin absorption of the fluorescent dye Nile red by around fourfold, whereas nanostructured carriers enhanced the absorption by twofold (Lombardi Borgia et al., 2005). A clinical study has demonstrated an improved efficacy of glucocorticoid therapy with clobetasol propionate-loaded SLNs when compared to conventional cream in the treatment of eczema (Kalariya et al., 2005). SLNs have demonstrated enhanced skin penetration and uptake of cyproterone acetate in topical acne treatment (Stecova et al., 2007).

Solid lipid nanoparticles have been extensively investigated for topical treatment of acne. Retinol-loaded solid lipid nanoparticles of Compritol® released the drug more rapidly and to a higher extent when compared with conventional vehicles (Jenning et al., 2000a, b). This rapid release was attributed to the burst effect from the nanoparticles following water evaporation from the skin surface, which may induce polymorphic transition of the lipid lattice. Moreover, SLNs have been reported to be an attractive carrier for a number of sunscreening agents (Wissing and Muller, 2001; Wissing et al., 2001).

1.4.10 Liquid crystalline systems

The liquid crystalline state has the properties of both solid and liquid states. The liquid state has the ability to flow while the solid state displays an ordered crystalline structure. The liquid crystals represent an intermediate state between solid and liquid states, and hence they are termed mesophases (Muller-Goymann, 2004). A number of polar surfactants hydrate in water to form spherical or ellipsoidal micelles. At high surfactant concentrations these associates assume a densely packed arrangement to form cubic liquid crystals (Schuetze and Mueller-Goymann, 1992). Permeation studies with excised stratum corneum have shown that the amount of ibuprofen permeated was much higher from cubic liquid crystalline microstructures than from a micellar solution of the drug (Mueller-Goymann, 2001). The high permeation rates were mainly attributed to the surfactant content and the alcohol present in these formulations.

1.4.11 Chemical permeation enhancers

Substances that are known to improve the diffusion of drugs through stratum corneum and epidermis are called penetration enhancers (PEs), accelerants, or sorption enhancers (Pfister et al., 1990). Penetration enhancers are known to improve drug transport across skin by reducing the resistance of the stratum corneum to drug permeation.

Ideally a PE should have the following properties (Williams and Barry, 2004):

- It must be pharmacologically inert, nontoxic, non-irritating and non-allergenic
- It must work rapidly and the duration of action should be predictable and reproducible
- It must permit the actives to penetrate into the body and at the same time prevent the loss of body fluids, electrolytes, and other endogenous materials from the body
- It should allow the skin to regain barrier properties on removal
- It should be compatible with the drug and excipients.

None of the PEs reported so far possesses all of the above-mentioned properties, though many enhancers do have most of the attributes. Due to the diverse physicochemical properties and mechanisms of action, it has been difficult to have a simple classification scheme for PEs. However, a classification system based on a conceptual diagrammatic approach has been proposed for the enhancers (Hori et al., 1990). The approach involves determining the organic and inorganic characteristics of the PEs, and the plot of the organic versus inorganic characteristics classifies them into three distinct areas on the diagram. Area I includes PEs that are solvents, area II contains PEs for hydrophilic compounds, while area III consists of PEs for lipophilic substances. PEs are also classified based on their polarity into either polar or nonpolar compounds (Pfister et al., 1990).

Lambert et al. (1993) classified PEs into three groups:

1. Solvents and hydrogen bond acceptors (e.g. dimethyl formamide, dimethyl sulfoxide, and dimethylacetamide)
2. Simple fatty acids and alcohols
3. Weak surfactants containing moderately sized polar compounds (e.g. azones).

1.4.11.1 Water as a drug permeability enhancer

Human stratum corneum contains water to an extent of 15–20% of the total tissue dry weight, which may vary depending on the environment to which it is exposed. The water content of the stratum corneum may attain equilibrium with the underlying tissues when the skin is soaked in water, exposed to higher humidity, or occluded, as in most clinical conditions (Williams and Barry, 2004). The water content of the stratum corneum may approach 400% of the tissue dry weight in most occluded conditions. Many topical and transdermal products such as ointments and patches are found to be occlusive, and the numerous patch formulations are known to deliver drugs at higher rates due to higher stratum corneum water content resulting from occlusion. Generally, increased tissue hydration is said to increase transdermal permeation of both hydrophilic and lipophilic drugs (Williams and Barry, 2004). Contrary to this, some reports have stated that occlusion may not enhance the percutaneous transport and transdermal delivery of polar permeants *in vivo* (Bucks and Maibach, 1999).

About 25–35% of the water present in the skin is in the bound form in association with structural elements within the tissue, while the remaining water is in the free form and acts as a solvent for the hydrophilic permeants (Walkley, 1972). Human stratum corneum contains a hygroscopic mixture of amino acids, amino acid derivatives, and salts known as natural moisturizing factor, which is known to maintain tissue flexibility (Williams and Barry, 2004). Moreover, the keratin-rich corneocytes containing the hydroxyl and carboxylic functional groups hold sufficient water molecules within the tissue.

The mechanism by which water enhances transdermal permeation is not clear. Water within the tissue would alter the solubility of the permeants in the stratum corneum and thereby modify their skin partitioning (Williams and Barry, 2004). Though the elevated fluxes of polar molecules under occlusive conditions could be explained by the above mechanism, it fails to explain hydration-induced enhancement of lipophilic permeants like steroids. Since the major barrier to transdermal permeation is the stratum corneum lipids, it has been thought that high water content due to occlusion would hydrate the polar head groups of the bilayer, causing them to swell and disrupt the lipid domains. However, no strong evidence exists to support this hypothesis.

Some authors report the presence of an aqueous pore pathway in the stratum corneum containing lacunar domains embedded within the bilayer lipids (Elias et al., 2002). Though they remain scattered and discontinuous under normal physiological conditions, the lacunae may expand, interconnect, and form a continuous pore pathway under high-stress conditions like extensive hydration, iontophoresis, or ultrasound. These pathways are expected to markedly enhance drug permeation through the stratum corneum.

1.4.11.2 Sulfoxides and similar solvents

Dimethyl sulfoxide (DMSO) is a powerful aprotic solvent that is colorless, odorless, and hygroscopic. Studies have demonstrated that DMSO is effective in promoting the permeation of a number of hydrophilic and lipophilic permeants. It has been found to enhance the permeation of beta blockers (Kai et al., 1993) and ephedrine hydrochloride (Singh et al., 1993). The enhancer effect is reported to be concentration dependent, with concentrations greater than 60% required to show optimum efficacy (Williams and Barry, 2004). However, erythema and wheals of the stratum corneum have been reported at these relatively high concentrations of DMSO, which have also resulted in protein denaturation (Anigbogu et al., 1995). Human volunteer studies have demonstrated erythema, scaling, contact urticaria, stinging and burning sensations, and some systemic symptoms (Kligman, 1965). The skin penetration mechanisms of the aprotic solvents, and DMSO in particular, are complex. Upon application to human skin, DMSO is known to denature proteins and change the intracellular keratin conformation from helical to β sheet (Oertel, 1997). DMSO is said to interact with the head group of the lipid bilayers to distort the packing geometry. Further partitioning of the drug from the formulation into DMSO present within the tissue is favored.

Another aprotic solvent, dimethylformamide (DMF), is known to enhance absorption through the polar pathway by increasing diffusion and partitioning (Sinha and Kaur, 2000). A 12-fold increase in the flux of caffeine across DMF-treated human skin was observed, though the authors reported irreversible skin damage (Southwell and Barry, 1983). DMF is reported to increase the permeation of ephedrine hydrochloride across rat skin and human epidermis (Singh et al., 1993), and also promote the *in vivo* bioavailability of betamethasone-17-benzoate (Barry et al., 1984; Bennett et al., 1984).

Other aprotic solvents employed as permeation enhancers include dimethylacetamide, dimethyloctanamide, and dimethyldecanamide. Dimethylacetamide was found to increase the permeation of indomethacin from ointments and creams in rats (El-Faham and Safwat, 1992). Dimethyloctanamide and dimethyldecanamide were found to enhance the permeation of ibuprofen and naproxen from aqueous propylene glycol solutions across rat skin (Irwin et al., 1990). Decylmethylsulfoxide, a structural analog of DMSO, is known to act reversibly on human skin and promote the permeation of hydrophilic permeants (Williams and Barry, 2004). It is reported to increase the flux of oxymorphone hydrochloride through guinea-pig skin and human skin (Aungst et al., 1990a) and 5-fluorouracil across human skin (Goodmann and Barry, 1988).

1.4.11.3 Azone

Azone (1-dodecylazacycloheptan-2-one) is a hybrid of two potent permeation enhancers, namely pyrrolidone and decylmethylsulfoxide, but lacks the aprotic sulfoxide group (Williams and Barry, 2004). Azone is a colorless, odorless liquid having a melting point of $-7°C$. It is a highly lipophilic substance, having a log *P* value of 6.2, and is miscible with most organic solvents. Azone has been found to be effective typically at concentrations between 0.1% and 5%, and exhibits low irritancy, very low toxicity, and little pharmacological activity.

Azone is reported to enhance the permeation of both hydrophilic and lipophilic molecules belonging to different classes (Williams and Barry, 2004). Azone in combination with propylene glycol enhanced the flux of methotrexate and edatrexate across hairless mouse skin in the presence of propylene glycol and isopropyl alcohol respectively (Chatterjee et al., 1997). Azone was found to be the most effective enhancer for low-molecular-weight heparin across human skin when compared with terpenes (Xiong et al., 1996). Though azone has no direct effect on the stratum corneum proteins, it is reported to increase the moisture content in the stratum corneum (Sugibayashi et al., 1992). Azone is likely to exert its effect by interaction with the lipid domains of the stratum corneum, where it may partition into the lipid bilayers and disrupt their packing arrangement (Williams and Barry, 2004). The azone molecules may remain dispersed within the barrier lipids or as separate domains within the bilayers. This mechanism is supported by the "soup spoon" model that has been proposed to explain the confirmation of azone within the stratum corneum (Hoogstraate et al., 1991). Moreover, electron diffraction studies have revealed that azone existed as a distinct phase within the lipids isolated from human stratum corneum (Pilgram et al., 2001).

Azone analogs with varying azacyclo ring were investigated for their penetration-enhancing effects on six drugs (5-fluorouracil, caffeine, salicyluric acid, salicylic acid, triamcinolone acetonide, and ibuprofen) that differed in their lipophilicities (log *P* ranged from -0.95 to 3.51) across hairless mouse skin. *N*-Dodecyl-2-pyrrolidone and *N*-dodecyl-2-piperdinone were found to be the most effective enhancers for hydrophilic drugs (Philips and Michniak, 1995).

1.4.11.4 Pyrrolidones

The two most widely studied penetration enhancers among pyrrolidones are *N*-methyl-2-pyrrolidone (NMP) and 2-pyrrolidone (2-P). The permeation of ibuprofen and flurbiprofen through cadaver skin was increased by 16-fold and threefold respectively from films containing NMP (Akhter and Barry, 1985). 2-P and NMP increased the bioavailability of topical model steroid betamethasone 17-benzoate from aqueous dimethylisosorbide used as a solvent (Bennett et al., 1984). The enhancement produced by structural modification of the pyrrolidone ring has also been investigated. 1-Propyl- and 1-butyl-3-dodecyl-2-pyrrolidone were found to be effective enhancers for indomethacin through skin from 60% ethanolic solution *in vitro* (Aoyagi et al., 1991).

However, clinical use of pyrrolidones has been restricted because of their adverse effects. Pyrrolidones have been reported to cause erythema in some volunteers, though the effect was found to be short-lived (Williams and Barry, 2004). A toxic hygroscopic contact reaction has been observed on exposure to *N*-methylpyrrolidone (Jungbauer et al., 2001).

1.4.11.5 Fatty acids

A number of long-chain fatty acids and their esters have been found to enhance the percutaneous absorption of drugs. Extensive literature exists that indicates that fatty acids have been successfully employed to enhance the transport of hydrophilic and lipophilic permeants (Sinha and Kaur, 2000; Williams and Barry, 2004).

Fatty acids with saturated alkyl chain lengths of around C_{10}–C_{12} attached to the polar head group and those with unsaturated alkyl chain lengths of C_{18} were found to be potent enhancers (Aungst et al., 1986; Aungst, 1989). A 6.5- to 17.5-fold increase in the permeation of flurbiprofen through rat skin has been reported with unsaturated fatty acids, whereas no significant enhancement was observed with saturated fatty acids (Chi et al., 1995). The flux of ketoprofen from carbopol patches across full-thickness human skin was found to be maximum when oleic acid was used as a penetration enhancer (Singh et al., 1996). *In vitro* permeation of flurbiprofen through rat skin has been enhanced by oleic acid in combination with urea (Chi et al., 1995). Oleic acid at concentrations of 5% in propylene glycol was found to enhance the permeation of estradiol by 10-fold, while it was moderately successful in increasing the permeation of 5-fluorouracil (Goodmann and Barry, 1988). Transdermal permeation of tetrapeptide melanotropin-hisetal across hairless mouse skin was increased by oleic acid by a factor of 28 (Ruland et al., 1994). By using a combination of oleic and propionic acids, transdermal delivery of physostigmine is reported to greatly increase in both *in vitro* (in isolated human skin) and *in vivo* (in guinea-pigs) studies (Jenner et al., 1995). A synergistic effect has been observed in the delivery of nicardipine hydrochloride through hairless guinea-pig skin when oleic acid was used with propylene glycol or dimethyl isosorbide was used with propylene glycol (Aboofazeli et al., 2002). *In vitro* permeation studies through hairless rat skin revealed that sodium oleate greatly increased the flux of indomethacin and urea (Ogiso et al., 1995).

Lauric acid enhanced the flux of metaproterenol (Elyan et al., 1996) and highly lipophilic antiestrogens (Funke et al., 2002). Lauric acid was also found to increase the *in vitro* and *in vivo* permeation of tegafur, alclofenac, and ibuprofen from an ethanol–water binary system (60:40) (Lee et al., 1993). *In vitro* studies have shown that capric acid, lauric acid, and neodecanoic acid increased the transport of naloxone, testosterone, indomethacin, methotrexate, and fluorouracil through human skin (Aungst et al., 1990b).

Most of the long-chain fatty acids are known to interact and modify the lipid domains of the stratum corneum (Williams and Barry, 2004). The fact that oleic acid exists as a separate phase within stratum corneum lipid bilayers has been indicated by spectroscopic (Ongipattanakul et al., 1991) and electron microscopic (Tanojo et al., 1997) studies. The presence of such domains in the stratum corneum would induce permeability defects in stratum corneum bilayer lipids, which in turn would facilitate the permeation of hydrophilic permeants (Williams and Barry, 2004).

1.4.11.6 Alcohols, fatty alcohols, and glycol

Ethanol is a commonly used penetration enhancer in many transdermal formulations and patch systems. It is reported to permeate through human skin with a steady state flux of approximately 1 $mg/cm^2/h$ (Berner et al., 1989). *In vitro* studies have demonstrated that ethanol increased the transdermal transport of levonorgesterol, estradiol, hydrocortisone, and 5-fluorouracil through rat skin (Friend et al., 1988). *In vivo* studies have confirmed that ethanol has been able to enhance estradiol permeation through human skin (Pershing et al., 1990). Ethanol increased the percutaneous permeation of ketoprofen lysine salt across rat skin from topical gel–spray formulation (Porzio et al., 1998). The permeability coefficient of methyl paraben has been increased by 16-fold by 1% menthol in combination with 15% ethanol (Kitagawa et al., 1997). An optimum enhancement of salicylate permeation was obtained with ethanol (63% v/v), while higher ethanol concentrations decreased the permeation (Kurihara-Bergstrom et al., 1990). The permeation of zalcitabine, didanosine, and zidovudine increased with ethanol concentration and reached a maximum at 50–60% of ethanol (Kim and Chien, 1996). Dehydration of the membranes at higher ethanol concentrations was probably the cause for the reduced skin permeation (Williams and Barry, 2004).

Various mechanisms have been proposed to explain the permeation-enhancing effect of ethanol. Primarily ethanol acts by increasing the solubility of the permeant in the vehicle. With poorly soluble drugs that tend to deplete in the donor phase during steady state, ethanol enhances the drug solubility in the donor phase (Pershing et al., 1990). Further, the permeation of ethanol into the stratum corneum alters the solubility properties of the tissue so as to favor drug partitioning into the membrane (Megrab et al., 1995b). In addition, it is also likely that the rapid permeation of ethanol or evaporative solvent loss from the donor phase alters the thermodynamic activity of the drug in the formulation (Williams and Barry, 2004). Moreover, when ethanol permeates into the skin, the solvent drag may carry the permeant into the tissue (Morimoto et al., 2002). Ethanol may also extract some of the stratum corneum lipids when used at high concentrations for prolonged periods, which in turn would enhance the drug flux (Williams and Barry 2004).

Fatty alcohols (n-alkanols) have been investigated for their penetration-enhancing activity. The flux of nicotinamide across hairless mouse skin increased with increase in the number of carbon chain length of the fatty alcohols (Kai et al., 1990). The absorption of propranolol hydrochloride through skin was increased by a factor of 8.2 in the presence of 1-nonanol (Hori et al., 1991). However, the enhancement was saturable, indicating a maximally perturbed stratum corneum. Octyl alcohol was found to increase the permeation of urea (Ogiso et al., 1995), whereas lauryl achohol was successful in increasing the flux of acetaminophen, ibuprofen (Bhatt et al., 1991), and 5-fluorouracil (Turunen et al., 1993). Lauryl alcohol was found to be an efficient permeation enhancer for more hydrophilic and large molecular size permeants (Itoh et al., 1992). A structure–activity relationship has recently been proposed for fatty alcohols using melatonin as a permeant across porcine and human skin *in vitro.* A parabolic relationship was observed with a maximum enhancement for decanol (Andega et al., 2001). The enhancement activity was found to increase when two unsaturated bonds were introduced into the alcohols, but dropped on introducing three double bonds.

Pretreatment with propylene glycol promoted the flux of heparin sodium through excised human skin *in vitro* (Bonina and Montenegro, 1992). Saturated solutions of terpenes in a propylene glycol–water cosolvent system enhanced the transport of 5-fluorouracil (Yamane et al., 1995). The terpene activity has been shown to be dependent on the propylene glycol content and maximum flux was observed with formulations containing 80%

propylene glycol (Sinha and Kaur, 2000). In combination with azone, propylene glycol increased the flux of methotrexate (Chatterjee et al., 1997), cyclosporine A (Duncan et al., 1990), and 5-fluorouracil (Goodmann and Barry, 1988). The flux of estradiol was increased by 10-fold when propylene glycol was used in conjunction with 5% oleic acid (Goodmann and Barry, 1988). Urea analogs were found to be effective penetration enhancers when propylene glycol was used as a vehicle (Williams and Barry, 1989a).

The mechanisms of action of propylene glycol are known to be similar to that of ethanol. Propylene glycol is said to permeate the stratum corneum, altering the thermodynamic activity of the drug that in turn may favor drug diffusion. Propylene glycol is suspected to partition into the tissue, favoring drug uptake by the skin, and it may also disturb intercellular lipid packing within the stratum corneum bilayers (Williams and Barry, 2004).

1.4.11.7 Surfactants

Some surfactants are known to enhance the flux of materials through biological membranes. Surfactants are made up of a lipophilic hydrocarbon chain and a hydrophilic polar head group. They are categorized into anionic, cationic, non-ionic, and zwitterionic surfactants. Anionic and cationic surfactants have the potential to damage human skin, while non-ionic surfactants are considered to be safe (Williams and Barry, 2004). Significant enhancement in the permeation of chromaphenicol across hairless mouse skin was observed with sodium lauryl sulphate (Sarpotdar and Zatz, 1986a). Tween surfactants were found to enhance the permeation of lidocaine through hairless mouse skin in the presence of various concentrations of propylene glycol (Sarpotdar and Zatz, 1986b). A sixfold increase in the enhancement of 5-fluorouracil permeation was observed across hairless mouse skin using Tween 20 in normal saline (Rigg and Barry, 1990). Surfactants are included in the formulations to solubilize the lipophilic actives and hence they are expected to act as enhancers by solubilizing stratum corneum lipids (Williams and Barry, 2004).

1.4.11.8 Urea

Urea finds its application as a hydrating agent in the treatment of scaling conditions like psoriasis, ichthyosis, and other hyper keratotic skin conditions (Williams and Barry, 2004). Human volunteer studies demonstrate significant stratum corneum hydration and improved barrier properties on application of urea alone or in combination with ammonium lactate (Gloor et al., 2001). The penetration enhancement effect of urea was related to increased stratum corneum hydration and its keratolytic activity. Cyclic urea analogs were found to be as effective as azone in enhancing the permeation of indomethacin through shed snake skin and hairless mouse skin (Wong et al., 1989). The alkyl and aryl analogs were moderately effective in promoting the permeation of 5-fluorouracil across human skin *in vitro* when applied in propylene glycol (Williams and Barry, 1989a).

1.4.11.9 Terpenes, terpenoids, and essential oils

Terpenes are important constituents of essential oils, having a chemical structure consisting of repeated isoprene (C_5H_8) units. They are classified into monoterpenes (C_{10}), sesquiterpenes (C_{15}), and diterpenes (C_{20}) based on whether they are made up of two, three, or four isoprene units respectively. Based on the number of cyclic rings in the structure, terpenes can also be classified as acyclic, monocyclic, and bicyclic.

The essential oils of eucalyptus, chenopodium, and ylang ylang effectively enhanced the penetration of 5-fluorouracil, with eucalyptus oil increasing the permeability coefficient by 34-fold (Williams and Barry, 1989b). Pretreatment of human epidermis with the terpene 1,8-cineole was found to increase the permeability coefficient of 5-fluorouracil by around 100-fold (Williams and Barry, 1991). In contrast, a moderate enhancement of 10- to 40-fold was observed with the model lipophilic drug estradiol across human skin with the same terpene (Williams and Barry, 1991).

L-Menthol is known to increase the skin permeation of testosterone by forming a eutectic mixture with it (Kaplun-Frischoff and Touitou, 1997). Formation of eutectic mixture lowered the melting point (from 153.7 to 39.9°C) which increased the drug solubility and enhanced the skin permeation. The permeation of the hydrophobic drug diclofenac and its hydrophilic sodium salt was studied through ethanol-treated and untreated silicone membranes using L-menthol as a permeation enhancer (Maitani et al., 1996). The results showed that L-menthol promoted the permeation of the hydrophobic drug and its hydrophilic salt by lipid and pore pathways respectively. The addition of L-menthol to 40% ethanol was found to improve transport across the stratum corneum by increasing the diffusion coefficients of morphine hydrochloride, atenolol, nifedipine, and vinpocetine (Kobayashi et al., 1994).

Pretreatment of human skin samples with α-bisabolol increased the permeation of 5-fluorouracil by 17-fold and triamcinolone by 73-fold compared to untreated skin (Kadir and Barry, 1991). Smaller terpenes are known to provide better penetration enhancement than the larger sesquiterpenes (Williams and Barry, 2004). Terpenes with a nonpolar group, like limonene, were found to be better enhancers for lipophilic substances, while terpenes containing polar groups provide better enhancement for hydrophilic permeants (Williams and Barry, 2004).

One proposed mechanism by which the terpenes are known to act is by modifying the solvent nature of stratum corneum, thereby improving drug partitioning into the tissue. In addition, terpenes are said to modify drug diffusivity through the membrane, thereby reducing the lag time for permeation. X-ray diffraction studies have revealed that D-limonene and 1,8-cineole disrupt stratum corneum lipids while nerolidol, a long-chain sesquiterpene, reinforces the bilayers, probably by orienting along the stratum corneum lipid bilayers (Cornwell et al., 1996).

1.5 CONCLUSIONS

The human stratum corneum acts as a barrier for the permeation of a number of active substances. This has limited the number of molecules commercially available as transcutaneous delivery systems. Various strategies have been employed over the last few decades to optimize drug delivery across the skin. Approaches based on supersaturation, eutectic mixtures, prodrugs, and ion pairing have been used with success to improve the transdermal delivery of several poorly permeable compounds. Liposomes have been used as carriers of steroids, antifungals, and vaccines for cutaneous delivery. A substantial database has been generated on chemical penetration enhancers over the last few years and the search for the ideal enhancers has been the prime focus of many research groups. The efficacy of a passive approach to enhance transdermal drug delivery is known to diminish with increasing size of the active substance. Therefore, the limited potential of these passive techniques to facilitate the delivery of macromolecules has given way to various device-based physical enhancement techniques such as iontophoresis, electoporation, sonophoresis, and microneedles. Iontophoresis is a technique that involves application of low-intensity electric current to skin to enhance the permeation of therapeutic agents. Electroporation uses high-voltage pulses over short time durations to facilitate percutaneous delivery of active agents. The use of ultrasound or ultrasonic energy to increase the transport of solutes across skin is referred to as sonophoresis. Microneedle-based devices consisting of a drug reservoir with plurality of projections have the ability to penetrate the stratum corneum to deliver the drug systemically. In spite of the emergence of several aggressive active enhancement methods, the demand for passive transdermal technologies to deliver smaller molecules continues to prevail. The prime reason for this is their safety and efficacy during long-term treatment, resulting in improved patient acceptance and compliance compared to the physical enhancement techniques.

REFERENCES

Aboofazeli, R., Zia, H., Needham, T.E., 2002. Transdermal delivery of nicardipine: an approach to *in vitro* permeation enhancement. Drug Deliv. 9, 239–247.

Agrawal, G.P., Juneja, M., Agrawal, S., Jain, S.K., Pancholi, S.S., 2004. Preparation and characterization of reverse micelle based organogels of piroxicam. Pharmazie. 59, 191–193.

Ainbinder, D., Touitou, E., 2005. Testosterone ethosomes for enhanced transdermal delivery. Drug Deliv. 12, 297–303.

Akhter, S.A., Barry, B.W., 1985. Absorption through human skin of ibuprofen and flurbiprofen; effects of dose variation, deposited drug films, occlusion and penetration enhancer N-methyl pyrrolidone. J. Pharm. Pharmacol. 37, 27–37.

Andega, S., Kanikkannan, N., Singh, M., 2001. Comparison of the effect of fatty alcohols on the permeation of melatonin between porcine and human skin. J. Control. Release 77, 17–25.

Anigbogu, A.N.C., Williams, A.C., Barry, B.W., Edwards, H.G.M., 1995. Fourier transform Raman spectroscopy of interaction between the penetration enhancer dimethyl sulphoxide and human stratum corneum. Int. J. Pharm. 125, 265–282.

Aoyagi, T., Yamamura, M., Suzuki, N., Matsui, K., Nagase, Y., 1991. Preparation of alkyl-substituted pyrrolidone derivatives and their evaluation as transdermal penetration enhancers. Drug Des. Discov. 8, 37–46.

Aungst, B.J., 1989. Structure–effects studies of fatty acid isomers as skin penetration enhancers and skin irritants. Pharm. Res. 6, 244–247.

Aungst, B.J., Rogers, N.J., Shefter, E., 1986. Enhancement of naloxone penetration through human skin *in vitro* using fatty acids, fatty alcohols, surfactants, sulfoxides and amides. Int. J. Pharm. 33, 225–234.

Aungst, B.J., Blake, J.A., Rogers, N.J., Hussain, M.A., 1990a. Transdermal oxymorphone formulation development and methods for evaluating flux and lag times for two skin permeation enhancing vehicles. J. Pharm. Sci. 79, 1072–1076.

Aungst, B.J., Blake, J.A., Hussain, M.A., 1990b. Contributions of drug solubilization, partitioning, barrier disruption and solvent permeation to the enhancement of skin permeation of various compounds with fatty acids and amines. Pharm. Res. 7, 712–718.

Barry, B.W., 1983. Dermatological Formulations: Percutaneous Absorption, vol 18. Marcel Dekker, New York.

Barry, B.W., 2001. Novel mechanisms and devices to enable successful transdermal drug delivery. Eur. J. Pharm. Sci. 14, 101–114.

Barry, B.W., Southwell, D., Woodford, R., 1984. Optimisation of bioavailability of topical steroids: penetration enhancers under occlusion. J. Invest. Dermatol. 82, 49–52.

Beall, H.D., Sloan, K.B., 1996. Transdermal delivery of 5-fluorouracil (5-FU) by 1-alkylcarbonyl-5 FU prodrugs. Int. J. Pharm. 129, 203–210.

Beall, H.D., Prankerd, R., Sloan, K., 1994. Transdermal delivery of 5-fluorouracil (5-FU) through hairless mouse skin by 1-alkyloxycarbonyl-5-FU prodrugs: physicochemical characterization of prodrugs and correlation with transdermal delivery. Int. J. Pharm. 111, 223–233.

Bennett, S.L., Barry, B.W., Woodford, R., 1984. Optimisation of bioavailability of topical steroids: non-occluded penetration enhancers under thermodynamic control. J. Pharm. Pharmacol. 37, 298–304.

Benson, H.A.E., 2005. Transdermal drug delivery: penetration enhancement techniques. Current Drug Deliv. 2, 23–33.

Bentley, M.V., Vianna, R.F., Wilson, S., Collett, J.H., 1997. Characterization of the influence of some cyclodextrins on the stratum corneum from the hairless mouse. J. Pharm. Pharmacol. 49, 397–402.

Berner, B., Mazzenga, G.C., Otte, J.H., Steffens, R.J., Juang, R.-H., Ebert, C.D., 1989. Ethanol:water mutually enhanced transdermal therapeutic system: II. Skin permeation of ethanol and nitroglycerin. J. Pharm. Sci. 78, 402–407.

Bhatnagar, S., Vyas, S.P., 1994. Organogel-based system for transdermal delivery of propranolol. J. Microencapsul. 11, 431–438.

Bhatt, P.P., Rytting, J.H., Topp, E.M., 1991. Influence of Azone® lauryl alcohol on the transport of acetaminophen and ibuprofen through shed snake skin. Int. J. Pharm. 72, 219–226.

Bonina, F.P., Montenegro, L., 1992. Permeation enhancer effects on *in vitro* absorption of heparin sodium salt. Int. J. Pharm. 82, 171–177.

Bouwstra, J.A., Honeywell-Nguyen, P.L., 2002. Skin structure and mode of action of vesicles. Adv. Drug Deliv. Rev. 54, S41–S55.

Brown, M.B., Martin, G.P., Jones, S.A., Akomeah, F.K., 2006. Dermal and transdermal drug delivery systems: current and future prospects. Drug Deliv. 13, 175–187.

Bucks, D., Maibach, H.I., 1999. Occlusion does not uniformly enhance penetration *in vivo.* In: Bronaugh, R.L., Maibach, H.I. (Eds), Percutaneous Absorption; Drugs, Cosmetics, Mechanisms, Methodology, 3rd ed. Marcel Dekker, New York, Chapter 4, pp. 81–106.

Buhse, L., Kolinski, R., Westenberger, B., Wokovich, A., Spencer, J., Chen, C.W., Turujman, S., Mamta, G.-B., Kang, G.J., Kibbe, A., Heintzelman, W.E., 2005. Topical drug classification. Int. J. Pharm. 295, 101–112.

Calpena, A.C., Blanes, C., Moreno, J., Obach, R., Domenech, J., 1994. A comparative *in vitro* study of transdermal absorption of anti-emetics. J. Pharm. Sci. 83, 29–33.

Cevc, G., 1996. Transfersomes, liposomes and other lipid suspensions on the skin: permeation enhancement, vesicle penetration, and transdermal drug delivery. Crit. Rev. Ther. Drug Carrier Syst. 13, 257–388.

Cevc, G., Blume, G., 1992. Lipid vesicles penetrate into intact skin owing to the transdermal osmotic gradients and hydration force. Biochim. Biophys. Acta. 1104, 226–232.

Cevc, G., Gebauer, D., 2003. Hydration-driven transport of deformable lipid vesicles through fine pores and skin barrier. Biophys. J. 84, 1010–1024.

Cevc, G., Blume, G., Schatzlein, A., 1995. Transdermal drug carriers: basic properties, optimization and transfer efficiency in the case of epicutaneously applied peptides. J. Control. Release 36, 3–16.

Cevc, G., Blume, G., Schatzlein, A., Gebauer, D., Paul, A., 1996. The skin: a pathway for systemic treatment with patches and lipid-based agent carriers. Adv. Drug Deliv. Rev. 18, 349–378.

Cevc, G., Gebauer, D., Stieber, J., Schatzlein, A., Blume, G., 1998. Ultraflexible vesicles, Transfersomes, have an extremely low pore penetration resistance and transport therapeutic amounts of insulin across the intact mammalian skin. Biochim. Biophys. Acta 1368, 201–215.

Cevc, G., Schätzlein, A., Richardsen, H., 2002. Ultradeformable lipid vesicles can penetrate skin and other semi-permeable membrane barriers unfragmented. Evidence from double label CLSM experiments and direct size measurement. Biochim. Biophys. Acta 1564, 21–30.

Chatterjee, D.J., Li, W.Y., Koda, R.T., 1997. Effect of vehicles and penetration enhancers on the *in vitro* and *in vivo* percutaneous absorption of methotrexate and edatrexate through hairless mouse. Pharm. Res. 14, 1058–1065.

Chen, H., Chang, X., Weng, T., Zhao, X., Gao, Z., Yang, Y., Xu, H., Yang, X., 2004. A study of microemulsion system for transdermal delivery of triptolide. J. Control. Release. 98, 427–436.

Chi, S.-C., Park, E.-S., Kim, H., 1995. Effect of permeation enhancers on flurbiprofen permeation through rat skin. Int. J. Pharm. 126, 267–274.

Chiang, C.M., Flynn, G.L., Weiner, N.D., Szpunar, G.J., 1989. Bioavailability assessment of topical delivery systems; effect of vehicle evaporation upon *in vitro* delivery of minoxidil from solution formulations. Int. J. Pharm. 55, 229–236.

Choi, M.J., Maibach, H.I., 2005. Liposomes and niosomes as topical drug delivery systems. Skin Pharmacol. Physiol. 18, 209–219.

Coldman, M.F., Poulsen, B.J., Higuchi, T., 1969. Enhancement of percutaneous absorption by the use of volatile/nonvolatile systems as vehicles. J. Pharm. Sci. 58, 1098–1102.
Cornwell, P.A., Barry, B.W., Bouwstra, J.A., Gooris, G.S., 1996. Modes of action of terpene penetration enhancers in human skin; differential scanning calorimetry, small angle X-ray diffraction and enhancer uptake studies. Int. J. Pharm. 127, 9–26.
Davis, A.F., Hadgraft, J., 1991. Effect of supersaturation on membrane transport: 1. Hydrocortisone acetate. Int. J. Pharm. 76, 1–8.
Dayan, N., Touitou, E., 2000. Carriers for skin delivery of trihexyphenidyl HCl: ethosomes vs. liposomes. Biomaterials 21, 1879–1885.
Dreher, F., Walde, P., Walter, P., Wehrli, E., 1997. Interaction of a lecithin microemulsion gel with human stratum corneum and its effect on transdermal transport. J. Control. Release 45, 131–140.
Dubey, V., Mishra, D., Asthana, A., Jain, N.K., 2006. Transdermal delivery of a pineal hormone: melatonin via elastic liposomes. Biomaterials 27, 3491–3496.
Duncan, J.I., Payne, S.N., Winfield, A.J., Ormerod, A.D., Thomson, A.W., 1990. Enhanced percutaneous absorption of a novel topical cyclosporine-A formulation and assessment of its immunosuppressive activity. Br. J. Dermatol. 123, 631–640.
Egbaria, K., Ramachandran, C., Kittayanond, D., Weiner, N., 1990. Topical delivery of liposomally encapsulated interferon evaluated by *in vitro* diffusion studies. Antimicrob. Agents Chemother. 34, 107–110.
Egbaria, K., Ramachandran, C., Kittayanond, D., Weiner, N., 1991. Topical application of liposomally entrapped cyclosporine evaluated by *in vitro* diffusion studies with human skin. Skin Pharmacol. 4, 21–28.
El-Faham, T.H., Safwat, S.M., 1992. Comparative study of release and anti-inflammatory activity of indomethacin from topical formulations. Pharmazeutische Industrie 54, 82–86.
Elias, P.M., 1983. Epidermal lipids, barrier function and desquamation. J. Invest. Dermatol. 80, 44–49.
Elias, P.M., Feingold, K.R., 1988. Lipid based barriers and gradients in the epidermis. Ann. New York Acad. Sci. 548, 4–13.
Elias, P.M., Friend, D.S., 1975. The permeability barrier in mammalian epidermis. J. Cell Biol. 65, 180–191.
Elias, P.M., Williams, M.L., Maloney, M.E., Bonefas, J.A., Brown, B.E., Grayson, S., Epstein Jr., E.H., 1984. Stratum corneum lipids in disorders of cornification: steroidal sulphatase and cholesterol sulphate in normal desquamation and the pathogenesis of recessive X-linked ichthyosis. J. Clin. Invest. 74, 1414–1421.
Elias, P.M., Tsai, J., Menon, G.K., Holleran, W.M., Feingold, K.R., 2002. The potential of metabolic interventions to enhance transdermal drug delivery. JID Symp. Proc. 7, 79–85.
El Maghraby, G.M., Williams, A.C., Barry, B.W., 1999. Skin delivery of oestradiol from deformable and traditional liposomes: mechanistic studies. J. Pharm. Pharmacol. 51, 1123–1134.
El Maghraby, G.M., Williams, A.C., Barry, B.W., 2000a. Oestradiol skin delivery from ultradeformable liposomes: refinement of surfactant concentration. Int. J. Pharm. 196, 63–74.
El Maghraby, G.M., Williams, A.C., Barry, B.W., 2000b. Skin delivery of oestradiol from lipid vesicles: importance of liposome structure. Int. J. Pharm. 204, 159–169.
El Maghraby, G.M., Williams, A.C., Barry, B.W., 2001. Skin hydration and possible shunt route penetration in controlled skin delivery of estradiol from ultradeformable and standard liposomes *in vitro*. J. Pharm. Pharmacol. 53, 1311–1322.
El Maghraby, G.M., Williams, A.C., Barry, B.W., 2006. Can drug-bearing liposomes penetrate intact skin? J. Pharm. Pharmacol. 58, 415–429.
El Maghraby, G.M., Barry, B.W., Williams, A.C., 2008. Liposomes and skin: from drug delivery to model membranes. Eur. J. Pharm. Sci. 34, 203–222.
Elsayed, M.M.A., Abdallah, O.Y., Naggar, V.F., Khalafallah, N.M., 2007a. Lipid vesicles for skin delivery: reviewing three decades of research. Int. J. Pharm. 332, 1–16.
Elsayed, M.M.A., Abdallah, O.Y., Naggar, V.F., Khalafallah, N.M., 2007b. Deformable liposomes and ethosomes as carriers for skin delivery of ketotifen. Pharmazie 62, 133–137.

Elyan, B.M., Sidhom, M.B., Plakogiannis, F.M., 1996. Evaluation of effect of different fatty acids on the percutaneous absorption of metaproterenol sulphate. J. Pharm. Sci. 85, 101–105.
Escribano, E., Calpena, A.C., Queralt, J., Obach, R., Doménech, J., 2003. Assessment of diclofenac permeation with different formulations: anti-inflammatory study of a selected formula. Eur. J. Pharm. Sci. 19, 203–210.
Fresta, M., Puglisi, G., 1996. Application of liposomes as potential cutaneous drug delivery systems. *In vitro* and *in vivo* investigation with radioactively labelled vesicles. J. Drug Target. 4, 95–101.
Fresta, M., Puglisi, G., 1997. Coticosteroid dermal delivery with skin-lipid liposomes. J. Control. Release 44, 141–151.
Friend, D., Cartz, P., Heller, J., Reid, J., Baker, R., 1988. Transdermal delivery of levonorgestrel. II. Effect of prodrugs structure on skin permeability *in vitro*. J. Control. Release 7, 251–261.
Funke, A.P., Schiller, R., Motzkus, H.W., Gunther, C., Muller, R.H., Lipp, R., 2002. Transdermal delivery of highly lipophilic drugs: *in vitro* fluxes of antiestrogens, permeation enhancers and solvents from liquid formulations. Pharm. Res. 19, 661–668.
Ganesan, M.G., Weiner, N.D., Flynn, G.L., Ho, N.F.H., 1984. Influence of liposomal drug entrapment on percutaneous absorption. Int. J. Pharm. 20, 139–154.
Garg, M., Mishra, D., Agashe, H., Jain, N.K., 2006. Ethinylestradiol-loaded ultraflexible liposomes: pharmacokinetics and pharmacodynamics. J. Pharm. Pharmacol. 58, 459–468.
Gloor, M., Fluhr, J., Wasik, B., Gehring, W., 2001. Clinical effect of salicylic acid and high dose urea applied in standardized NRF formulations. Pharmazie 56, 810–814.
Goodmann, M., Barry, B.W., 1988. Action of penetration enhancers on human skin as assessed by the permeation of model drugs 5-fluorouracil and estradiol. I. Infinite dose technique. J. Invest. Dermatol. 91, 323–327.
Gregoriadis, G., Florence, A.T., 1993. Liposomes in drug delivery: clinical, diagnostic and ophthalmic potential. Drugs 45, 15–28.
Guo, J., Ping, Q., Sun, G., Jiao, C., 2000. Lecithin vesicular carriers for transdermal delivery of cyclosporin A. Int. J. Pharm. 194, 201–207.
Gupta, P., Garg, S., 2002. Recent advances in semisolid dosage forms for dermatological application. Pharm. Tech. 26, 144–162.
Gupta, P.N., Mishra, V., Rawat, A., Dubey, P., Mahor, S., Jain, S., Chatterji, D.P., Vyas, S.P., 2005a. Non-invasive vaccine delivery in transfersomes, niosomes and liposomes: a comparative study. Int. J. Pharm. 293, 73–82.
Gupta, P.N., Mishra, V., Singh, P., Rawat, A., Dubey, P., Mahor, S., Vyas, S.P., 2005b. Tetanus toxoid-loaded transfersomes for topical immunization. J. Pharm. Pharmacol. 57, 295–301.
Hadgraft, J., 1999. Passive enhancement strategies in topical and transdermal drug delivery. Int. J. Pharm. 184, 1–6.
Hadgraft, J., 2004. Skin deep. Eur. J. Pharm. Biopharm. 58, 291–299.
Higuchi, T., 1960. Physical chemical analysis of percutaneous absorption process. J. Soc. Cosmetic Chemists 11, 85–97.
Honeywell-Nguyen, P.L., Bouwstra, J.A., 2005. Vesicles as a tool for transdermal and dermal delivery. Drug Discov. Today: Technol. 2, 67–74.
Hoogstraate, A.J., Verhoef, J., Brussee, J., Ijzerman, A.P., Spies, F., Bodde, H.E., 1991. Kinetics, ultrastructural aspects and molecular modelling of transdermal peptide flux enhancement by *N*-alkylazacycloheptanones. Int. J. Pharm. 76, 37–47.
Hori, M., Satoh, S., Maibach, H.I., 1990. Classification of penetration enhancers: a conceptual diagram. J. Pharm. Pharmacol. 42, 71–72.
Hori, M., Satoh, S., Maibach, H.I., Guy, R.H., 1991. Enhancement of propranolol hydrochloride and diazepam skin absorption *in vitro*: effect of enhancer lipophilicity. J. Pharm. Sci. 80, 32–35.
Iervolino, M., Raghavan, S.L., Hadgraft, J., 2000. Membrane penetration enhancement of ibuprofen by super saturation. Int. J. Pharm. 198, 229–238.

Irwin, W.J., Sanderson, F.D., WanPo, A.L., 1990. Percutaneous absorption of ibuprofen and naproxen: effect of amide enhancer on transport through rat skin. Int. J. Pharm. 66, 243–252.

Itoh, T., Wasinger, L., Turunen, T.M., Rytting, J.H., 1992. Effects of transdermal permeation enhancers on the permeability of shed snake skin. Pharm. Res. 9, 1168–1172.

Jansen, C.J., Hopsu-Havu, V.K., 1969. Proteolytic enzymes in the skin. 3. Studies on the extractability, stability and modifier characteristics of the caseinolytic enzymes in the rat skin. Acta Derm-Venereol. 49, 525–535.

Jenner, J., Saleem, A., Swanston, D., 1995. Transdermal delivery of physostigmine: a pretreatment against organophosphate poisoning. J. Pharm. Pharmacol. 47, 206–212.

Jenning, V., Gysler, A., Schafer-Korting, M., Gohla, S.H., 2000a. Vitamin A loaded solid lipid nanoparticles for topical use: occlusive properties and drug targeting to the upper skin. Eur. J. Pharm. Biopharm. 49, 211–218.

Jenning, V., Schäfer-Korting, M., Gohla, S., 2000b. Vitamin A loaded solid lipid nanoparticles for topical use: drug release properties. J. Control. Release 66, 115–126.

Jona, J.A., Dittert, L.W., Crooks, P.A., Milosovich, S.M., Hussain, A.A., 1995. Design of novel prodrugs for enhancement of transdermal permeation of indomethacin. Int. J. Pharm. 123, 127–136.

Jungbauer, F.H.W., Coenraads, P.J., Kardaun, S.H., 2001. Toxic hygroscopic contact reaction to *N*-methyl-2-pyrrolidone. Contact Dermatitis 45, 303–304.

Kadir, R., Barry, B.W., 1991. α-Bisabolol, a possible safe penetration enhancer for dermal and transdermal therapeutics. Int. J. Pharm. 70, 87–94.

Kai, T., Mak, V.H.W., Potts, P.O., Guy, R.H., 1990. Mechnasim of percutaneous penetration enhancement: effect of n-alkanols on the permeability barrier of hairless mouse skin. J. Control. Release 12, 103–112.

Kai, T., Nakazona, M., Kurosaki, Y., Nakayama, T., Kimura, T., 1993. Keratinized epithelial transport of beta-blocking agents. III. Evaluation of enhancing effect on percutaneous absorption using model lipid liposomes. Biol. Pharm. Bull. 16, 801–805.

Kalariya, M., Padhi, B.K., Chougule, M., Misra, A., 2005. Clobetasol propionate solid lipid nanoparticles cream for effective treatment of eczema: formulation and clinical implications. Indian J. Exp. Biol. 43, 233–240.

Kang, L.S., Jun, H.W., McCall, J.W., 2000. Physicochemical studies of lidocaine–menthol binary systems for enhanced membrane transport. Int. J. Pharm. 206, 35–42.

Kaplun-Frischoff, Y., Touitou, E., 1997. Testosterone skin permeation enhancement by menthol through formation of eutectic with drug and interaction with skin lipids. J. Pharm. Sci. 86, 1394–1399.

Katahira, N., Murakami, T., Kugai, S., Yata, N., Takano, M., 1999. Enhancement of topical delivery of a lipophilic drug from charged multilamellar liposomes. J. Drug Target. 6, 405–414.

Kemken, J., Ziegler, A., Muller, B.W., 1992. Influence of supersaturation on the thermodynamic effect of bupranolol after dermal administration using microemulsions as vehicle. Pharm. Res. 9, 554–558.

Kim, D.D., Chien, Y.W., 1996. Transdermal delivery of dideoxynucleoside-type anti HIV drugs. 2. The effect of vehicle and enhancer on skin permeation. J. Pharm. Sci. 85, 214–219.

Kirjavainen, M., Urtti, A., Jaaskelainen, I., Suhonen, T.M., Paronen, P., Koskela, R.V., Kiesvaara, J., Monkkonen, J., 1996. Interaction of liposomes with human skin *in vitro* – the influence of lipid composition and structure. Biochim. Biophys. Acta 1304, 179–189.

Kitagawa, S., Li, H., Sato, S., 1997. Skin permeation of parabens in excised guinea pig dorsal skin, its modification by penetration enhancers and their relationship with n-octanol/water coeffecients. Chem. Pharm. Bull. 45, 1354–1357.

Kligman, A.M., 1965. Topical pharmacology and toxicology of dimethylsulfoxide. J. Am. Med. Assoc. 193, 796–804.

Kobayashi, D., Matsuzawa, T., Sugibayashi, K., Morimoto, Y., Kimura, 1994. Analysis of the combined effects of 1-menthol and ethanol as skin penetration enhancers based on two layer skin model. Pharm. Res. 11, 96–103.

Kogan, A., Garti, N., 2006. Microemulsions as transdermal drug delivery vehicles. Adv. Colloid Interface Sci. 123–126, 369–385.
Kondo and Sugimoto, 1987. Enhancement of transdermal delivery by superfluous thermodynamic potential. I. Thermodynamic analysis of nifedipine transport across the lipoidal barrier. J. Pharmacobio-Dyn, 10, 587–594.
Kondo, S., Yamasaki-Konishi, H., Sugimoto, I., 1987a. Enhancement of transdermal delivery by superfluous thermodynamic potential. II. In vitro in vivo correlation of percutaneous nifedipine transport. J. Pharmacobio-Dyn, 10, 662–680.
Kondo, S., Yamanaka, C., Sugimoto, I., 1987b. Enhancement of transdermal delivery by superfluous thermodynamic potential. III. Percutaneous absorption of nifedipine in rats. J. Pharmacobio-Dyn, 10, 743–749.
Kreilgaard, M., Pedersen, E.J., Jaroszewski, J.W., 2000. NMR characterization and transdermal delivery potential of microemulsion systems. J. Control. Release 69, 421–433.
Kurihara-Bergstrom, T., Knutson, K., De Noble, L.J., Goates, C.Y., 1990. Percutaneous absorption enhancement of an ionic molecule by ethanol–water systems in human skin. Pharm. Res. 7, 762–766.
Lambert, W.J., Kudlar, R.J., Hollard, J., Curry, J.T., 1993. A biodegradable transdermal penetration enhancer based on N-(2-hydroxyethyl)-2-pyrrolidone. I. Synthesis and characterization. Int. J. Pharm. 45, 181–192.
Landmann, L., 1986. Epidermal permeability barrier transformation of lamellar granule-disks into intercellular sheets by a membrane-fusion process, a freeze fracture study. J. Invest. Dermatol. 87, 202–209.
Langer, R., 2004. Transdermal drug delivery: past present, current status and future prospects. Adv. Drug Deliv. Rev. 56, 557–558.
Lee, C.K., Uchida, T., Kitagawa, K., Yagi, A., Kim, N.S., Gota, S., 1993. Effect of hydrophilic and lipophilic vehicles on skin permeation of tegafur, alclofenac and ibuprofen with or without permeation enhancers. Biol. Pharm. Bull. 16, 1264–1269.
Legendre, J.Y., Rault, I., Petit, A., Luijten, W., Demuynck, I., Horvath, S., Ginot, Y.M., Cuine, A., 1995. Effects of β-cyclodextrins on skin: implications for transdermal delivery of piribedil and a novel cognition enhancing drug, S-9977. Eur. J. Pharm. Sci. 6, 311–322.
Liu, H., Pan, W.S., Tang, R., Luo, S.D., 2004. Topical delivery of different acyclovir palmitate liposome formulations through rat skin *in vitro*. Pharmazie 59, 203–206.
Lodzki, M., Godin, B., Rakou, R., Mechoulam, R., Gallily, R., Touitou, E., 2003. Cannabidiol-transdermal delivery anti-inflammatory effect in a murine model. J. Control. Release 93, 377–387.
Loftsson, T., Brewster, M.E., 1996. Pharmaceutical applications of cyclodextrins. 1. Drug solubilization and stabilization. J. Pharm. Sci. 85, 1017–1025.
Loftsson, T., Masson, M., 2001. Cyclodextrins in topical drug formulations: theory and practice. Int. J. Pharm. 225, 15–30.
Loftsson, T., Vogensen, S.B., Brewster, M.E., Konradsdottir, F., 2007. Effect of cyclodextrins on drug delivery through biological membranes. J. Pharm. Sci. 96, 2532–2546.
Lombardi Borgia, S., Regehly, M., Sivaramakrishnan, R., Mehnert, W., Korting, H.C., Danker, K., Röder, B., Kramer, K.D., Schäfer-Korting, M., 2005. Lipid nanoparticles for skin penetration enhancement – correlation to drug localization within the particle matrix as determined by fluorescence and parelectric spectroscopy. J. Control. Release 110, 151–163.
Lopez-Pinto, J.M., Gonzalez-Rodriguez, M.L., Rabasco, A.M., 2005. Effect of cholesterol and ethanol on dermal delivery from DPPC liposomes. Int. J. Pharm. 298, 1–12.
Maitani, Y., Shimada, K., Nagai, T., 1996. l-Menthol, oleic acid lauricidin in absorption enhancement of free and sodium salt of diclofenac using ethanol treated silicone membrane as model for skin. Chem. Pharm. Bull. 44, 403–408.
Manosroi, A., Kongkaneramit, L., Manosroi, J., 2004. Stability and transdermal absorption of topical amphotericin B liposome formulations. Int. J. Pharm. 270, 279–286.

Megrab, N.A., Williams, A.C., Barry, B.W., 1995a. Estradiol permeation through human skin and silastic membrane: effects of propylene glycol and supersaturation. J. Control. Release 36, 277–294.

Megrab, N.A., Williams, A.C., Barry, B.W., 1995b. Oestradiol permeation across human skin, silastic and snake skin membranes: the effects of ethanol water cosolvent systems. Int. J. Pharm. 116, 101–112.

Megwa, S.A., Cross, S.E., Benson, H.A.E., Roberts, M.S., 2000a. Ion pair formation as a strategy to enhance topical delivery of salicylic acid. J. Pharm. Pharmacol. 52, 919–928.

Megwa, S.A., Cross, S.E., Whitehouse, M.W., Benson, H.A.E., Roberts, M.S., 2000b. Effect of ion pairing with alkylamines on the in vitro dermal penetration and local tissue deposition of salicylates. J. Pharm. Phamacol. 52, 929–940.

Menon, G.K., 2002. New insights into skin structure: scratching the surface. Adv. Drug Deliv. Rev. 54 (Suppl. 1), S3–S17.

Menon, G.K., Grayson, S., Elias, P.M., 1986. Cytochemical and biochemical localization of lipase and sphingomyelinase activity in mammalian epidermis. J. Invest. Dermatol. 86, 591–597.

Mezei, M., 1985. Liposomes as skin drug delivery system. In: Breimer, D.D., Speiser, P. (Eds), Topics in Pharmaceutical Sciences. Elsevier, Amsterdam, pp. 345–358.

Mezei, M., Gulasekharam, V., 1980. Liposomes, a selective skin drug delivery system for the topical route of administration: lotion dosage form. Life Sci. 26, 1473–1477.

Mezei, M., Gulasekharam, V., 1982. Liposomes, a selective drug delivery system for the topical route of administration: gel dosage form. J. Pharm. Pharmacol. 34, 473–474.

Mier, P., van den Hurk, J., 1975. Lysosomal hydrolases of the epidermis. Br. J. Dermatol. 93, 509–517.

Mishra, D., Dubey, V., Asthana, A., Saraf, D.K., Jain, N.K., 2006. Elastic liposomes mediated transcutaneous immunization against hepatitis. B. Vaccine 24, 4847–4855.

Morimoto, H., Wada, Y., Seki, T., Sugibayashi, K., 2002. *In vitro* skin permeation of morphine hydrochloride during the finite application of penetration-enhancing system containing water, ethanol and l-menthol. Biol. Pharm. Bull. 25, 134–136.

Moser, K., Kriwet, K., Naik, A., Kalia, Y.N., Guy, R.H., 2001a. Passive penetration enhancement and its quantification in vitro. Eur. J. Pharm. Biopharm. 54, 103–112.

Moser, K., Kriwet, K., Froehlich, C., Kalia, Y.N., Guy, R.H., 2001b. Supersaturation: enhancement of skin penetration and permeation of lipophilic drugs. Pharm. Res. 18, 1006–1011.

Mueller-Goymann, C.C., 2001. Liquid crystals in drug delivery. In: Swarbrick, J., Boylan, J.C. (Eds), Encyclopedia of Pharmaceutical Technology. Marcel Dekker, New York, pp. 117–146.

Muller-Goymann, C.C., 2004. Physicochemical characterization of colloidal drug delivery systems such as micelles, vesicles, liquid crystals and nanoparticles for topical administration. Eur. J. Pharm. Biopharm. 58, 343–356.

Müller, R.H., Radtke, M., Wissing, S.A., 2002. Solid lipid nanoparticles (SLN) and nanostructured lipid carriers (NLC) in cosmetic and dermatological preparations. Adv. Drug Deliv. Rev. 54, S131–S155.

Najjar, T.A., Sleeper, H.R., Calabresi, P., 1969. The use of 5-iodo-2′-deoxyuridine (IUDR) in Orabase and plastibase for treatment of oral herpes simplex. J. Oral Med. 24, 53–57.

Nastruzzi, C., Gambari, R., 1994. Antitumor activity of transdermally delivered aromatic tetra-amidines. J. Control. Release 29, 53–62.

Oertel, R.P., 1997. Protein conformational changes induced in human stratum corneum by organic sulphoxides: an infrared spectroscopic investigation. Biopolymers 16, 2329–2345.

Ogiso, T., Iwaki, M., Paku, T., 1995. Effect of various enhancers on transdermal penetration of indomethacin and urea and relationship between penetration parameters and enhancement factors. J. Pharm. Sci. 84, 482–488.

Ogiso, T., Yamaguchi, T., Iwaki, M., Tanino, T., Miyake, T., 2001. Effect of positively and negatively charged liposomes on skin permeation of drugs. J. Drug Target. 9, 49–59.

Oh, Y.K., Kim, M.Y., Shin, J.Y., Kim, T.W., Yun, M.O., Yang, S.J., Choi, S.S., Jung, W.W., Kim, J.A., Choi, H.G., 2006. Skin permeation of retinol in Tween 20-based deformable liposomes: *in-vitro* evaluation in human skin and keratinocyte models. J. Pharm. Pharmacol. 58, 161–166.
Ongipattanakul, B., Burnette, R.R., Potts, R.O., Francoeur, M.L., 1991. Evidence that oleic acid exists in a separate phase within stratum corneum lipids. Pharm. Res. 7, 350–354.
Osborne, D.W., 2008. Review of changes in topical drug classification. Pharm. Tech. 32, 66–74.
Padula, C., Nicole, C., Colombo, P., Santi, P., 2007. Single layer transdermal film containing lidocaine: modulation of drug release. Eur. J. Pharm. Biopharm. 66, 422–428.
Pellett, M.A., Davis, A.F., Hadgraft, J., 1994. Effect of supersaturation on membrane transport. 2. Piroxicam. Int. J. Pharm. 111, 1–6.
Pellett, M.A., Castellano, S., Hadgraft, J., Davis, A.F., 1997. The penetration of supersaturated solutions of piroxicam across siloxane membranes and human skin *in vitro*. J. Control. Release 46, 205–214.
Penzes, T., Blazso, G., Aigner, Z., Falkay, G., Eros, I., 2005. Topical absorption of piroxicam from organogels – *in vitro* and *in vivo* correlations. Int. J. Pharm. 298, 47–54.
Pershing, L.K., Lambert, L.D., Knutson, K., 1990. Mechanism of ethanol-enhanced estradiol permeation across human skin *in vivo*. Pharm. Res. 7, 170–175.
Pfister, W.R., Dean, S., Hsieh, S.T., 1990. Permeation enhancers compatible with transdermal drug delivery systems. I. Selection and formulation considerations. Pharm. Tech. 8, 132–139.
Philips, C.A., Michniak, B.B., 1995. Transdermal delivery of drugs with differing lipophilicities using azone analogs as dermal penetration enahancers. J. Pharm. Sci. 84, 1427–1433.
Pilgram, G.S.K., van der Meulen, J., Gooris, G.S., Koerten, H.K., Bouwstra, J.A., 2001. The influence of two azones and sebaceous lipids on the lateral organization of lipids isolated from human stratum corneum. Biochim. Biophys. Acta Biomembr. 1511, 244–254.
Porzio, S., Caselli, G., Pellegrini, L., Pallottini, V., Del Rosario, M., Coppola, A., Boltri, L., Gentile, M., Clavenna, G., Malillo, G., 1998. Efficacy of a new topical gel–spray formulation of ketoprofen lysine in the rat: percutaneous permeation *in vitro* and *in vivo* and pharmacological activity. Pharmacol. Res. 37, 41–47.
Potts, R.O., Guy, R.H., 1992. Predicting skin permeability. Pharm. Res. 9, 663–669.
Raghavan, S.L., Trividie, A., Davis, A.F., Hadgraft, J., 2000. Effect of cellulose polymers on supersaturation and *in vitro* membrane transport of hydrocortisone acetate. Int. J. Pharm. 193, 231–237.
Ren, C., Fang, L., Li, T., Wang, M., Zhao, L., He, Z., 2008. Effect of penetration enhancers and organic acids on the skin permeation of indapamide. Int. J. Pharm. 350, 43–47.
Rigg, P.C., Barry, B.W., 1990. Shed snake skin and hairless mouse skin as model membranes for human skin during permeation studies. J. Invest. Dermatol. 94, 235–240.
Ruland, A., Kreuter, J., Rytting, J.H., 1994. Transdermal delivery of tetrapeptide hisetal (melanotropin 6–9). 1. Effect of various enhancers: *in vitro* study across hairless mouse skin. Int. J. Pharm. 101, 57–61.
Santos Maia, C., Mehnert, W., Schaller, M., Korting, H., Gysler, A., Haberland, A., Schäfer-Korting, M., 2002. Drug targeting by solid lipid nanoparticles for dermal use. J. Drug Target. 10, 489–495.
Sarpotdar, P.P., Zatz, J.L., 1986a. Percutaneous absorption enhancement by non-ionic surfactants. Drug Dev. Ind. Pharm. 12, 1625–1647.
Sarpotdar, P.P., Zatz, J.L., 1986b. Evaluation of penetration enhancement of lidocaine by non-ionic surfactants through hairless mouse skin in vitro. J. Pharm. Sci. 75, 176–181.
Scartazzini, R., Luisi, P.L., 1988. Organogels from lecithins. J. Phys. Chem. 92, 829–833.
Schaefer, H., Redelmeier, T.E., 1996. Skin Barrier Principles of Percutaneous Absorption. Karger, Basel, pp. 235–237.

Schafer-Korting, M., Mehnert, W., Korting, H.-C., 2007. Lipid nanoparticles for improved topical application of drugs for skin diseases. Adv. Drug Deliv. Rev. 59, 427–443.
Scheuplein, R.J., Blank, I.H., 1971. Permeability of the skin. Physiol. Rev. 51, 702–747.
Schuetze, W., Mueller-Goymann, C.C., 1992. Mutual interactions between nonionic surfactants and gelatin – investigations in cubic liquid crystalline systems and micellar systems. Colloid Polym. Sci. 269, 85–90.
Schwarb, F.P., Imanidis, G., Smith, E.W., Haigh, J.M., Surber, C., 1999. Effect of concentration and degree of saturation of topical flucinonide formulations on *in vivo* membrane transport and *in vivo* bioavailability on human skin. Pharm. Res. 16, 909–915.
Shaikh, I.M., Jadhav, K.R., Gide, P.S., Kadam, V.J., Pisal, S.S., 2006. Topical delivery of aceclofenac from lecithin organogels: preformulation study. Curr. Drug Deliv. 3, 417–427.
Simeoni, S., Scalia, S., Benson, H.A.E., 2004. Influence of cyclodextrins on *in vitro* skin absorption of the sun screen butyl methoxydibenzoylmethane. Int. J. Pharm. 280, 163–171.
Singh, J., Tripathi, K.P., Sakya, T.R., 1993. Effect of permeation enhancer on the *in vitro* transport of ephedrine through rat skin and human epidermis from matrix based transdermal formulations. Drug Dev. Ind. Pharm. 19, 1623–1628.
Singh, S.K., Roane, D.S., Reddy, I.K., Durrani, M.J., Khan, M.A., 1996. Effect of additives on diffusion of ketoprofen through human skin. Drug Dev. Ind. Pharm. 22, 471–474.
Sinha, V.R., Kaur, M.P., 2000. Permeation enhancers for transdermal drug delivery. Drug Dev. Ind. Pharm. 26, 1131–1140.
Sinico, C., Manconi, M., Peppi, M., Lai, F., Valenti, D., Fadda, A.M., 2005. Liposomes as carriers for dermal delivery of tretinoin: *in vitro* evaluation of drug permeation and vesicle–skin interaction. J. Control. Release 103, 123–136.
Sintov, A.C., Shapiro, L., 2004. New microemulsion vehicle facilitates percutaneous penetration *in vitro* and cutaneous bioavailability *in vivo*. J. Control. Release 95, 173–183.
Sloan, K.B., 1989. Prodrugs for dermal delivery. Adv. Drug Deliv. Rev. 3, 67–101.
Southwell, D., Barry, B.W., 1983. Penetration enhancers for human skin: mode of action of 2-pyrrolidone and dimethylformamide on partition and diffusion of model compounds water, *n*-alcohols and caffeine. J. Invest. Dermatol. 80, 507–515.
Stecova, J., Mehnert, W., Blaschke, T., Kleuser, B., Sivaramakrishnan, R., Zouboulis, C.C., Seltmann, H., Korting, H.C., Kramer, K.D., Schäfer-Korting, M., 2007. Cyproterone acetate loading to lipid nanoparticles for topical acne treatment: particle characterization and skin uptake. Pharm. Res. 24, 991–1000.
Stott, P.W., Williams, A.C., Barry, B.W., 1998. Transdermal delivery for eutectic systems: enhanced permeation of model drug, ibuprofen. J. Control. Release 50, 297–308.
Stott, P.W., Williams, A.C., Barry, B.W., 2001. Mechanistic study into the enhanced transdermal permeation of a model β blocker propranolol by fatty acids: a melting point depression effect. Int. J. Pharm. 219, 161–176.
Sugibayashi, K., Nakayama, S., Seki, T., Hosoya, K., Morimoto, Y., 1992. Mechanism of skin permeation enhancing effect of laurocapram. J. Pharm. Sci. 81, 58–64.
Tanojo, H., Bosvan Geest, A., Bouwstra, J.A., Junginger, H.E., Bodde, H.E., 1997. *In vitro* human skin barrier perturbation by oleic acid: thermal analysis and freeze fracture electron microscopy studies. Thermochim. Acta 293, 77–85.
Touitou, E., Dayan, N., Bergelson, L., Godin, B., Eliaz, M., 2000a. Ethosomes – novel vesicular carriers for enhanced delivery: characterization and skin penetration properties. J. Control. Release 65, 403–418.
Touitou, E., Godin, B., Weiss, C., 2000b. Enhanced delivery of drugs into and across the skin by ethosomal carriers. Drug Dev. Res. 50, 406–415.
Trotta, M., Peira, E., Carlotti, M.E., Gallarate, M., 2004. Deformable liposomes for dermal administration of methotrexate. Int. J. Pharm. 270, 19–125.
Turunen, T.M., Buyuktimkin, S., Urtti, A., Peronen, P., Rytting, J.H., 1993. Enhanced delivery of 5-fluorouracil through shed snake skin by two new transdermal permeation enhancers. Int. J. Pharm. 92, 89–95.

Valenta, C., Siman, U., Kratzel, M., Hargraft, J., 2000. The dermal delivery of lignocaine: influence of ion pairing. Int. J. Pharm. 197, 77–85.
Vintiloiu, A., Leroux, J.-C., 2008. Organogels and their use in drug delivery – a review. J. Control. Release 125, 179–192.
Vollmer, U., Muller, B.W., Peeters, J., Mesens, J., Wilffert, B., Peters, T., 1994. A study of percutaneous absorption enhancing effects of cyclodextrin derivatives in rats. J. Pharm. Pharmacol. 46, 19–22.
Walkley, K., 1972. Bound water in stratum corneum measured by differential scanning calorimetry. J. Invest. Dermatol. 59, 225–227.
Weiner, N., Williams, N., Birch, G., Ramachandran, C., Shipman, G.J.R., Flynn, G., 1989. Topical delivery of liposomally encapsulated interferon evaluated in cutaneous herpes guinea pig model. Antimicrob. Agents Chemother. 33, 1217–1221.
Wertz, P.W., Downing, D.T., 1989. Stratum corneum: biological and biochemical considerations. In: Hadgraft, J., Guy, R.H. (Eds), Transdermal Drug Delivery. Marcel Dekker, New York, pp. 1–17.
Williams, A.C., Barry, B.W., 1989a. Urea analogues in propylene glycol as penetration enhancers in human skin. Int. J. Pharm. 56, 43–50.
Williams, A.C., Barry, B.W., 1989b. Essential oils as novel human skin penetration enhancers. Int. J. Pharm. 57, R7–R9.
Williams, A.C., Barry, B.W., 1991. The enhancement index concept applied to terpene penetration enhancers for human skin and model lipophilic (oestradiol) and hydrophilic (5-fluorouracil) drugs. Int. J. Pharm. 74, 157–168.
Williams, A.C., Barry, B.W., 2004. Penetration enhancers. Adv. Drug Deliv. Rev. 56, 603–618.
Williams, A.C., Shatri, S.R., Barry, B.W., 1998. Transdermal permeation modulation by cyclodextrins: a mechanistic study. Pharm. Dev. Tech. 3, 283–296.
Willimann, H., Walde, P., Luisi, P.L., Gazzaniga, A., Stroppolo, F., 1992. Lecithin organogel as matrix for transdermal transport of drugs. J. Pharm. Sci. 81, 871–874.
Wissing, S.A., Muller, R.H., 2001. A novel sunscreen system based on tocopherol acetate incorporated in solid lipid nanoparticles (SLN). Int. J. Cosmetic Sci. 23, 233–243.
Wissing, S.A., Lippacher, A., Muller, R.H., 2001. Investigations on the occlusive properties of solid lipid nanoparticles (SLNTM). J. Cosmetic Sci. 52, 313–323.
Wokovich, A.M., Prodduturi, S., Doub, W.H., Hussain, A.S., Buhse, L.F., 2006. Transdermal drug delivery system (TDDS) adhesion as a safe and quality attribute. Eur. J. Pharm. Biopharm. 64, 1–8.
Wong, O., Tsuzuki, N., Nghiem, B., Kuehnhoff, J., Itoh, T., Masaki, K., Huntingdon, J., Konishi, R., Rytting, J.H., Higuchi, T., 1989. Unsaturated cyclic ureas as new non-toxic biodegradable transdermal penetration enhancers: II. Evaluation study. Int. J. Pharm. 52, 191–192.
Woolfson, A.D., Malcolm, R.K., Campbell, K., Jones, D.S., Russell, J.A., 2000. Rheological, mechanical, and membrane penetration properties of novel dual drug systems for percutaneous delivery. J. Control. Release 67, 295–408.
Xiong, G.L., Quan, D., Maibach, H.I., 1996. Effect of permeation enhancers on *in vitro* percutaneous absorption of low molecular weight heparin through human skin. J. Control. Release 42, 289–296.
Yamane, M.A., Williams, A.C., Barry, B.W., 1995. Terpene penetration enhancers in propylene glycol/water cosolvent systems: effectiveness and mechanism of action. J. Pharm. Pharmacol. 47, 978–989.
Yu, H.Y., Liao, H.M., 1996. Triamcinolone permeation from different liposome formulations through rat skin *in vitro*. Int. J. Pharm. 127, 1–7.

CHAPTER 2

Micellar Nanoparticles: Applications for Topical and Passive Transdermal Drug Delivery

Robert W. Lee [1*], **Dinesh B. Shenoy** [2] *and* **Rajiv Sheel** [3]
[1] Particle Sciences Inc., 3894 Courtney Street, Bethlehem, PA 18017, USA; email: rlee@particlesciences.com
[2] Cadila Pharmaceuticals Ltd (CPL), India
[3] Orgenus Pharma Inc., 116 Village Blvd Ste 200, Princeton, NJ 08540, USA

Contents

2.1 Introduction 37
  2.1.1 MNP composition and structures 39
  2.1.2 Physicochemical characterization 42
  2.1.3 Antimicrobial properties 44
2.2 Transdermal drug delivery applications of MNP technology 45
  2.2.1 Estrasorb™ – commercial validation of MNP technology 45
  2.2.2 Raloxifene MNP product 47
  2.2.3 Nicotine MNP product 51
2.3 Topical drug delivery applications of MNP technology 53
2.4 Conclusion 54
References 56
Further reading 58

2.1 INTRODUCTION

Nanotechnology has evolved to be an integral part of the twenty-first century. Nanotech-enabled products find applicability in almost everything we touch on a day-to-day basis, such as medicine, pharmaceuticals, chemicals, biologics, and information technology. In particular, the pharmaceutical industry has been energized with breakthroughs in nano-engineering, especially in the fields of drug delivery and formulation development. Over the last few decades, there has been an explosion of research – at both academic and industrial levels – pertaining to nano-formulations: liposomes

* Author to whom correspondence should be addressed.

Handbook of Non-Invasive Drug Delivery Systems ISBN 9780815520252

(El Maghraby et al., 2006; Kazakov and Levon, 2006; Mainardes et al., 2006; Minko et al., 2006; Sharma et al., 2006; Torchilin, 2006; Tyagi et al., 2006; Weissig et al., 2006; Dutta, 2007; Elsayed et al., 2007; Karanth and Murthy, 2007; Koning and Krijger, 2007; Letchford and Burt, 2007; Malik et al., 2007; Vyas and Khatri, 2007), nanoparticles (Asghar and Chandran, 2006; Conti et al., 2006; Mainardes et al., 2006; Singhvi, 2006; Caruthers et al., 2007; Emerich and Thanos, 2007; Garg and Saraf, 2007; Goldberg et al., 2007; Illum, 2007; Kohane, 2007; Koning and Krijger, 2007; Navalakhe and Nandedkar, 2007; Patel and Vavia, 2007; Silva, 2007; Torchilin, 2007; Wang et al., 2007), nanoemulsions (Sarker, 2005; Pattani et al., 2006; Saupe et al., 2006; Tiwari and Amiji, 2006; Date and Nagarsenker, 2007; de Araujo et al., 2007; Fatouros et al., 2007; Khandavilli and Panchagnula, 2007; Schaffazick et al., 2007), dendrimers (Duncan and Izzo, 2005; Kitchens et al., 2005; Koo et al., 2005; Bai et al., 2006; Gupta et al., 2006; Najlah and D'Emanuele, 2006; Qiu and Bae, 2006; Reddy et al., 2006; Yang and Kao, 2006), supramolecular assemblies (Giraud-Guille et al., 2003; Lukin and Vogtle, 2005; Hamacek et al., 2006; Perez-Garcia and Amabilino, 2007), and surface nano-engineered products, to name a few.

Transdermal delivery involves application of a pharmacologically active compound on to the skin to achieve therapeutic blood levels in order to treat diseases remote from the site of application. Ever since the approval of Transderm-Scop®, the first transdermal drug delivery system (TDDS) in 1981, there has been explosive research in the field of transdermal therapeutics for treatment of a variety of clinical conditions (Gordon and Peterson, 2003). Unmatched clinical benefits (Gordon and Peterson, 2003), profound industry interest, existence of strong and *niche* markets, and regulatory precedence show why the TDDS has become a flourishing and viable dosage form. The current transdermal therapeutics market is segmented into traditional formulations (gels), advanced delivery systems (patches), and novel physical technologies (microporation, iontophoresis, and sonophoresis). Transdermal delivery is particularly advantageous for those drugs having significant hepatic first-pass metabolism or degradation in the gastrointestinal tract. Over the years, the US Food and Drug Administration (FDA) has approved more than 40 transdermal products, spanning about 15 molecules with sales of nearly $2.5 billion.

Micellar nanoparticle (MNP) technology was invented in the mid-1990s (Wright, 1997; Simon, 2006; Singhvi, 2006). Scientists at Novavax developed and patented MNP technology and subsequently rolled out the first nano-engineered transdermal lotion product (Estrasorb™) in 2003. Estrasorb is commercially manufactured on a kiloton scale and the manufacturing process is economical. The ingredients used in Estrasorb are all generally recognized as safe (GRAS).

MNP is a nanotechnology-based formulation that has achieved a breakthrough in transdermal therapeutics. The formulation represents a robust and versatile delivery system that can accommodate a range of therapeutic compounds having varying physicochemical properties. MNP-based emulsions (lotions) are attractive alternatives for systemic drug delivery via topical application. The technology allows high concentrations of drug to penetrate the skin and functionally create a drug depot in the stratum corneum and epidermis. This route of delivery provides similar advantages of patch technology in avoiding both contact with the gastrointestinal tract and hepatic first-pass effects, and is cosmetically more acceptable to many patients. MNP drug delivery offers a potentially fast and inexpensive pharmaceutical development model by using drugs already proven safe and effective to create new proprietary formulations.

2.1.1 MNP composition and structures

In broad terms, MNP is a multiphasic nanoemulsion. MNP technology presents the active pharmaceutical ingredient (API) in a more readily bioavailable form. There are five basic components of an MNP system: (i) one or more APIs; (ii) solvent; (iii) stabilizer; (iv) oil; and (v) aqueous medium. When these components are mixed together and subjected to a milling process (assisted by high-shear or high-pressure mixing), the API presents in one or more composite fractions (Figure 2.1):

- Solid particulates (micro/nanoparticles)
- Micelle-associated
- Oil-associated
- Solubilized (in aqueous and/or solvent medium).

MNPs can accommodate both water-soluble and poorly water-soluble APIs. While the technology can accommodate more traditional crystalline compounds, surprisingly it can also be used with amorphous drugs. Depending upon the physicochemical properties of the API and the dose requirements, drug loading up to 20% (w/w) can be achieved. The range of APIs that can be formulated in MNP technology is quite broad – from physicochemical and therapeutic perspectives. This aspect will be elaborated upon in the following sections.

A solvent is generally used to assist solubilization of the API during processing – though it is not a prerequisite. The typical solvent used in MNP

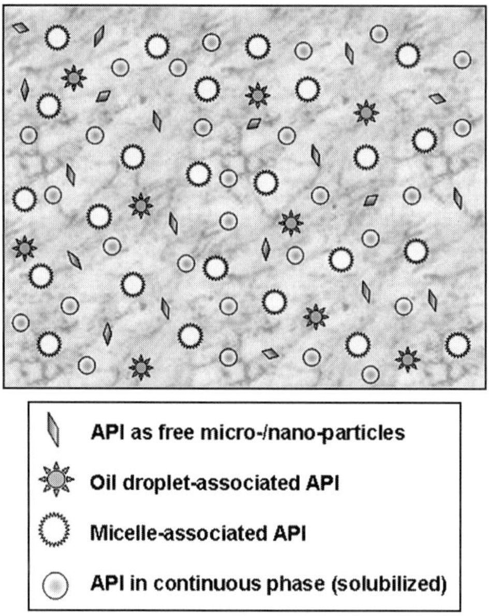

Figure 2.1 Schematic representation of the micro/nanostructures within an MNP formulation showing the different API components.

preparation is ethanol. In addition, stable MNPs can be obtained using other solvents such as propylene glycol, low-molecular-weight polyethylene glycols, triacetin, and *N*-methylpyrrolidinone. The solvent plays an important role in controlling the solubilized fraction of the drug, which is a key facilitator for rapid transmembrane permeation of the API.

The stabilizers used are generally non-ionic surfactants. Stable MNP preparations have been prepared using both hydrophilic and lipophilic stabilizers that encompass a wide hydrophilic–lipophilic balance (HLB) range. The surfactants include such classes as sorbitan esters, glycerol esters, block copolymers, polyethylene glycol esters, and ethoxylated fatty esters. The surfactant helps to sterically stabilize the micro/nanoparticles and the oil droplets, besides contributing to formation of the micellar phase.

The oil forms the internal phase of the emulsion. Depending upon the properties of the drug, the oil phase can accommodate an API fraction in soluble form. Some of the oils used are mineral oil, vegetable oils (soybean, corn, etc.), medium-chain triglycerides, and squalane. The aqueous medium used is generally purified water. A buffering agent may be included to maintain the pH and maximize stability of the API. The product-specific composition of the MNP formulation is dependent on the physicochemical properties of the API, the therapeutic need, intended site of action (local or systemic), and target product profile.

For topical or transdermal administration, MNPs can be classified as a type of microreservoir-dissolution-controlled system that can be tailored to deliver drugs topically (skin being the site of action) or transdermally (systemic availability). The physicochemical properties of MNP formulations can be tailored for a given route of administration. This may encompass adjusting the viscosity appropriately for topical or transdermal formulations, incorporating a mucoadhesive for vaginal or rectal administration, changing the particle/droplet size, tuning the formulation composition and components, adjusting the zeta potential, or tailoring the fraction of drug in solution versus in suspension. In a highly fragmented transdermal drug delivery market, MNP is the only passive, nanotechnology-enabled, cosmetically appealing, lotion-like topical dosage form offering a tunable delivery profile for a wide range of APIs. Figure 2.2 depicts a model for deposition of the MNP formulation within skin layers. The composite structures within the MNP preparation are complementary to the skin architecture and we hypothesize that this facilitates stratified positioning of the API within different skin layers.

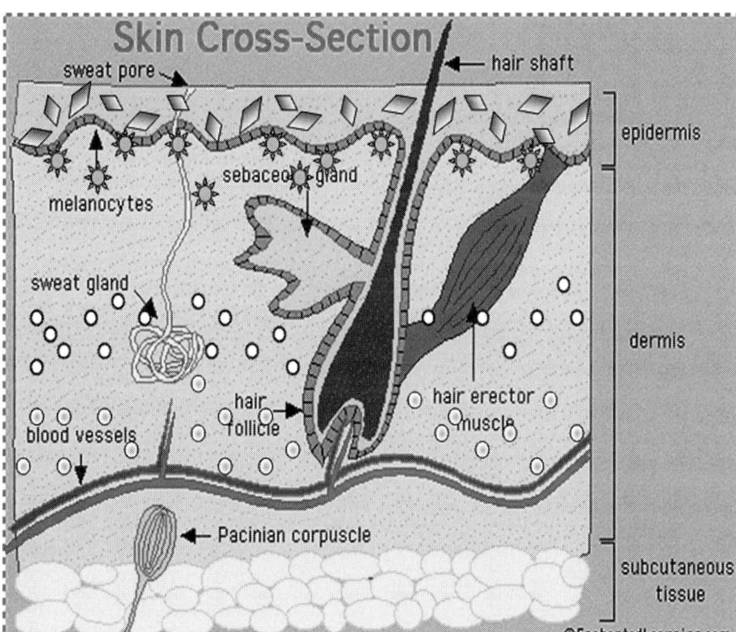

Figure 2.2 Schematic representation (hypothesis) of deposition and disposition of MNP structures within skin layers showing stratification of API.

2.1.2 Physicochemical characterization

The composite and multiphasic nature of the MNP formulation makes it difficult to capture the complete picture with a single characterization tool. Being the first of its kind, Estrasorb underwent tight scrutiny by the US FDA during the approval process, which led to state-of-the-art quality control test procedures (including particle sizing, crystal quantity, crystal number, *in vitro* release test).

The physical appearance of MNPs has been captured using transmission electron microscopy (TEM; negative staining with phosphotungstic acid – Figure 2.3) as well as freeze-fracture electron microscopy (FFEM – Figure 2.4). The TEM image captures the internal emulsion droplets, while FFEM shows oil droplets as well as micelles (tiny nubbles spread throughout the image – approximate size 8 nm). Particle size analysis (using dynamic light scattering) confirms the coexistence of micelles and oil-in-water (o/w) emulsion droplets (Figure 2.5). It is worth noting that particles fall into two fairly tight peaks of uniform size, with a peak at 8 nm representing the micelles and the second peak at 110 nm representing the oil droplets. Furthermore, particle size can be modulated via alterations to the manufacturing process or the formulation.

The amount of drug present as solid particulates varies and correlates with drug loading and the solubility of the API. Typically, the particle size of a poorly water-soluble API (i.e. acyclovir) can be effectively reduced to less

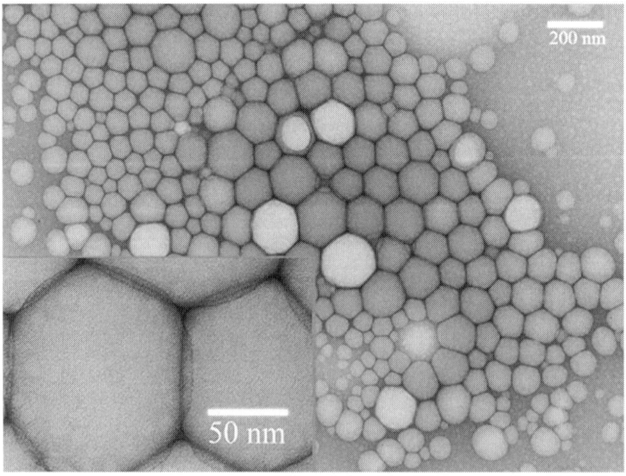

Figure 2.3 Transmission electron microscopic (TEM) image of a representative MNP formulation manufactured using a high-pressure process.

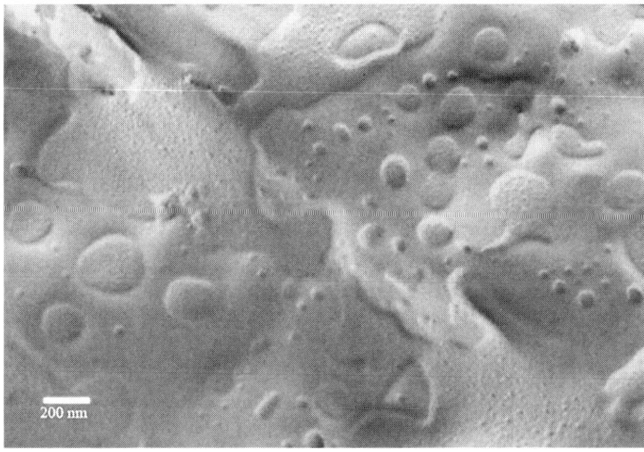

Figure 2.4 Freeze-fracture electron microscopic (FFEM) image of a representative MNP formulation manufactured using a high-pressure process.

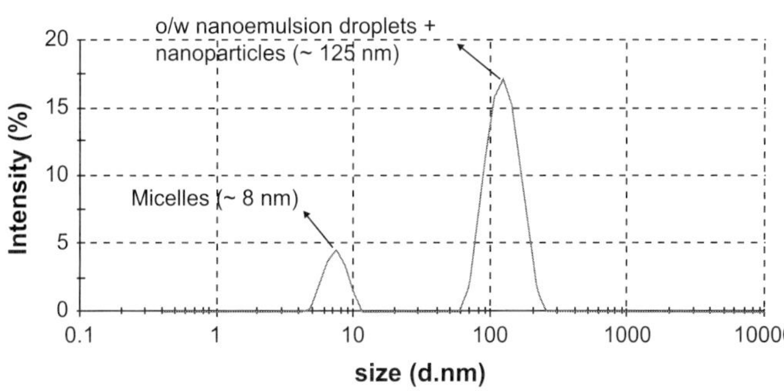

Figure 2.5 Particle size data for a representative MNP formulation showing co-existence of o/w emulsion droplets and micelles.

than 3 μm in the MNP formulation. Figure 2.6a shows the particle size of the API (raw material – before incorporation into MNP formulation) and Figure 2.6b shows the size of the o/w emulsion (approximately 200 nm) and the solid particulate drug upon formulating into the MNP product. The particle size of the droplet and particulates can be tailored by choosing the appropriate manufacturing technique (high-shear or high-pressure).

The viscosity of the MNP formulation can be tuned depending upon the intended application. For a lotion-like appearance, a viscosity of about 80–350 mPa s can be achieved. A stiffer preparation (like a semisolid) can be obtained by incorporating a suitable thickening agent (i.e. carbopol, xanthan gum, stearyl alcohol). The nature and amount of oil and stabilizer also play a role in modulating the viscosity of the final product.

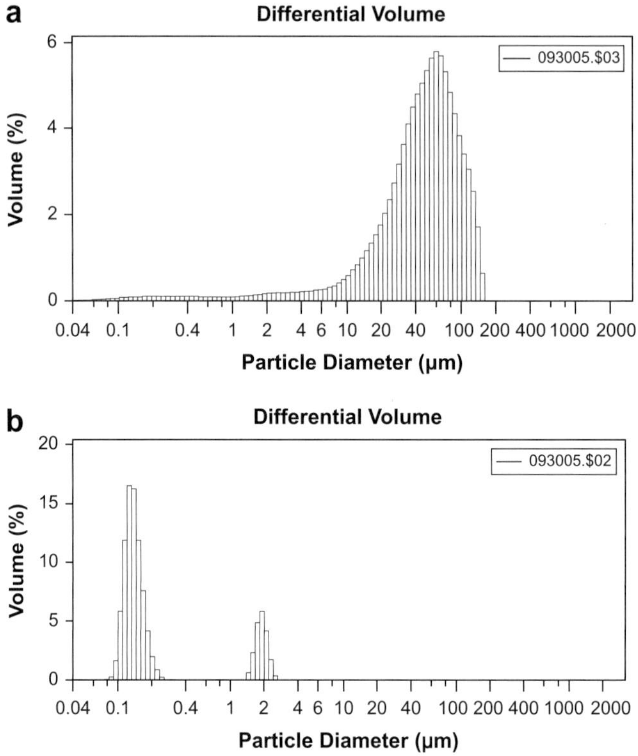

Figure 2.6 Particle size data for acyclovir. (a) Raw material. (b) Upon formulating as MNP product.

Optimized MNPs are highly stable products. The validated shelf-life claim for the commercial product based on MNP technology (Estrasorb™) is 3 years at room temperature with excursions allowed to 40°C. It has been demonstrated that there is no Oswald ripening phenomenon occurring in the MNP product, and that both the number of crystals per unit volume and crystal quantity of the API (estradiol) remain stable during room temperature and accelerated stability storage conditions. The MNP vehicle composition can be altered to offer enhanced thermal stability and some preparations can withstand a standard terminal heat sterilization cycle (i.e. 120°C for 25 min at 15 psi).

2.1.3 Antimicrobial properties

The MNP composition is inherently antimicrobial. Figure 2.7 depicts the results of the United States Pharmacopeia (USP) antimicrobial effectiveness test (AET) for a representative placebo MNP formulation. The results indicate that the MNPs are not only microbistatic, but are essentially

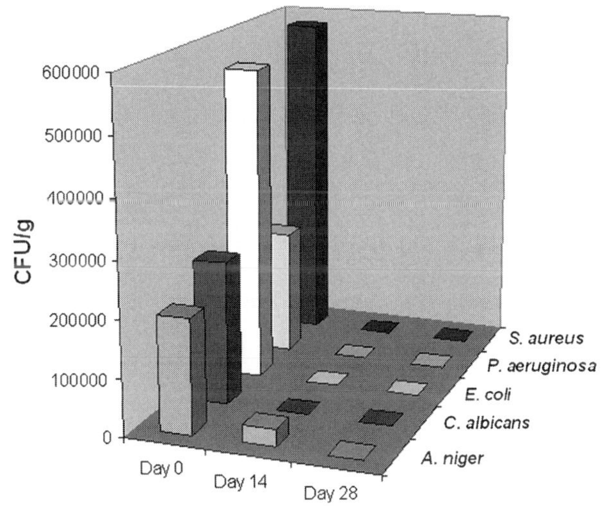

Figure 2.7 Antimicrobial effectiveness testing (USP) data for a representative placebo MNP preparation.

microbicidal. This can be attributed to the nano-size of the preparation and the nature of the composition (i.e. the high concentration of non-ionic surfactant). However, MNPs exhibit good safety profiles and they are relatively non-irritating dermally. This property offers commercial benefits such as the possible elimination of an antimicrobial preservative (especially for product filled in a multi-dose container), or the possible synergy of the microbicidal effect of an MNP preparation when formulated with an antibacterial, antifungal, and/or antiviral API.

2.2 TRANSDERMAL DRUG DELIVERY APPLICATIONS OF MNP TECHNOLOGY

2.2.1 Estrasorb™ – commercial validation of MNP technology

MNP technology was originally developed for transdermal delivery of APIs. MNP technology has been applied for estrogen replacement therapy with 17β-estradiol in Estrasorb (Estrasorb Package Insert, http://www.estrasorb.com/EstrasorbBrief.pdf) – Novavax's first internally developed FDA-approved product and the only emulsion-based formulation in the topical estrogen replacement market (primary indication being moderate-to-severe vasomotor symptoms associated with menopause). Estrasorb is the world's first nano-engineered topical dosage form that is approved by the US FDA for hormone replacement therapy, and represents commercial validation of the MNP technology.

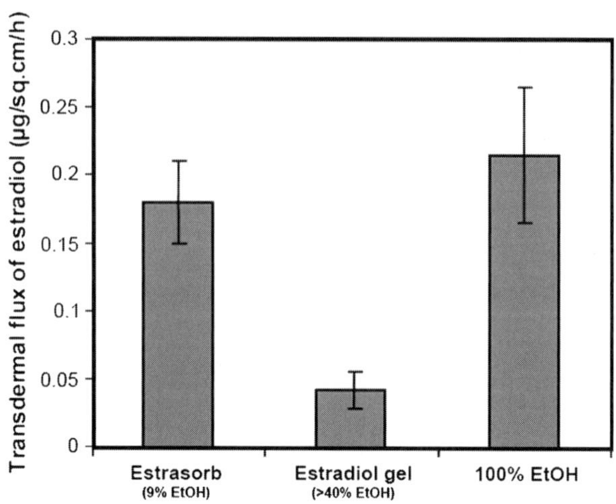

Figure 2.8 *In vitro* (Franz cell–cadaver skin assembly) data comparing transdermal flux rates for estradiol preparations.

The principal driving force behind transdermal flux is the concentration gradient. An *in vitro* Franz cell study was conducted using human cadaver skin to compare the relative flux rates obtained for Estrasorb (containing about 9% w/w ethanol), a commercial estradiol gel (containing about 40% w/w ethanol), and a 100% ethanolic solution of estradiol (Figure 2.8). The three formulations were applied on the skin at equivalent estradiol concentrations and the concentration of drug that permeated across the skin into the donor compartment was measured as a function of time. The results indicated that there was no significant difference between the 100% ethanolic solution and Estrasorb – while the gel exhibited one-fourth the rate of drug transfer of Estrasorb. This supports the claim that the composition of the MNP formulation promotes improved product–skin interactions and drives the API more efficiently across the skin – in a comparable fashion to a pure drug solution.

MNPs behave like a pseudo-patch or patchless-patch. The data from human clinical trials are shown in Figure 2.9 (once-daily application of 3.45 g of Estrasorb containing 0.25% w/w estradiol in 100 patients). It took approximately 2 weeks to attain steady-state plasma levels. A constant and controlled infusion of the drug from the topically applied estradiol emulsion maintained the drug at therapeutic levels for prolonged periods of time. Such a "depot effect" can be attributed to the multiphasic composition of the MNP preparation and stratified skin deposition upon dermal application, which results in establishment and maintenance of a concentration gradient across the skin. As the API from deeper skin layers becomes

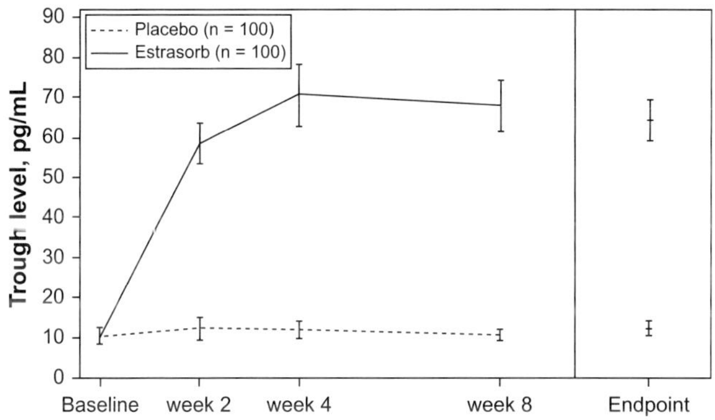

Figure 2.9 Mean trough serum estradiol concentrations following daily topical application of 3.45 g of Estrasorb™ containing 2.5 mg/g estradiol for 12 weeks.

depleted (through absorption into systemic circulation), more API dissolves from the solid particulate drug reservoir (deposited in superficial skin layers), maintaining a steady drug infusion. The effective plasma half-life for estradiol in Estrasorb (57.6 h) is significantly higher as compared to the commercial estradiol gel (36 h) or oral tablet (16.5 h). This provides strong evidence of the patch-like delivery profile for the MNPs.

Several small-molecular-weight compounds have been evaluated to prove the versatility and expandability of the MNP technology. A testosterone MNP formulation (Androsorb™) has completed phase I clinical evaluation for two indications: hormone replacement therapy in hypogonadal males, and to treat sexual dysfunction in females. A brief list of APIs that have been successfully formulated as MNP products and have completed key proof-of-concept (PoC) investigation has been compiled in Table 2.1.

Two case studies are presented, which will help define the MNP technology in terms of delivering a nontraditional transdermal API (raloxifene), or tuning the delivery profile for a classical transdermal candidate (nicotine).

2.2.2 Raloxifene MNP product

Raloxifene is a selective estrogen receptor modulator that belongs to the benzothiophene class of compounds. It is commercially available in tablet form. Approximately 60% of the oral dose is absorbed, but extensive hepatic conjugation to a number of inactive glucuronides results in an absolute bioavailability of 2%. The rationale for developing a transdermal delivery system for raloxifene was based on two considerations: (i) if therapeutic concentrations of raloxifene could be delivered to the systemic circulation transdermally, high hepatic concentrations would be avoided – thereby

Table 2.1 Proof-of-concept studies for various APIs formulated using MNP technology

API and indication	PoC investigation model	Key outcomes
Traditional transdermal APIs		
Testosterone (hormone replacement therapy in males or female sexual dysfunction)	• Phase I completed for both indications	• Pharmacokinetic end-points have been met in phase I • Dose-dependent blood levels seen in males and females • Same strength formulation can be used for both indications – simply by varying the amount applied
Nicotine (smoking cessation)	• *In vitro* Franz cell–cadaver skin study for formulation screening • Preclinical pharmacokinetic evaluation in rabbits	• Not a replacement product for patch • Product ideally suited for intermediate duration of action (3–6 h) to address withdrawal symptoms • Delivery profile can be tuned to fit quick onset of action (to address craving)
Oxybutynin (urinary incontinence)	• Preclinical pharmacokinetic evaluation in rabbits	• Data showed clinically exploitable transdermal delivery profile • Ideal formulation for treating urinary incontinence considering the drawbacks of the commercial patch • Can be tuned to create once-a-day application product
Fentanyl (severe pain)	• Preclinical pharmacokinetic evaluation in rabbits	• Data demonstrate a product with a rapid onset of action (and pain relief) • Opportunity to create an abuse-resistant product through formulation engineering
Clonidine (hypertension)	• Preclinical pharmacokinetic evaluation in rabbits	• Data showed clinically exploitable transdermal delivery profile • Can be tuned to create once-a-day application product

Nontraditional transdermal APIs		
Raloxifene (osteoporosis in postmenopausal women)	• *In vitro* Franz cell–cadaver skin study for formulation screening	• Data showed clinically exploitable transdermal delivery profile – unlike the hydro-alcoholic gel (50% ethanol), which showed zero transdermal delivery
Alprostadil (erectile dysfunction)	• Preclinical pharmacodynamic evaluation in rabbits (vasodilatation in ears)	• Positive data showed improved efficacy over pure ethanolic solution • Significant degree (visual scoring) and extent (vein diameter) of vasodilatation suggest an increased probability of success for use as localized treatment for erectile dysfunction
Cetirizine (antihistaminic)	• Preclinical pharmacokinetic evaluation in rabbits	• Data showed clinically exploitable transdermal delivery profile • In addition, the product is likely to offer significant benefits at the site of action (i.e. local inflammation)
Naltrexone (narcotic antagonist)	• Preclinical pharmacokinetic evaluation in rabbits • *In vitro* Franz cell–cadaver skin study for formulation screening	• Data showed clinically exploitable transdermal delivery profile
Cyclobenzaprine (antispasmodic)	• Preclinical pharmacokinetic evaluation in rabbits	• Data showed clinically exploitable local and transdermal drug delivery profile

reducing or avoiding adverse effects on coagulation factors and the consequent risk of thromboembolism; and (ii) by avoiding extensive first-pass metabolism to inactive metabolites, the total amount of raloxifene required to achieve therapeutic concentrations is reduced – with an expected result of a reduction in the adverse effects of metabolites. The physicochemical properties of raloxifene (molecular weight 473, melting point 145°C, log *P* 5.7, water solubility 0.25 mg/L) make it challenging to formulate using conventional transdermal technologies, and difficult to create an elegant topical formulation.

An MNP-based raloxifene formulation was prepared at 3% w/w of raloxifene base. The inactive ingredients were ethanol, benzyl alcohol, soybean oil, poloxamer 188, and water. A control formulation was prepared at the same drug loading using 50% w/w ethanol in 4% w/w hydroxypropylmethylcellulose (HPMC) aqueous gel. The rate and extent of drug transportation across human cadaver skin was evaluated *in vitro* using a Franz cell assembly. A known quantity (20 or 80 mg) of each test article was applied per cell (area of exposure = 0.64 cm^2) and the quantity of raloxifene diffusing across the skin into the receptor medium (phosphate-buffered saline (PBS), pH 7.4–ethanol, 60:40 v/v) was measured as a function of time.

The results are summarized in Figure 2.10. There was no recorded transdermal drug transportation across the skin with the control formulation (gel) up through 48 h – even though the formulation contained 50% w/w ethanol. The MNP formulation produced significant passive transdermal

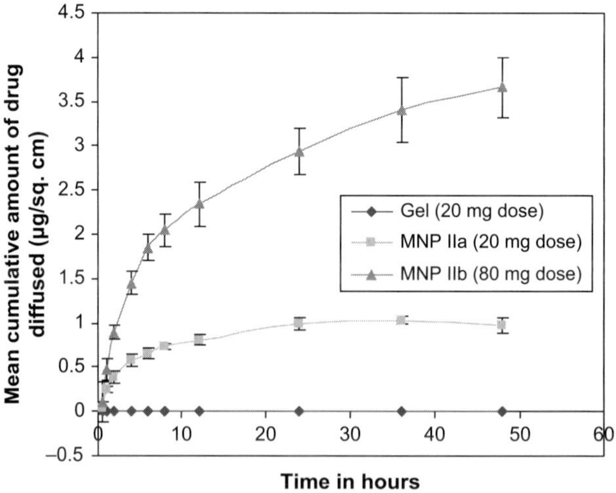

Figure 2.10 *In vitro* (Franz cell–cadaver skin assembly) data comparing transdermal flux rates for MNP compared to gel raloxifene preparations.

drug flux and showed a linear dose–response (MNP IIa and IIb) relationship at two doses – a fourfold increase in the amount of raloxifene applied resulted in a corresponding fourfold increase in transdermal drug flux. Based on these data, 2 g of 3% w/w MNP product (60 mg raloxifene) applied on two thighs (~375 cm^2 each) could provide the targeted 1.2 mg/day dose (equivalent to a single oral dose of 60 mg).

It is evident from this study that the MNP technology can facilitate transdermal transportation of APIs that may not be considered as the ideal candidates for transdermal delivery, or that cannot be formulated using conventional dosage forms.

2.2.3 Nicotine MNP product

Nicotine is an alkaloid found in the nightshade family of plants, predominantly in tobacco. It functions as an antiherbivore chemical, being a potent neurotoxin with particular specificity to insects. In low concentrations (an average cigarette yields about 1 mg of absorbed nicotine), the substance acts as a stimulant in mammals and is one of the main factors responsible for the dependence-forming properties of tobacco smoking. The pharmacologic and behavioral characteristics that determine tobacco addiction are similar to those that determine addiction to drugs such as heroin and cocaine.

Current nicotine products include chewing gum, oral lozenge, nasal spray, oral inhalant, and transdermal patch. The objective of the nicotine MNP product is to provide controlled and continuous delivery of nicotine through the skin. The MNP formulation can be customized with respect to onset and duration of action. Based on *in vitro* data (given below), it is expected that the MNP product will have an intermediate pharmacokinetic profile falling between that of a gum and a patch.

The physicochemical properties of nicotine (molecular weight 162.26, melting point −79°C, log P 1.305, oily liquid but miscible in water) make it an ideal fit for transdermal therapeutics – this fact is evident through the multitude of transdermal formulations commercially available. However, for a nonpolymer-based topical lotion technology like MNP, the challenge is to deliver the drug upon dermal application over longer periods of time. The objective of the *in vitro* proof-of-concept study (Franz cell–cadaver skin assembly) was to demonstrate the capability of the MNP technology to modulate the rate and extent of nicotine delivery through variations in the formulation.

Four different MNP formulations were prepared with different excipients and at different drug-loading levels. Nicoderm® patch was used as the standard, and a hydro-alcoholic gel (HPMC based) as the control (Table 2.2). The first-generation MNPs were prepared using essentially the same components and process as used in Estrasorb, while the second-generation MNPs used different components and a different manufacturing process. In addition to the formulation variables given in Table 2.2, the nicotine MNPs were further optimized with varying quantities of inactive ingredients and pH (data not presented).

Forty milligrams of each test article (or a single Nicoderm patch per cell) were applied to the cadaver skin (donor side of the Franz cell) with 0.64 cm^2 area of exposure. The quantity of nicotine diffusing across the skin into the receptor medium (PBS, pH 7.4–ethanol 90:10 v/v) was measured as a function of time (Figure 2.11).

The Nicoderm patch exhibited a greater rate (flux) of drug transfer across the skin (refer to Table 2.2 for flux rates; graph for Nicoderm not shown due to significant difference in scale). The drug–polymer matrix within the patch not only ensured a continual concentration gradient, but a larger one when

Table 2.2 Formulation, dose, and flux data (*in vitro*) for nicotine MNP products

Formulation	Transdermal flux ($mg/15\ cm^2/24\ h$)
Nicoderm patch (78 mg/patch)	42.03
Nicotine gel (14 mg/g) (ethanol, HPMC, citric acid, water)	6.95
Nicotine MNP, first gen. (30 mg/g) (ethanol, soybean oil, polysorbate 80, citric acid, water)	4.67
Nicotine MNP, second gen. I (14 mg/g) (ethanol, squalane, poloxamer 188, citric acid, water)	7.82
Nicotine MNP, second gen. II (14 mg/g) (squalane, poloxamer 188, citric acid, water)	1.22
Nicotine MNP, second gen. III (50 mg/g) (ethanol, soybean oil, poloxamer 188, tartaric acid, water)	13.49

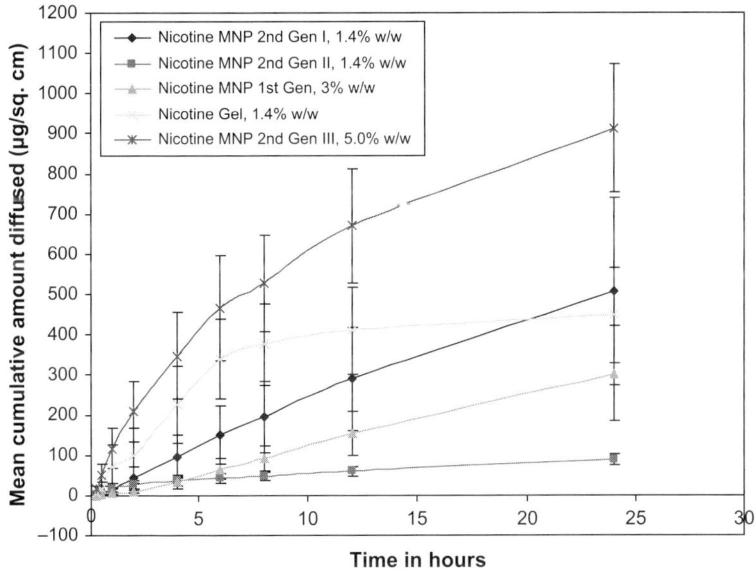

Figure 2.11 *In vitro* (Franz cell–cadaver skin assembly) data comparing transdermal flux rates for MNP compared to gel nicotine preparations.

compared to the MNPs or gel (0.64 cm^2 of patch contained approximately 3.33 mg nicotine vs. 0.56, 1.2, and 2.0 mg of nicotine for 14, 30, and 50 mg/g strengths respectively). The nicotine gel had no effective control over drug disposition as it followed a first-order release during the initial 8-h release phase and reached a plateau within about 8 h. The two second-generation MNP formulations (I and II) containing the same loading of nicotine as the gel (14 mg/g) not only exhibited prominent control over drug transportation rates (zero-order release), but also a significant difference in the rate or drug transport (sixfold difference in flux). This was essentially because of the difference in the composition of the two MNPs, which were tailored to release the drug at faster or slower rates. In spite of having higher drug loading (30 mg/g), the first-generation MNP showed a lower drug transportation pattern than the second-generation MNP I, which can be attributed to the difference in formulation components.

2.3 TOPICAL DRUG DELIVERY APPLICATIONS OF MNP TECHNOLOGY

Although the transdermal drug delivery field has enjoyed a significant amount of research effort and technological breakthrough, there has not been much corresponding innovation taking place in the field of topical drug delivery. The majority of the dosage forms are limited to traditional creams, ointments, and gels. Some of the new additions have been sprays, foams, and patches. MNP technology can be exploited to design improved topical dosage forms that deliver the API locally (at site of application) in an efficient and effective manner. It is possible to tailor drug deposition, disposition, and permeation kinetics through formulation engineering (altered composition, drug loading, droplet size, etc.). This concept has been demonstrated using acyclovir as the model drug. Commercially, a topical acyclovir product is available (Zovirax®) and is indicated for the treatment of recurrent herpes labialis (cold sores), for genital herpes, and in limited nonlife-threatening mucocutaneous herpes simplex viral infections in immunocompromised patients. The product needs to be applied topically five to seven times a day for 4–7 days. A comparative investigation (*in vitro* using Franz cell–cadaver skin assembly) was carried out with two MNP formulations (designed to differentiate topical and transdermal delivery) and Zovirax cream. All the formulations had a drug loading of 5% w/w. The product was applied to the skin (donor compartment), and drug that permeated across the skin as well as that retained within skin layers was estimated. The results are captured in Figure 2.12a and b. MNP I was designed to retain the API preferentially in the skin layers, while MNP II was engineered to facilitate transdermal permeation of the same drug. It is clear from the data that the transdermal flux rate for MNP I was comparable to that for Zovirax – but the amount of drug retained within skin was about twofold higher for the former. If this effect can be translated into clinical use (or *in vivo* systems), one can expect to witness:

- Increased rate of skin permeation, resulting in faster onset of action
- Greater degree of skin deposition, leading to higher local drug concentrations
- Skin depot effects, leading to longer drug residence time
- Potential reductions of total dose, frequency of application, or both.

In addition, the inherent antimicrobial nature of the MNP vehicle would be beneficial from a therapeutic and packaging perspective. Based on these benefits, the MNP technology could offer a novel perspective to the field of topical drug delivery – especially for nonsteroidal anti-inflammatory drugs (NSAIDs), antifungals, antibacterials, antivirals, antispasmodics, and vasodilatory drugs.

2.4 CONCLUSION

Transdermal drug delivery is not suited or clinically justified for all drugs, yet it is viewed to be much more limited than is warranted. MNP technology helps to incorporate and deliver many therapeutic compounds

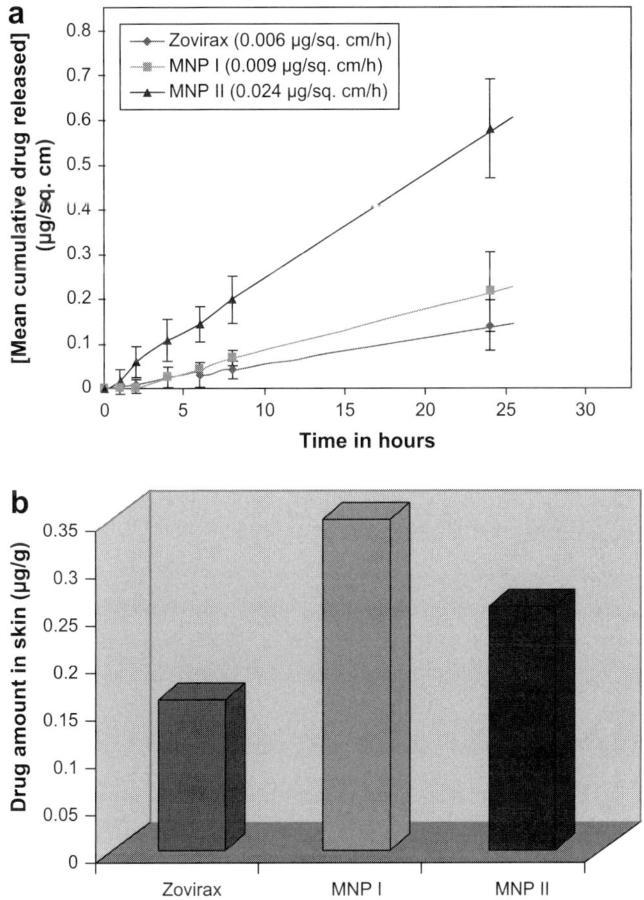

Figure 2.12 *In vitro* (Franz cell–cadaver skin assembly) data comparing acyclovir preparations. (a) Transdermal flux rates. (b) Amount retained within skin layers after 23 h.

that are otherwise viewed as unsuitable for transdermal delivery. MNP technology allows fast, low-cost product development compared with the typical development of new chemical entities. From proof of principle in a validated preclinical model through beginning a phase I study in humans requires approximately 12 months to complete. The data from the preclinical studies described here show a high probability of clinical success within a shorter development time frame, and a lower cost than a typical NDA (New Drug Application).

Research to date has shown the MNP drug delivery platform to be a versatile technology for multiple routes of administration (data not presented for ophthalmic, vaginal, and oral routes). Understanding the basic physicochemical properties of MNPs has enabled a degree of control over pharmacokinetic parameters that may provide an attractive option for pharmaceutical formulators. The technology is validated for transdermal delivery and a commercial MNP product, Estrasorb, is manufactured on a kiloton scale. The Estrasorb ingredients are GRAS and the manufacturing process is attractive from a cost of goods perspective. The heterogeneous, multiphasic nanoemulsion that comprises the MNPs is surprisingly stable and, in some cases, amenable to terminal heat sterilization. Experiments have shown the potential for MNPs to be used for intranasal, vaginal, rectal, and parenteral routes of administration – for poorly water-soluble drugs in particular. Better commercial exploitation of the MNP technology, to its fullest potential, is expected in coming years.

REFERENCES

Asghar, L.F., Chandran, S., 2006. Multiparticulate formulation approach to colon specific drug delivery: current perspectives. J. Pharm. Sci. 9 (3), 327–338.

Bai, S., Thomas, C., Rawat, A., Ahsan, F., 2006. Recent progress in dendrimer-based nanocarriers. Crit. Rev. Ther. Drug Carrier Syst. 23 (6), 437–495.

Caruthers, S.D., Wickline, S.A., Lanza, G.M., 2007. Nanotechnological applications in medicine. Curr. Opin. Biotechnol. 18 (1), 26–30. Epub 2007, Jan 24.

Conti, M., Tazzari, V., Baccini, C., Pertici, G., Serino, L.P., De Giorgi, U., 2006. Anti-cancer drug delivery with nanoparticles. In Vivo 20 (6A), 697–701.

Date, A.A., Nagarsenker, M.S., 2007. Design and evaluation of self-nanoemulsifying drug delivery systems (SNEDDS) for cefpodoxime proxetil. Int. J. Pharm. 329 (1–2), 166–172. Epub 2006, Sep 1.

de Araujo, S.C., de Mattos, A.C., Teixeira, H.F., Coelho, P.M., Nelson, D.L., de Oliveira, M.C., 2007. Improvement of in vitro efficacy of a novel schistosomicidal drug by incorporation into nanoemulsions. Int. J. Pharm. 337 (1–2), 307–315. Epub 2007, Jan 14.

Duncan, R., Izzo, L., 2005. Dendrimer biocompatibility and toxicity. Adv. Drug Deliv. Rev. 57 (15), 2215–2237. Epub 2005, Nov 16.

Dutta, R.C., 2007. Drug carriers in pharmaceutical design: promises and progress. Curr. Pharm. Des. 13 (7), 761–769.

El Maghraby, G.M., Williams, A.C., Barry, B.W., 2006. Can drug-bearing liposomes penetrate intact skin? J. Pharm. Pharmacol. 58 (4), 415–429.

Elsayed, M.M., Abdallah, O.Y., Naggar, V.F., Khalafallah, N.M., 2007. Lipid vesicles for skin delivery of drugs: reviewing three decades of research. Int. J. Pharm. 332 (1–2), 1–16. Epub 2006, Dec 8.

Emerich, D.F., Thanos, C.G., 2007. Targeted nanoparticle-based drug delivery and diagnosis. J. Drug Target. 15 (3), 163–183.

Fatouros, D.G., Deen, G.R., Arleth, L., Bergenstahl, B., Nielsen, F.S., Pedersen, J.S., Mullertz, A., 2007. Structural development of self nano emulsifying drug delivery systems (SNEDDS) during in vitro lipid digestion monitored by small-angle X-ray scattering. Pharm. Res. Epub ahead of print, Apr 26.

Garg, G., Saraf, S., 2007. Cubosomes: an overview. Biol. Pharm. Bull. 30 (2), 350–353.

Giraud-Guille, M.M., Besseau, L., Martin, R., 2003. Liquid crystalline assemblies of collagen in bone and in vitro systems. J. Biomech. 36 (10), 1571–1579.

Goldberg, M., Langer, R., Jia, X., 2007. Nanostructured materials for applications in drug delivery and tissue engineering. J. Biomater. Sci. Polym. Ed. 18 (3), 241–268.

Gordon, R.D., Peterson, T.A., 2003. 4 myths about transdermal drug delivery. Drug Deliv. Tech. 3 (4), 44–50.

Gupta, U., Agashe, H.B., Asthana, A., Jain, N.K., 2006. A review of in vitro–in vivo investigations on dendrimers: the novel nanoscopic drug carriers. Nanomedicine 2 (2), 66–73.

Hamacek, J., Borkovec, M., Piguet, C., 2006. Simple thermodynamics for unraveling sophisticated self-assembly processes. Dalton Trans. 12, 1473–1490. Epub 2006, Mar 1.

Illum, L., 2007. Nanoparticulate systems for nasal delivery of drugs: a real improvement over simple systems? J. Pharm. Sci. 96 (3), 473–483.

Karanth, H., Murthy, R.S., 2007. pH sensitive liposomes – principle and application in cancer therapy. J. Pharm. Pharmacol. 59 (4), 469–483.

Kazakov, S., Levon, K., 2006. Liposome–nanogel structures for future pharmaceutical applications. Curr. Pharm. Des. 12 (36), 4713–4728.

Khandavilli, S., Panchagnula, R., 2007. Nanoemulsions as versatile formulations for paclitaxel delivery: peroral and dermal delivery studies in rats. J. Invest. Dermatol. 127 (1), 154–162. Epub 2006, Jul 20.

Kitchens, K.M., El-Sayed, M.E., Ghandehari, H., 2005. Transepithelial and endothelial transport of poly (amidoamine) dendrimers. Adv. Drug Deliv. Rev. 57 (15), 2163–2176. Epub 2005, Nov 11.

Kohane, D.S., 2007. Microparticles and nanoparticles for drug delivery. Biotechnol. Bioeng. 96 (2), 203–209.

Koning, G.A., Krijger, G.C., 2007. Targeted multifunctional lipid-based nanocarriers for image-guided drug delivery. Anticancer Agents Med. Chem. 7 (4), 425–440.

Koo, O.M., Rubinstein, I., Onyuksel, H., 2005. Role of nanotechnology in targeted drug delivery and imaging: a concise review. Nanomedicine 1 (3), 193–212.

Letchford, K., Burt, H., 2007. A review of the formation and classification of amphiphilic block copolymer nanoparticulate structures: micelles, nanospheres, nanocapsules and polymersomes. Eur. J. Pharm. Biopharm. 65 (3), 259–269. Epub 2006, Nov 23.

Lukin, O., Vogtle, F., 2005. Knotting and threading of molecules: chemistry and chirality of molecular knots and their assemblies. Angew. Chem. Int. Ed. Engl. 44 (10), 1456–1477.

Mainardes, R.M., Urban, M.C., Cinto, P.O., Chaud, M.V., Evangelista, R.C., Gremiao, M.P., 2006. Liposomes and micro/nanoparticles as colloidal carriers for nasal drug delivery. Curr. Drug Deliv. 3 (3), 275–285.

Malik, D.K., Baboota, S., Ahuja, A., Hasan, S., Ali, J., 2007. Recent advances in protein and peptide drug delivery systems. Curr. Drug. Deliv. 4 (2), 141–151.

Minko, T., Pakunlu, R.I., Wang, Y., Khandare, J.J., Saad, M., 2006. New generation of liposomal drugs for cancer. Anticancer Agents Med. Chem. 6 (6), 537–552.

Najlah, M., D'Emanuele, A., 2006. Crossing cellular barriers using dendrimer nanotechnologies. Curr. Opin. Pharmacol. 6 (5), 522–527. Epub 2006, Aug 4.

Navalakhe, R.M., Nandedkar, T.D., 2007. Application of nanotechnology in biomedicine. Indian J. Exp. Biol. 45 (2), 160–165.

Patel, A.R., Vavia, P.R., 2007. Nanotechnology and pharmaceutical inhalation aerosols. Indian J. Exp. Biol. 45 (2), 166–174.

Pattani, A.S., Mandawgade, S.D., Patravale, V.B., 2006. Development and comparative anti-microbial evaluation of lipid nanoparticles and nanoemulsion of Polymyxin B.J. Nanosci. Nanotechnol. 6 (9–10), 2986–2990.

Perez-Garcia, L., Amabilino, D.B., 2007. Spontaneous resolution, whence and whither: from enantiomorphic solids to chiral liquid crystals, monolayers and macro- and supramolecular polymers and assemblies. Chem. Soc. Rev. 36 (6), 941–967. Epub 2007, Feb 5.

Qiu, L.Y., Bae, Y.H., 2006. Polymer architecture and drug delivery. Pharm. Res. 23 (1), 1–30. Epub 2006, Jan 11.

Reddy, S.T., Swartz, M.A., Hubbell, J.A., 2006. Targeting dendritic cells with biomaterials: developing the next generation of vaccines. Trends Immunol. 27 (12), 573–579. Epub 2006, Oct 16.

Sarker, D.K., 2005. Engineering of nanoemulsions for drug delivery. Curr. Drug Deliv. 2 (4), 297–310.

Saupe, A., Gordon, K.C., Rades, T., 2006. Structural investigations on nanoemulsions, solid lipid nanoparticles and nanostructured lipid carriers by cryo-field emission scanning electron microscopy and Raman spectroscopy. Int. J. Pharm. 314 (1), 56–62. Epub 2006, Mar 30.

Schaffazick, S.R., Pohlmann, A.R., Guterres, S.S., 2007. Nanocapsules, nanoemulsion and nanodispersion containing melatonin: preparation, characterization and stability evaluation. Pharmazie 62 (5), 354–360.

Sharma, G., Anabousi, S., Ehrhardt, C., Ravi Kumar, M.N., 2006. Liposomes as targeted drug delivery systems in the treatment of breast cancer. J. Drug Target. 14 (5), 301–310.

Silva, G.A., 2007. Nanotechnology approaches for drug and small molecule delivery across the blood brain barrier. Surg. Neurol. 67 (2), 113–116.

Simon, J.A., 2006. Estrasorb Study Group. Estradiol in micellar nanoparticles: the efficacy and safety of a novel transdermal drug-delivery technology in the management of moderate to severe vasomotor symptoms. Menopause 13 (2), 222–231.

Singhvi, R., 2006. Micellar nanoparticles: a new drug delivery platform. Drug Deliv. Tech. 6 (2), 72–75.

Tiwari, S.B., Amiji, M.M., 2006. Improved oral delivery of paclitaxel following administration in nanoemulsion formulations. J. Nanosci. Nanotechnol. 6 (9–10), 3215–3221.

Torchilin, V.P., 2006. Multifunctional nanocarriers. Adv. Drug Deliv. Rev. 58 (14), 1532–1555. Epub 2006, Sep 28.

Torchilin, V.P., 2007. Micellar nanocarriers: pharmaceutical perspectives. Pharm. Res. 24 (1), 1–16. Epub 2006, Nov 16.

Tyagi, P., Wu, P.C., Chancellor, M., Yoshimura, N., Huang, L., 2006. Recent advances in intravesical drug/gene delivery. Mol. Pharm. 3 (4), 369–379.

Vyas, S.P., Khatri, K., 2007. Liposome-based drug delivery to alveolar macrophages. Expert Opin. Drug Deliv. 4 (2), 95–99.

Wang, M.D., Shin, D.M., Simons, J.W., Nie, S., 2007. Nanotechnology for targeted cancer therapy. Expert Rev. Anticancer Ther. 7 (6), 833–837.

Weissig, V., Boddapati, S.V., Cheng, S.M., D'Souza, G.G., 2006. Liposomes and liposome-like vesicles for drug and DNA delivery to mitochondria. J. Liposome Res. 16 (3), 249–264.

Wright, C., 1997. Micellar nanoparticles. US patent, number 5,620.021.

Yang, H., Kao, W.J., 2006. Dendrimers for pharmaceutical and biomedical applications. J. Biomater. Sci. Polym. Ed. 17 (1–2), 3–19.

FURTHER READING

Biorise® technology from Eurand. <http://www.eurand.com/Technologies/Bioavailability-enhancement/Biorise/>.

Burgess, D. (Ed.), 2005. Injectable Dispersed Systems: Formulation, Processing and Performance. Taylor & Francis, Florida.

Insoluble Drug Delivery (IDD®) technology from SkyePharma. <http://www.skyepharma.com/solubilization_idd.html>.

Liu, R. (Ed.), 2000. Water-Insoluble Drug Formulations. Interpharm Press, Colorado.

NanoCrystal® technology from élan Drug Delivery Inc. <http://www.elan.com/EDT/nanocrystal_technology/>.

NanoEdge® technology from Baxter BioPharma Solutions. <http://www.baxterbiopharmasolutions.com/content/formulation_tech/nanoedge.html>.

CHAPTER 3

Emulsions and Microemulsions for Topical and Transdermal Drug Delivery

Guang Wei Lu [1] *and* **Ping Gao** [2]

[1] Research Fellow, Parenteral Center of Emphasis, Global Research & Development, Pfizer Inc., Eastern Point Road, Groton, CT 06340, USA; email: guang.w.lu@pfizer.com

[2] Director, Life Cycle Technology, Global Pharmaceutical & Analytical Sciences, Abbott Lab., 100 Abbott Park Rd., Abbott Park, IL 60064, USA; email: ping.gao@abbott.com

Contents

3.1 Overview of basic aspects of emulsions and microemulsions 59
 3.1.1 Introduction 59
 3.1.2 Methods of characterization 63
 3.1.3 Stability evaluation 69
3.2 Application of emulsions for topical and transdermal drug delivery 71
 3.2.1 Drug transport from an emulsion to the skin 71
 3.2.2 Investigation of topical and transdermal delivery in human subjects 78
 3.2.3 Formulation strategies for emulsion-based products 80
3.3 Advances in using microemulsions as topical and transdermal drug delivery systems 86
References 91

3.1 OVERVIEW OF BASIC ASPECTS OF EMULSIONS AND MICROEMULSIONS

3.1.1 Introduction

3.1.1.1 Emulsions

Emulsions are heterogeneous systems composed of at least two immiscible liquids, water and oil, one of which is usually uniformly dispersed as fine droplets throughout the other liquid phase by a mechanical agitation process. Emulsions are considered as a type of liquid–liquid colloid. The phase existing as small droplets is called the dispersed phase and the surrounding liquid is known as the continuous phase. Emulsions are commonly classified

Handbook of Non-Invasive Drug Delivery Systems
ISBN 9780815520252

as oil-in-water (O/W) or water-in-oil (W/O) depending on whether the continuous phase is water or oil.

Emulsions are thermodynamically unstable as the dispersed and continuous phases can revert back as separate phases, oil and water, by fusion or coalescence of droplets. However, emulsions are commonly stabilized by an emulsifying agent, often referred to as a surfactant. In general, after vigorous agitation of the two immiscible phases, the more rapidly coalescing droplets form the continuous phase. This is usually the liquid that is present in the larger amount – the greater the number of droplets, the higher the probability of collision and coalescence. Therefore, emulsification can be considered as the result of two competing processes that occur simultaneously, namely the disruption of bulk liquids to produce fine droplets and the recombination of the dispersed droplets back to the bulk liquids. Theoretically, the dispersed phase of an emulsion can occupy up to 74% of the phase volume, and such high internal phase O/W emulsions have been produced with suitable surfactants (Eccleston, 2007). It is more difficult to formulate a W/O emulsion with greater than 50% dispersed phase because of the steric mechanisms involved in the physical stability. Although the droplet diameters of emulsions may vary enormously depending on their composition and manufacturing process, pharmaceutical and cosmetic emulsions are typically polydispersed with droplet sizes ranging from 0.1 to 100 μm.

With the application of surfactants and/or cosurfactants, the resulting emulsion is dictated by the relative solubility of these surfactants in oil and water. Usually, the phase in which the surfactant is most soluble becomes the continuous phase. Thus, hydrophilic surfactants promote O/W emulsions whereas lipophilic surfactants promote W/O emulsions. A range of different types of emulsifying agents are used in emulsion formulations, the individual emulsifying agents functioning on different principles to achieve a stable product. As a result, it is not easy to develop a universal theory that would account for emulsification and emulsion stability.

Emulsions are commonly used for topical pharmaceutical and cosmetic products, such as lotions and creams. The largest group of emulsions commercially available as medicines are dermatological products for topical application. Emulsions can be designed to facilitate drug penetration into and/or through the skin. Both O/W and W/O emulsions have been extensively used to deliver drugs and cosmetic agents to the skin, depending on the property of active agents and the indications of the medicines. Although the microstructure of many of these complex emulsions is now better characterized and understood, the underlying mechanisms by which the structure of an emulsion and the function of individual excipient (e.g. penetration enhancing agent) can influence drug bioavailability are far from clear. Droplet size of the emulsion may influence the drug penetration through the skin, but the effect often is not clinically significant. The evaporation of volatile excipients can occur and so affect the drug permeation across the skin. Judicious selection of an appropriate emulsifying agent and additional stabilizer is a critical factor in the design and development of emulsions. These factors will be discussed in more detail later.

3.1.1.2 Microemulsions

Microemulsions are thermodynamically stable dispersions of oil and water. A great deal of debate surrounding the definition of microemulsions originated from their different attributes. The definition of microemulsions as a single-phase system containing water, oil, and an amphiphile(s), which is a thermodynamically stable and isotropic solution, is widely accepted.

The term microemulsion appears to imply a close relationship to the ordinary or conventional emulsions, and this could be misleading. Microemulsions differ significantly from conventional emulsions by their physical properties (transparency, low viscosity, small particle size of <100–200 nm, etc.) and an ability to form spontaneously. Microemulsions embrace several different microstructures that have little in common with conventional emulsions. While a microemulsion is thermodynamic stable, emulsion droplets, which may be as small as those of microemulsion, will undergo coalescence, leading to phase separation.

It is worth noting that the term microemulsion in a broad sense includes kinetically stable, but thermodynamically unstable, emulsions that are similar in droplet size. Often, it is unclear whether a microemulsion described in a publication is truly a thermodynamically stable or unstable system. In many applications, an emulsion system having two phases in which the droplet sizes are extremely small appears transparent or translucent, and it could be considered as a microemulsion. The distinction between a conventional emulsion and a microemulsion is sometimes difficult to determine from empirical observation. Although an ultrafine emulsion is a true emulsion that is thermodynamically unstable, a markedly stable ultrafine emulsion with a long shelf-life can be obtained with optimized compositions and manufacturing processes. Accordingly, they can be considered pseudo-equivalent to microemulsions in a broad sense and could be applied in the same manner.

Homogeneous microemulsion systems, which can be prepared over a wide range of surfactant concentrations and oil-to-water ratios (10–80%), are all fluids of low viscosity.

In some pharmaceutical and cosmetic applications, both microemulsions and emulsions could be the dosage form of choice. Microemulsions have some advantages over emulsions. In particular, a low energy input in the preparation is involved due to its spontaneous formation and physical stability. The major attributes of microemulsions are summarized as follows:

- Microemulsions form spontaneously without the aid of high-shear equipment and the preparation is independent of the order of addition of the excipients. Optical transparency and low viscosity of microemulsions ensure cosmetic elegance, and easiness of handling and packaging. The indefinite stabilities of microemulsions ensure a long shelf-life. Moreover, microemulsions are stable to phase separation over a wide temperature range.
- The stability of a microemulsion is a consequence of the ultralow interfacial tension between the oil and water phases. The surfactants and cosurfactants used in the composition contribute to the low interfacial tension. The cosurfactant molecules intercalate with the surfactant molecules at the oil–water interface, thus affecting the curvature of the droplet. The molecular structure of the surfactants and cosurfactants, as well as their concentrations, determines the microstructure. Additionally, the properties and concentration of the oil in the formulation play an important role.
- The dispersed phase, either lipophilic (O/W) or hydrophilic (W/O), can behave as a potential reservoir of lipophilic or hydrophilic drugs. Surfactants and cosurfactants usually are excellent solubilizers for a variety of compounds. When an emulsion system comes into contact with the skin, the drug can be readily transported from either the oil or the water phase to the skin surface. The mean diameter of the droplets in microemulsions is typically below 100 nm. Such a small particle size yields a very large interfacial area, from which the drug can quickly be released into the external phase where *in vivo* absorption takes place. The rapid release of the drug from microemulsions at the absorption surface (e.g. skin) possibly permits the desired rapid absorption kinetics. Drug release with pseudo-zero-order kinetics is likely to be obtained, depending on the volume of the dispersed phase, the partitioning of the drug between the interphases, and the diffusion rate of the drug.

Nagarajan and Ruckenstein (2000) developed a molecular thermodynamic approach to predict the structural and compositional characteristics of microemulsions. The theory can be applied not only to O/W- and

W/O-type microemulsions, but also to bicontinuous microemulsions. This treatment constituted an extension of the theoretical consideration of solubilization with micelles and mixed micelles, and also took into account the self-association of short-chain alcohols in the oil phase and the excluded-volume interactions among droplets. Microstructural features including the droplet radius, the thickness of the surfactant layer at the interface, the number of molecules of various species in a droplet, the size and composition of the droplets, and the distribution of the surfactant, oil, alcohol, and water molecules in the various microdomains were all taken into consideration. Further, these authors included chain packing, which depended on the curvature of the aggregates, the self-association of alcohols in the oil, the penetration of the oil molecules through the interfacial layer region of the microemulsion, the effects on chain packing and interfacial energies, and the interactions among the droplets. Two aspects included in this theoretical model were self-association of alcohols in oil phases and hard-sphere repulsive interactions among droplets. Further, the model allowed the identification of the transition from a two-phase droplet-type microemulsion system to a three-phase microemulsion system involving a bicontinuous microemulsion. The persistence length of the bicontinuous microemulsion was also predicted by the model. Finally, the model permitted calculation of the interfacial tension between a microemulsion and the coexisting phase.

3.1.2 Methods of characterization

Emulsions and microemulsions have been characterized using a variety of analytical techniques over the past 50 years. This implies that the characterization of emulsions and microemulsions is a rather complicated task. This is due to their complexity, namely the variety of structures and components involved in the systems, as well as the limitations associated with each analytical technique. Therefore, complementary studies using a combination of techniques are usually required to obtain a comprehensive view of the physicochemical properties and structures of emulsions and microemulsions. In this section, some of the analytical methods and techniques used to identify and characterize emulsions and microemulsions are briefly described.

3.1.2.1 Phase behavior

Phase diagrams have been used extensively in the study of emulsions and microemulsions. This approach has afforded comprehensive knowledge of how to characterize the domains of microemulsions and other systems (such as liquid crystals and coarse emulsions), although many times an unambiguous distinction between a microemulsion and an ultrafine emulsion was unclear or could not readily be made.

Phase behavior studies, with phase diagram determinations, are essential in the study of emulsion and microemulsion systems. They provide information on the boundaries of the different phases as a function of composition variables, temperature, and structural organization. In addition, phase behavior studies allow a comparison of the efficiency of different surfactants for a given application. The boundaries of one-phase regions can be assessed directly by visual observation of samples of known composition. The experiments for preparing a phase diagram need to be carefully designed and conducted, particularly for a system in which supersaturation is likely to occur. The slow equilibration in multiple regions, especially if liquid crystalline phases are involved, can make these determinations long, tedious, and subjective.

The phase behavior of interest for microemulsion and emulsion studies includes at least three components: water, surfactant, and oil. Although most of the formulations of practical interest (e.g. product development) consist of more than three components, the study of a simple system containing the basic three or four components is a prerequisite for understanding the behavior of such complex compositions. The phase behavior of three-component systems at a fixed temperature is best represented by a ternary diagram. A triangular prism would be suitable if temperature is considered as a variable. Another common approach representing the phase behavior are to keep either the concentration of one component or the ratio of two components constant. As the number of components increases in such characterizations, the experimental runs needed to define the complete phase behavior increase significantly, and the characterization becomes highly complex. Combining more than one component in the vertices of the ternary diagrams (pseudo-ternary diagrams) can make the characterization of these multiple component systems simpler. Most of the phase studies concerning microemulsions have been focused on determining one-liquid-isotropic phase boundaries. However, information about the number and compositions of the coexisting phases in equilibrium is of utmost interest in characterizing these systems.

3.1.2.2 Rheology

In general, the rheological properties of emulsions are influenced by a number of interacting factors, the nature of the continuous and dispersed phases, the phase volume ratio, and to a lesser extent particle size distributions.

A variety of products ranging from running liquids to thick semisolids can be formulated by altering the dispersed phase volume and/or the nature and concentration of surfactants. For emulsions of low internal phase volume, the overall consistency of the emulsion is generally similar to that of the continuous phase. Thus, W/O emulsions are usually thicker than O/W emulsions. Within the dispersed phases, waxy components such as cetyl alcohol and stearyl alcohol significantly increase the viscosity of the emulsions at room temperature. The rheological properties of emulsions can be controlled by the addition of natural or synthetic polymers that import plastic or pseudoplastic flow properties. Some mixed emulsifiers interact in water to form a viscoelastic continuous phase to give a semisolid O/W cream.

A recent study on the rheology of high internal phase emulsions (HIPEs) suggested trends that were consistent with shear deformation and the formation of highly elongated drops that subsequently ruptured to produce uniform, small droplets (Welch et al., 2006). In the search for a predictive tool to aid in the manufacture and use of HIPEs, the rheological data of these shear-thinning HIPEs were compared to the data from different models in the literature. Existing models did not correctly account for the effect of a high internal phase viscosity on the rheological properties of the HIPE. These authors concluded that the internal phase viscosity did not seem to affect the shear-thinning property and the surfactant apparently played an important role, possibly through its modification of the interfacial tension and continuous phase rheology.

3.1.2.3 Particle size and dispersion

Droplet size distributions in pharmaceutical and cosmetic emulsions are important from both stability and biopharmaceutical considerations. An emulsion containing a large particle size usually shows a strong tendency to coalescence. Droplet size distributions are substantially influenced by the type and concentration of surfactants used in emulsions and microemulsions, and by the preparation methods. From a biological point of view, fine emulsions may enhance drug absorption through the skin by rapid release and increase of the contact area of the internal phase with the skin.

Dynamic light scattering, also referred to as photon correlation spectroscopy (PCS), is used to analyze the fluctuations in intensity of scattering by droplets due to Brownian motion. The self-correction function is measured and gives information on the dynamics of the system. Dynamic light scattering has been widely applied in the study of emulsions and microemulsions to determine emulsion and microemulsion droplet size and shape. In these experiments, the intensity of scattered light is measured at various angles for different concentrations of the emulsion or microemulsion droplets. At sufficiently low concentrations, provided that the particles are small enough, the Rayleigh approximation can be applied. Droplet size can be estimated by plotting the intensity as a function of droplet volume fraction. Although dynamic light scattering measurements are relatively easy and quick, extrapolation of the results to infinite dilution is problematic in most emulsion and microemulsion systems.

3.1.2.4 Zeta potential

Zeta potential, or ζ potential, is an abbreviation for electrokinetic potential in colloidal systems. From a theoretical viewpoint, zeta potential is the electric potential in the interfacial double layer of a dispersed particle or droplet versus a point in the continuous phase away from the interface. In other words, zeta potential is the potential difference between the mobile dispersion medium and the stationary layer of the dispersion medium attached to the dispersed particle. The most important factor that affects zeta potential is the pH of the medium. Other factors include ionic strength, the concentration of any additives, and temperature.

The significance of zeta potential is that its value can be related to the short- and long-term stability of emulsions. Emulsions with high zeta potential (negative or positive) are electrically stabilized while emulsions with low zeta potentials tend to coagulate or flocculate, possibly leading to poor physical stability. In general, when the zeta potential of an emulsion is high, the repulsive forces exceed the attractive forces, resulting in a relatively stable system.

It was reported by Losso et al. (2005) that the surface properties of most emulsions were associated with their half-life stability. In particular, the half-life stability of phosphatidylcholine-stabilized emulsions with a zeta potential of −11 mV was shorter than that of phosphatidylglycerol-stabilized emulsions with a zeta potential of −53 mV. It was suggested that the emulsions with zeta potential values of −11 to −20 mV were close to the threshold of agglomeration, while the emulsions with zeta potential values of −41 to −50 mV had good stability. In a study of tristearoylglycerol crystals in olive oil, Arts et al. (1994) found that phosphatidylcholine was the most effective additive for increasing the zeta potential when compared to oleic acid, stearic acid, and monostearoyglycerol – generating a positive zeta potential from about 10 to 60 mV. Additionally, the zeta potential was strongly influenced by the impurities. The zeta potential is one of multiple indications of physical stability. However, it is sometimes not a directly relevant parameter for assessing stability when the difference of zeta potentials among various emulsions is small. As was reported by Roland et al. (2003), the authors did not observe any correlation between zeta potential (−43.1 to −50.2 mV) and overall stability. In fact, the most visually stable emulsions in this case exhibited the lowest zeta potential. It appears that absolute differences in zeta potential values should be at least 10 mV to allow prediction of distinct stability. Other factors, such as size reduction, also played important roles in stabilizing the emulsions.

Methylparaben (MP) was used as a model compound for investigating the effect of surface charge on the rate of degradation in oil-in-water emulsions (Pongcharoenkiat et al., 2002). The surface charge was varied by adding small amounts of phosphatidylglycerol (anionic) or stearylamine (cationic) to a standard intravenous lipid emulsion stabilized by egg phospholipid. The rates of hydrolysis at pH 8.0 in the water phase, oil phase, interface, and aqueous micellar phase were determined by application of a four-phase kinetic model. The rate of hydrolysis in the aqueous phase was found to be dependent on the zeta potential. This was attributed to the effect of surface charge through the phenomenon known as surface acidity. MP in the aqueous phase hydrolyzed at a rate associated with the pH of the microenvironment, not the pH of the bulk. The effect of the partition coefficient of the solute was studied by substituting ethylparaben (EP), propylparaben (PP), and butylparaben (BP) for MP in the emulsions. The rate of hydrolysis was inversely related to the partition coefficient. The effect of surface charge on hydrolysis was observed in the emulsions containing MP and EP, while partitioning had the greatest effect on the emulsions containing PP and BP. Therefore, the effect of surface charge predominated when the partition coefficient was small.

3.1.2.5 Pulsed-field-gradient spin echo nuclear magnetic resonance

Self-diffusion measurements using the pulsed-field-gradient spin echo nuclear magnetic resonance (PFGSE NMR) technique allows simultaneous and rapid determination of the self-diffusion coefficients (in the range of 10^{-9} to 10^{-12} m^2/s) of many components. The PFGSE NMR technique has been used extensively to investigate the size distribution in a variety of emulsion systems by measuring the diffusion of fluid contained inside emulsion droplets (Callaghan et al., 1983; Van den Enden et al., 1990; Loennqvist et al., 1991; Li et al., 1992; Hakansson et al., 1999). A detailed review of the PFGSE NMR technique has been published by Price (1997).

The PFGSE NMR method has been applied to determine particle size polydispersity, as it is not subject to the usual limitations of standard techniques (Ambrosone et al., 2004). In particular, it does not require a prior hypothesis on the form of the particle size distribution, or the prerequisite that suitable experimental conditions exist to enclose the whole volume fraction distribution. Very good agreement was found for the size distribution of an O/W emulsion obtained via NMR and optical microscopy data analysis, allowing for a broad range of droplet sizes (0.01–300 μm).

In a report by Hindmarsh et al. (2005), it was demonstrated that monitoring the oil droplet size distribution could detect swelling and a certain level of rupture of the inner compartment water droplet. Therefore, this method provided a more valuable and direct indication of the stability of W/O/W emulsions. The inner droplet size distribution of each W/O/W emulsion showed an increased mean radius and spread (standard deviation) of their distributions from the primary emulsion. The increased mean radius was attributed to the swelling of the droplets, but the increased spread of the distribution was presumably associated with the product of coalescence and rupture of the water droplets during the secondary emulsification and aging of the W/O/W emulsion. The presence of coalescence and rupturing was considered as an indication of the formulation instability. It has been demonstrated that a PFGSE NMR two-compartment model with inter-compartmental exchange can be used to describe the diffusional behavior of species in a W/O/W emulsion. This technique with established methodology permits detection of changes in the properties of W/O/W emulsions as a result of variations in formulation compositions. It is possible to obtain a level of information with a single PFG NMR that otherwise would have had to be gained with a number of experiments using different analytical techniques.

Coarsening of surfactant-stabilized oil-in-water emulsions ($C_{12}E_5$–decane–D_2O) induced by a quick temperature drop from their microemulsion phase was studied using ^{1}H PFGSE NMR spectroscopy by Hedin and Furó (2001). The results showed that the tested emulsions exhibited regular Ostwald ripening. Both the actual amounts of the expelled oil and the obtained coarsening rates agreed well with theoretical estimates.

3.1.2.6 Interfacial tension measurements

Interfacial tension measurements are useful in the study of the formation and characterization of emulsions and microemulsions. Ultralow values of interfacial tensions are found to be correlated with phase behavior, particularly the existence of a surfactant phase, or middle-phase emulsion or microemulsion in equilibrium with aqueous and oil phases. Ultralow interfacial tensions can be measured with a spinning-drop apparatus. Interfacial tensions are derived from the measurement of the shape of a drop of the low-density phase.

A study by Forgiarini et al. (2001) indicated that low values of equilibrium interfacial tensions and phase equilibrium involving a lamellar liquid crystalline phase were probably required but not sufficient to obtain nano-emulsions in this system. The key factor for nano-emulsion formation has been attributed to the kinetics of the emulsification process. The change in the natural curvature of the surfactant during the emulsification process may play a major role in achieving emulsions with small droplet size.

Characterization of the interfacial tension of an emulsion with the use of oil-soluble ethylcellulose indicated that the O/W emulsions can be stabilized in the absence of conventional low-molecular-weight surfactants (Melzer et al., 2003). Measurements of the interfacial tension confirmed that ethylcellulose was positively adsorbed at the water–oil interface with a diverse range of lipids. The appearance of the tested model emulsions was dependent on the processing temperature. At low temperatures (15°C) cream-like O/W emulsions were obtained, while processing at 30°C yielded fluid W/O lotions. Investigation of the microstructure showed that the surface of the emulsion droplets was covered with fine particles that formed a mechanical barrier. These colloidal particles were precipitated when the ethylcellulose dissolved in the lipids came into contact with water. Thus, ethylcellulose was demonstrated to represent a new type of particulate polymeric emulsifier.

3.1.3 Stability evaluation

Probably the most important consideration with respect to pharmaceutical and cosmetic emulsions is the physical and chemical stability of the finished product, related to the shelf-life of such products. The stability of a pharmaceutical emulsion is characterized by the absence of coalescence of the internal phase, absence of creaming, and maintenance of the uniformity and elegance with respect to appearance, order, color, and other physical properties.

The physicochemical principles underlying emulsion formulation and stabilization are complex and will not be discussed in depth here. Several factors for stabilizing emulsions, such as emulsifiers, polymeric excipients, and particle size distribution, have been briefly discussed in the previous sections. In short, an emulsion is formed when two immiscible liquids, water and oil, are mechanically agitated. When this occurs in the absence of any form of interfacial stabilization during agitation, both liquids will form droplets that rapidly flocculate and coalesce into two separate phases on standing. Flocculation is the term used to describe the close accumulation of two or multiple dispersed droplets without loss of the interfacial film. The interaction among the droplets is largely the result of Van der Waals forces. The flocculated droplets may then coalesce into one large droplet to reduce the surface tension. In reality, there is a brief period where one of the phases becomes the continuous phase because the droplets of this liquid coalesce more rapidly than the droplets of the other. In general, physical stability of an emulsion is determined by the ability and rate of an additive, such as an emulsifier, to counteract the Van der Waals attractions, thereby retarding flocculation and coalescence of the dispersed phase. This may be achieved by the establishment of energy and/or mechanical barriers among the droplets. In addition, the emulsion stability can be enhanced by increasing the viscosity of the continuous phase. While it is true that increasing the viscosity of the continuous phase will reduce the rate of flocculation of droplets, a stable system is more likely to be obtained when the continuous phase is gelled and the droplet diameter is $<0.1\ \mu m$.

In pharmaceutical and cosmetic applications, a commonly used approach for improving the physical stability of emulsions is to enhance the energy barrier among the dispersed droplets. The basis of the energy barrier is that droplets experience repulsion when they approach each other. Repulsion can be generated either by the establishment of an electric double layer, or by sterically adsorbed non-ionic surfactants or polymeric excipients on the droplet surface. Electrostatic repulsion is provided by ionic components in the oil–water interface, which will orient such that the polar ionic group faces the water and the hydrophobic tail points inwards. Electrostatic repulsion occurs when two similarly charged droplets approach each other, preventing or reducing the rate of flocculation. It is obvious that this mechanism is only appropriate for O/W emulsions. In addition, an emulsion stabilized by electrostatic repulsion is sensitive to the presence of additional electrolytes that may disrupt or change the electrical double layer. Steric repulsion may be produced using polymeric substances in the formulation. Repulsion force is generated primarily by polymer adsorption on the droplet surface. For instance, polyacrylic acid polymers linked to hydrophobic chains (Pemulen™, BFGoodrich) have been used as primary emulsification systems in O/W formulations because the polyacrylic acid chain, anchored to the oil droplet, forms both strong steric and electrical barriers at the interface.

3.2 APPLICATION OF EMULSIONS FOR TOPICAL AND TRANSDERMAL DRUG DELIVERY

Emulsions are the most commonly used formulation platform for both dermatological and cosmetic products. As a topical drug delivery system, emulsions have been formulated in a variety of forms including lotions, creams, foams, sometimes emulsion-gels and emulsion-ointments. Emulsion-based formulations have physically, chemically, and biologically superior compatibility with the skin compared with other topical and transdermal dosage forms, such as solutions, ointments, and patches. The outermost layer of the skin, known as the stratum corneum, is comprised of 10–20% lipids and 70–80% proteins by dry weight. Under normal skin conditions, the stratum corneum contains 15–20% water. However, in fully hydrated skin, it could contain up to 300% water (Scheuplein and Blank, 1971; Wilkes et al., 1973). The surface of the skin, particularly in the facial and scalp region, is typically covered with a lipophilic sebum layer comprised of wax, wax esters, triglycerides, and fatty acids, as well as a small fraction of water resulting from normal perspiration. Therefore, upon topical application, emulsions comprised of well-dispersed oil and water phases closely resembling conditions at the surface of the skin have the greatest compatibility with the skin. Compared to ointments, emulsions do not block the normal evaporation of moisture, resulting in less discomfort. Additionally, emulsions are much less greasy and therefore easy to wash away with water, leaving no palpable residue. Compared to liquid formulations, the viscosity of emulsions can be easily adjusted to prevent running during and after application, and to better control the surface area of the application. Moreover, emulsions are cosmetically elegant and more appealing to consumers. As a delivery system for both cosmetic and pharmaceutical active ingredients, emulsions are compatible with compounds that are either water soluble or poorly water soluble, making the preparation of molecular dispersions simple. In the case of a suspension matrix, emulsions have the ability to mask the dispersed particles while maintaining the integrity and consistency of the formulations.

3.2.1 Drug transport from an emulsion to the skin

Upon application of the emulsion, the drug delivery process occurs in a series of steps. When the emulsion is applied to the skin, it first releases the drug from the formulation, then the drug partitions into the skin surface, and finally the drug diffuses through the skin surface into the underlying tissues. While the inherent physicochemical properties of the drug molecules are of utmost importance in the delivery process, the choice of excipients used as delivery vehicles can also substantially influence all phases of the drug delivery process. Roberts et al. (2002) provide a summary and illustration of the fundamentals and recent advances in drug transport across the skin from various topical and transdermal formulations.

3.2.1.1 Solubility and thermodynamic activity

The release of a drug from a formulation depends on its thermodynamic activity in the vehicle (Higuchi, 1960). Thermodynamic activity is a function of the drug concentration, drug solubility, percentage of drug saturation in the vehicle, and both drug–drug and drug–excipient interactions. Under certain conditions, drugs with the same level of saturation or fractional solubility in different vehicles have similar thermodynamic activity. Assuming that drug–skin and vehicle–skin interactions do not substantially alter normal skin transport, the drug-saturated formulations possess the highest thermodynamic activity, resulting in maximal drug release and skin delivery potential. The strength of a drug in a formulation does not always directly correlate to the potential of drug release from that formulation to the skin. That is, a higher strength of the drug does not necessarily mean that more medicine is delivered. Given the same strength of a drug, on the other hand, the release potential of the drug from a vehicle with low drug solubility is superior to that from a vehicle with high drug solubility (Flynn and Weiner, 1993). When drug loading in an emulsion is higher than the solubility, drug particles become dispersed. In this case, drug dissolution (which is also a function of drug particle size) may substantially impact drug release and possible delivery. Therefore, drug solubility in a topical formulation is one of the important indications of drug deliverability. Solubilization of a drug in a vehicle may be achieved through simple solvation or more complicated processes, such as partitioning into micelles, entrapping in micro/nanostructures (cyclodextrin, for example), binding with polymers, and forming complexes with excipients. An emulsion is a multiple-phase system containing hydrophilic components such as water, lipophilic components such as oils and fatty alcohols, and amphiphilic components such as surfactants. Throughout the various phases of emulsion application, drug solubility may vary. For example, the molecules of an emulsifier form a film with a certain microstructure on the interface of oil/water droplets, which are different from their orientation as micelles in an aqueous solution. Therefore, drug solubility in an emulsion is not the simple addition of drug solubilities in the aqueous solution and in the oil phase.

It is very challenging to predict drug solubility in a multi-excipient formulation based only on the known drug solubility of each individual excipient. Conventional methods for determining drug solubility in solvents include a process of separating excess drug solids from a saturated vehicle, through either filtration or centrifugation. This process is not directly applicable to separating drug particles from emulsions, because of the relatively high viscosity and multiphase property of emulsions. It is especially not applicable to semisolid emulsions. Therefore, several different techniques have been developed to measure drug solubility in semisolids, the most notable of which is for creams. Microscopy has been commonly used to identify the highest drug concentration that can be achieved in a formulation before saturation occurs. This technique gives a good indication of a drug's solubility, but requires a preliminary screening to find critical concentration ranges and appropriate dose increments for the drug loading in order to accurately estimate the solubility. Tanaka et al. (1985) obtained the solubility of dexamethasone acetate in various creams by first testing the drug solubility in the aqueous surfactant solutions and in the oil components separately, and then estimating the drug solubility in the creams based on the formulation composition and individual solubility data. To estimate the solubility in stearyl alcohol, a waxy excipient at room temperature, the same authors also determined the partition coefficient by mixing an aqueous drug solution with stearyl alcohol at high temperature and then cooling down to reach equilibrium. This method provides an approximate estimation of drug solubility in creams, subject to several assumptions. One assumption is that the drug solubility in creams is linearly related to the drug solubility in each component. This assumption, however, does not consider the potential effect of emulsion microstructure on solubility. More recently, Spangler and Harris (1999) reported a novel percolating method to determine drug solubility in petrolatum-based formulations. A series of formulations at incremental drug loadings were sheared to disrupt their internal microcrystalline structures and placed in taut cheesecloth pouches. The bleeds from the pouches were collected and the drug concentrations in the bleeds were analyzed. After plotting the drug concentrations in the bleeds versus the formulation potencies, the onset of the plateau region on the plotted curve was defined as the solubility. On comparison, this method showed a good correlation with the microscopic method.

Flynn's lab has developed two useful methods to determine the drug solubility in creams and other semisolids. Both use a silicone sheet loaded with a suspended drug as a donor and a placebo formulation as the receiver. In the first method, the placebo formulation is spread on the drug-loaded sheet, which is then rolled up and wrapped with foil. In the second method, a piece of drug-loaded sheet is placed in a metal well, again with a layer of placebo on the sheet, with the apparatus then sealed using a metal cap. Drug solubility can be determined by measuring the drug concentrations once drug diffusion into the placebo formulation reaches equilibrium. Depending on the drug and vehicle properties, it could take 1–4 weeks to reach equilibrium (Flynn and Weiner, 1993; Pillai et al., 2001). Despite all these efforts, a simple, rapid, and accurate method for solubility measurement in emulsions and other semisolids is yet to be developed.

3.2.1.2 Drug release from emulsions

The first stage of drug transport from a formulation to the skin is the drug release. As shown in equation (3.1), the first 30% of release from a solution matrix is a function of concentration (or thermodynamic activity) and diffusivity through the formulation matrix (Higuchi, 1963; Flynn et al., 1999):

$$M = 2C_0(Dt/p\pi)^{1/2} \tag{3.1}$$

In this equation, M is the amount of drug release into the skin, C_0 the drug concentration in the matrix, D the diffusion coefficient, and t the time. Equation (3.2) shows the drug release from a suspension matrix when an M vs. $t^{1/2}$ linear stage is achieved:

$$M = (2QDC_st)^{1/2} \tag{3.2}$$

where Q is the total amount of drug in solution and suspension and C_s is the drug solubility in the matrix. The time to reach the M vs. $t^{1/2}$ linear stage is dependent on the partition coefficient.

The methodology of an *in vitro* release test and its applications have been discussed and summarized in an AAPS–FDA sponsored workshop report (Flynn et al., 1999). The testing conditions were evaluated and established by Shah et al. (1999). Eros et al. (2003) recently investigated the factors governing the release of salicylic acid from creams, gels, and ointments. It was demonstrated that: (1) an increase of the drug loading from 1% to 5% (dissolved and suspended) exponentially accelerated the release rate; (2) the distribution conditions of salicylic acid in each phase had a significant effect on the release rate; (3) a decrease in the particle size enhanced the dissolution rate; (4) an increase in the solubility using non-ionic surfactants facilitated the release rate, but not in the case where complexes were formed using cyclodextrin; and (5) the effects were dependent upon the character and composition of the vehicles. This is an interesting and extensive study, which provided a quantitative insight into the optimization of semisolid formulations, particularly emulsions, for salicylic acid. It is known that most of these effects, if not all, are compound dependent, and that a release test is often not a reliable indicator of drug delivery because the use of an aqueous solution as the receptor phase in release tests differs significantly from use of a formulation–skin interface.

3.2.1.3 Drug partition from an emulsion to the skin

The second step of topical drug delivery is the partitioning of the drug from the emulsion to the skin surface. Here, the stratum corneum, which is the outermost layer of skin, is the primary barrier for drug delivery, though other skin layers also contribute to the barrier function. Therefore, the partition coefficient of the drug between a vehicle and the stratum corneum is another critical parameter. Since the partitioning of drugs between octanol and water (*P*) can be readily determined or calculated, the log *P* or *c*log *P* is often used as a surrogate for estimating the drug partitioning property from an aqueous phase to the skin surface (Potts and Guy, 1992). However, the partition coefficient of the drug between the stratum corneum and an aqueous solution or suspension (K_{sc}) is more relevant to drug delivery across the skin. The Hidebrand solubility parameter approach offers a powerful tool to understanding drug–excipient interactions, excipient–excipient interactions, and even drug/excipient–stratum corneum interactions. It has been used for predicting skin permeation (Martin et al., 1980; Sloan et al., 1986; Groning and Braun, 1996). Using this approach, and assuming a stratum corneum solubility parameter of 10 $cal^{1/2}\ cm^{3/2}$, Sloan et al. found that isopropyl myristate (IPM) was the preferable solvent for theophylline partitioning into the stratum corneum. However, this technique is seldom used to estimate K_{sc}, primarily due to technical complexity. Wiechers et al. (2004) recently proposed a unique approach to selecting suitable emollients and/or oily excipients to achieve desirable delivery rates of active ingredients. The relative polarity index (RPI) was applied to compare the polarity of the active ingredients to those of the stratum corneum and the emollients in formulations. The polarity of an active agent is expressed by log $P_{oct/water}$. For example, the polarity of 1-butanol was 0.88 and the polarity of the stratum corneum was set at 0.8. The difference between the polarity of the stratum corneum and the active ingredient (the RPI) determines the polarity of the formulation required to drive the active compound into the stratum corneum. Similar to the solubility parameter method, this approach is based on an assumption that the polarities of the stratum corneum and a testing formulation remain unchanged during permeation of the active ingredients. In fact, hydrophilic and lipophilic excipients, as well as surfactants from emulsions, are likely to partition into the stratum corneum and thereby alter the drug distribution between the stratum corneum and a vehicle. If we use K_{sc} as a predictive parameter for drug delivery, the possibility of errors resulting from real-time dynamic partitioning at a non-equilibrium state should be considered, as K_{sc} is measured under the condition that the stratum corneum is saturated by the vehicle. When applying a topical formulation to the skin surface using finger or applicator, a thin layer of film (10–30 μm) forms such that the vehicle interaction with the stratum corneum is much less than when the stratum corneum is saturated by the vehicle. Therefore, the dynamic partitioning of an active ingredient from a formulation to the stratum corneum could be somewhere between the estimations based on solubility parameters or RPI and the vehicle/stratum corneum partitioning coefficient, again depending upon the property of the active ingredients and the excipients.

3.2.1.4 Drug diffusion across the stratum corneum

The final primary step of topical drug delivery is diffusion of the drug across the stratum corneum, i.e. drug molecules passing through the skin surface and permeating into deeper tissues. Although the other layers of the skin affect the permeation of drug across the bulk skin – for example, the dermis presents a barrier to highly lipophilic compounds – it is well accepted that the stratum corneum is the principal barrier to drug transport. The stratum corneum is essentially metabolically inactive tissue comprised of acutely flattened cells stacked in hexagonal building blocks, formed from once-living cells. These structural building blocks comprising 15–25 layers cover most of the body's surface (Wilkes et al., 1973). Intercellular lipids function as a "mortar," sealing the block structure together. The densely packed keratin platelets caulked with intercellular lipids make the stratum corneum an incredibly resistive barrier to penetrate. The rate of diffusion of a molecule is proportional to its potential gradient, and the diffusion coefficient of a molecule in a given matrix can be experimentally determined using Fick's law. Commonly used methods to estimate the diffusion coefficient of an active substance involve the experimental measurement of flux and lag time, with calculations using equations (3.3) and (3.4):

$$D = h^2/6t_{lag} \tag{3.3}$$

$$D = Jh/KC \tag{3.4}$$

In these equations, D is the diffusion coefficient, h the diffusion distance or thickness of the membrane, t_{lag} the lag time, J the flux, K the partition coefficient, and C the drug concentration on the donor side. The diffusion coefficient is compound dependent, affected by temperature, matrix property, and sometimes drug concentrations. Therefore, a delivery vehicle that penetrates into the stratum corneum and affects the matrix property could substantially change the diffusion coefficient.

3.2.1.5 Dose applied and dosing frequency

To evaluate the drug delivery property of emulsions, a commonly used method is to measure the flux and tissue concentrations using diffusion cells and human skin. In general, the skin flux and tissue concentration of a drug reflect a collective result of the thermodynamic activity of the drug in the formulation, the drug release from its vehicle, the partitioning of the drug from the vehicle to the skin surface, and the diffusion into/through the skin tissues. There are two major concerns in conducting experiments and analyzing the results. The first is the dose (volume or weight) of the applied formulation. In order to understand the excipient effect, an infinite dose (>100 mg formulation per cm^2) is often applied as a donor phase. Application of an infinite dose provides a relatively constant driving force for the active and inactive ingredients, allowing it to quickly achieve and maintain a steady state. The results from these studies are better correlated to the thermodynamic activity, as well as the partitioning and diffusion properties of the tested compounds. However, in practice the actual dose of an emulsion is closer to 2–5 mg per cm^2, which is much smaller than the doses used in the infinite dose studies. Furthermore, due to evaporation, the percentage of volatile solvents in the emulsions, such as water, is quickly reduced. The absorption of the excipients that have a different skin permeability to the active ingredient also changes the composition of a delivery system. In other words, the driving force starts changing immediately after topical application. Due to the evaporation of the volatile solvents, the degree of saturation and thermodynamic activity may increase. As a consequence, the drug delivery can be enhanced. Or the opposite may occur – delivery could be reduced due to precipitation of the active ingredient as a result of rapid evaporation of a volatile solvent. The overall outcome and magnitude of this effect are associated with drug solubility in both the volatile and nonvolatile solvents. In some ways, finite dosing (<10 mg/cm^2) has advantages over infinite dosing for assessing skin delivery from a topical emulsion, especially in the product development stage.

The second concern relates to dosing frequency. It is questionable whether the evaluation of skin delivery obtained from a single application of either a finite dose or an infinite dose can provide sufficient information to relate to real-world conditions where the product may be applied twice a day for weeks or months. As mentioned previously, some of the excipients penetrate into the stratum corneum and the viable epidermis, resulting in hydration, solvation and/or disturbing of intercellular lipids, and so altering the partitioning and diffusion of an active ingredient in the skin. These effects often vary greatly between single-dosing studies and multiple-dosing studies. To better understand the dosing frequency effect, a study using multiple finite dosing is recommended. Although the target sites for topical and transdermal deliveries are radically different, not much can be done once the drug molecules cross the epidermis. Most of the excipient molecules that penetrate the epidermis are absorbed by capillary circulation between the epidermis and dermis. Consequently, the effect of the vehicle on delivery is diminished.

In summary, emulsions offer unique compatibility with both lipophilic and hydrophilic compounds – both drugs and excipients. When used as a topical or transdermal drug delivery system, the vehicle effect on delivery is dynamic because of composition changes after application. This effect is limited to the superficial layers of the skin, because dilution and absorption of the excipients occur after penetration through the epidermis.

3.2.2 Investigation of topical and transdermal delivery in human subjects

As a drug delivery system, emulsions consist of lipophilic, hydrophilic, and amphiphilic components with a great capacity to influence drug thermodynamic activity in vehicles, drug partitioning from vehicles into the stratum corneum, and drug diffusion in and through the skin tissues. In a randomized, three-way crossover study on 12 subjects, Seth (1992) found that an emulsion gel formulation containing 1% diclofenac delivered approximately half of the active ingredient to systemic circulation, compared with a solution gel formulation. This indicates that the systemic efficacy of the emulsion was likely lower than that of the solution, though the local effect was not necessarily relevant. In another three-way crossover study on six human subjects, the systemic exposure of ibuprofen after a single topical application was gel > cream > ointment – based on C_{max}, t_{max}, and area under the curve (AUC) (Seth, 1993).

While the majority of emulsion dosage forms are intended for localized delivery and treating skin conditions, there is an increasing interest in developing emulsion-type formulations for transdermal drug delivery. As shown in Table 3.1, an emulsion containing estradiol has been commercialized as a transdermal delivery system for hormone replacement therapy (HRT).

Table 3.1 Estradiol topical emulsions*

Product	API/formulation	Primary emulsifier	Oil phase	Stabilizer	Others
Estrasorb® lotion	2.5 mg estradiol/ 1.74 g	Polysorbate 80	Soybean oil		Ethanol, water
Premarin® vaginal cream	0.625 mg conjugated estrogens/g	Sodium lauryl sulfate	Cetyl alcohol, cetyl esters wax, white wax, mineral oil, methyl stearate	Glyceryl monostearate, propylene glycol monostearate	Benzyl alcohol, glycerin, water
Estrace® cream	0.1 mg estradiol/g	Sodium lauryl sulfate	White ceresin wax, stearyl alcohol	HPMC, mono- and diglycerides	Methylparaben, EDTA, TBHQ, propylene glycol, water

* Abbreviations: EDTA, ethylenediaminetetra-acetate; HPMC, hydroxypropylmethylcellulose; TBHQ, tertiary butylhydroquinone.

A locally applied estrogen cream was initially designed for treating vaginal atrophy, to avoid the undesirable systemic effects that occur with orally administered hormone treatment. However, systemic absorption of estrogens from vaginal creams was also found to be clinical significant (Schiff et al., 1977; Rigg et al., 1978). After this initial failure, the emulsion-based products Estrace® cream and Premarin® cream were developed to achieve more targeted delivery of estrogen hormones. In a separate study, Martin et al. (1979) measured the systemic absorption from these emulsion-based products in comparison with oral hormone therapy. They found that the serum levels of E1 and E2 at 12 hours after local application were the same as those at 15 days of daily local dosing. This indicated that a topical formulation could achieve the rapid, efficient, and sustained delivery of the hormone to the local target as well as the systemic targets without the undesirable systemic side-effects. These findings opened a door to a new administration route, which is transdermal delivery for HRT. Since then, various transdermal dosage forms have been developed and commercialized for estradiol. Among them, Estrasorb® is an estrogen emulsion particularly for transdermal delivery. Although a head-to-head comparison study has not been reported, it is possible that the emulsion-based formulation reaches a steady state of estradiol slower in human subjects when compared to the estradiol gel (Estrogel®) and even the transdermal patch (Climara®) (PDR, 2007). Since HRT is a long-term treatment, the fast onset associated with rapid absorption is not as critical a factor as comfort and mildness of the formulations. In this sense, an estrogen emulsion may have better acceptance with patients.

3.2.3 Formulation strategies for emulsion-based products

There are several informative reviews and book chapters illustrating the general strategies for developing dermatological formulations, including emulsion-based formulations (Barry, 1983; Davis et al., 2002; Walter and Brain, 2002). As a formulation scientist in the pharmaceutical or cosmetic industry, it is essential to have a well-organized formulation development strategy in the first place. Emulsions have been used as a basic topical dosage form for many decades, and tremendous efforts have been made to understand the mechanisms of emulsions and skin delivery. Yet, formulation development is still semi-empirical and relies heavily on experience. Throughout the product development stage, the formulation scientist should periodically ask a few key questions to make sure the project is on the right track:

1. Is the formulation physically, chemically, and biologically (i.e. antimicrobial activity) stable under the anticipated conditions throughout the product shelf-life – i.e. through manufacturing, transport, shelf storage, and normal consumer usage conditions?

2. Can the formulation deliver a therapeutically sufficient amount of the active ingredient(s) into or through the skin?
3. Is the formulation safe for its intended use?
4. Do the specifications and the information made available about the product adequately meet consumer regulations?
5. Are the formulation and manufacturing process robust enough to be scaled up to commercial production levels? What are the allowable deviations in the formulation composition, manufacturing conditions, and packaging process?
6. Does the product meet the product profile for clinical use and market positioning?
7. Is the product costly to produce?
8. Can the formulation or process be patented, and does the product infringe upon existing patents?
9. What are the potential issues with raw materials, active pharmaceutical ingredients (APIs), manufacturing, packaging, drug delivery, safety, and cost? Are there backup plans?
10. Can the project/product be delivered on time?

3.2.3.1 Project timeline

External timings often play a big role in developing a project plan. An experienced formulator is usually good at balancing between science/technology and timeline/resources. It is very common that a formulator has to deliver a pharmaceutically acceptable prototype formulation within 6 months, whereas developing an optimized formulation often requires additional time and resources. Therefore, knowing the difference between what is crucial for an acceptable emulsion product and what is desirable or nice to have allows a formulator to deliver a project in a timely manner. Table 3.2 provides a general checklist for the development of an emulsion dosage form.

3.2.3.2 Consideration of thermodynamic activity

As discussed before, the thermodynamic activity of an API is the inherent driving force for drug delivery from a formulation. In general, drug-saturated emulsions with or without suspended drug particles provide maximum delivery potential. However, most emulsions are formulated at drug potencies much lower than the maximum drug solubility in the vehicles, and this may reduce the efficiency of drug delivery. The low level of saturation for drugs in these formulations could result from the good solubilizing property of the vehicles and low drug strength needed. In these cases, a low degree of saturation of an API in the formulations may retard

Table 3.2 A general checklist for development of emulsion dosage forms

I. Literature search (technical reports, patents, publications, presentations, websites, databases)

1. Summarize the physiochemical properties of API and potential excipients (solubility, pK_a, log P, etc.)
2. Understand the physical and chemical stability of the ingredients
3. Be aware of the regulatory acceptance
4. Search the use of API and excipients in commercial products and select one or two benchmarks
5. Compare the safety profiles of excipients
6. Establish a product profile

II. Selecting primary excipients

1. Research the compendia excipients
2. If possible, start with the excipients that may have been used in the commercial products
3. Determine API solubility in the primary excipients
4. Estimate API compatibility with the excipients
5. Assess potential interaction of the ingredients with skin
6. Find the cost, grade, quality of the ingredients, and qualify suppliers

III. Preparing prototype formulations

1. Formulate emulsions with different types and levels of emulsifiers, oil phase, polymers, buffers, and preservatives at small scale
2. Prepare emulsions with different API loadings
3. Preliminary physical and chemical stabilities at 5, 30, and 40°C
4. Preliminary process for preparation (order and rate of addition, rate and duration of mixing, rate and temperature of heating/cooling, light and oxygen effect, filling and packaging, etc.)

IV. Developing analytical methods to quantify

1. API and degradation products
2. Key excipients
3. Impurities in API and key excipients

V. Monitoring and testing

1. Appearance: uniformity, color, fluidity, cosmetic elegancy
2. Potency, degradation products, and impurity of API and key excipients
3. Particle size and size distribution
4. Rheological property
5. Zeta potential
6. Solubility in emulsion
7. Packaging and manufacturing compatibility
8. Antimicrobial activity

VI. Optimize formulations

1. Select two to five relatively stable emulsions
2. Find the boundary of the acceptable compositions (for example, the upper and lower limits of an emulsifier)
3. Finalize the manufacturing process and the deviation of process parameters (for example, the upper and lower limits of temperature for mixing)
4. Test stability at various temperatures and with heating/cooling cycles
5. Monitor the antimicrobial activity at 70–80% of the preservative strength

delivery into the skin, although it is sometimes intentional to develop a less-saturated formulation that can achieve sustained delivery. On the other hand, if drug loading is too close to saturation, a small change of the composition may cause supersaturation or precipitation of an API, such that the physical stability of an emulsion could be compromised. In addition, emulsions that are saturated or near saturation are also likely to have adsorption issues during manufacturing and packaging as well as during storage, which could reduce the drug potency of the finished product.

The thermodynamic activity of an API starts changing once an emulsion is applied to the skin as a result of the evaporation of volatile solvents, and sometimes the different absorption rates of an API and the excipients. These changes can occur within minutes, if not seconds. A formulator is able to estimate the direction and relative rate of the change according to the solubility and volatility of solvents. The evaporation of the solvents can be experimentally observed from the rate of weight loss and phase change under the microscope after applying a layer of emulsion to a glass plate. Obviously, the change of driving force on the surface of the skin greatly depends upon the properties of the formulation. Therefore, an assessment of the effect of thermodynamic activity on skin delivery should include the original driving force in the formulations and also the dynamic change of the driving force on the skin.

3.2.3.3 Selection of excipients

Selecting the right emulsion excipients is vital to the efficiency and smoothness of formulation development. For example, if a prototype emulsion is not stable or causes skin irritation due to its excipients, and in the worst-case scenario the formulator does not have a backup formulation, it would be difficult to start over again to select alternative excipients without delaying the project timeline. Therefore, a relatively conservative approach in selecting excipients, including backups, is preferable. Using precedent excipients in commercial products for initial screening is highly recommended. Formulators should make an effort to understand the properties of the excipients and their conditions of use in commercial products, including concentration ranges, compatibility with other ingredients, interaction with the skin, and potential interaction with the API. Publications and suppliers are often good sources of this information. If a novel or unprecedented excipient is considered as a potential ingredient, a prudent risk/benefit analysis must be conducted. For example, azone is an innovative and excellent permeation enhancer for skin delivery. A number of scientific studies have demonstrated the function, safety, and potential applications of azone for transdermal and topical products since 1982

(Stoughton and McClure, 1983; Hadgraft et al., 1993). However, as yet there is no FDA-approved product containing azone on the market. Another example is transcutol (diethylene glycol monoethyl ether), which is a very good solvent for many compounds that are poorly soluble in either water or oil. Although studies of the use of transcutol for topical formulations started decades ago, transcutol has only just recently been accepted as a topical pharmaceutical excipient (Inactive Ingredient Search for Approved Drug Products, http://www.accessdata.fda.gov/cder/iig). Due to safety concerns, regulatory acceptance of a new excipient can take much longer than a project timeline can afford. Therefore, a formulator is better off to avoid selecting unapproved excipients unless it is absolutely necessary. To accelerate the application of "new" excipients in pharmaceutical products, the International Pharmaceutical Excipients Council (IPEC) of America has recently developed a New Excipient Evaluation Procedure (NEEP). The NEEP panel can provide independent and regulatory acceptable evaluation of "new" excipients, and the FDA has agreed to consider (but not guarantee) the conclusions of the panel. This is good news for formulators who are interested in the use of "new" excipients (DeMerlis et al., 2008).

The most critical excipient in an emulsion is the emulsifier. Commonly used emulsifiers, particularly in pharmaceutical products, are shown in Table 3.3. The concentration limits are obtained from the FDA database of inactive ingredients. In general, the hydrophilic–lipophilic balance (HLB) values of emulsifiers provide guidance in selecting emulsifiers for either O/W or W/O emulsions. A mixture of more than one emulsifier often works better than a single emulsifier to improve physical stability. The use of a secondary emulsifier is becoming more common both in cosmetic and pharmaceutical products, not only for improving stability, but also for enhancing solubilities. Some examples include methyl glucose sesquistearate and PEG-20 methyl glucose sesquistearate in an adapalene cream (Differin®), lecithin and polysorbate 80 in an azelaic acid emulsion gel (Finacea®), methyl gluceth-10 and PEG-100 stearate in Tri-luma® cream, and polysorbate 60 and sorbitan monostearate in imiquimod cream (Aldara®). The inherent chemical stability of emulsifiers is also a factor of selection. Surfactants in ether forms are more stable than those in ester forms, but they are also more stable in the human body – which may not be desirable because of safety concerns. In addition, impurities found in emulsifiers could become problematic under some circumstances. Polyoxyl surfactants often contain trace amounts of peroxides, which sometimes cause oxidation of both API and inactive excipients in formulations. To resolve this issue, suppliers have developed higher quality emulsifiers, such as super-refined or ultrapure polysorbate 80, containing much less trace

Table 3.3 Commonly used emulsifiers in current pharmaceutical emulsions

Excipient	Hydrophilic–lipophilic balance (HLB)	Concentration (%)
Polysorbate 20	16.7	<15
Polysorbate 60	14.9	<5
Polysorbate 80	15	<9.4
Sorbitan monolaurate	8.6	<4.74
Sorbitan monopalmitate	6.7	<2
Sorbitan monostearate	4.7	<8
Sorbitan mono-oleate	4.3	<7
Methyl glucose sesquistearate	6	3.5
Lecithin	2–10	<1.4
Ceteth-2	5.3	2.5
Ceteth-10	12.9	2.5
Ceteth-20	15.7	<4
Ceteareth-30	17	<5
Sodium lauryl sulfate	40	0.5–2.5
Polyoxyl 100 stearate	18.8	2.1
Polyoxyl 100 glyceryl stearate	11.2	5
Polyoxyl 40 stearate	16.9	8.8

peroxides, metal ions, and other impurities. Using a purer emulsifier may significantly improve the stability of emulsions.

3.2.3.4 Stabilization of emulsions

As previously mentioned, selection of the appropriate emulsifiers and their concentrations is essential for the formulation of physically and kinetically stable emulsions – even though the emulsions themselves are thermodynamically unstable. In addition to the choice of type and concentration of an emulsifier, the formulation process, including the temperature, time, and rate of agitation or homogenization of mixing, order of addition and mixing of multiple excipients can influence particle size distribution and uniformity of the dispersion, as well as the physical stability of the finished products. Increasing viscosity often effectively enhances the physical stability of an emulsion; hence, polymeric thickeners are commonly formulated in topical formulations. However, viscosity is restricted by the dosage form. For example, the acceptable viscosity range for a lotion is much lower than that for a cream. Polymeric excipients have been added to cosmetic and dermatological emulsions to control rheological properties and so improve physical stability. Commercial examples include xanthan gum in Aldara® cream, hydroxyethylcellulose/xanthan gum in Klaron® lotion, Carbomer 934P in Differin® cream, hydroxypropylmethycellulose (HPMC)/Carbomer 1342 and Poloxamer 407 in Elocon® cream. Carbomer 1342 (Pemulen®) is a surface-active copolymer functioning both as an emulsifier and stabilizer in O/W emulsions. Because of its surface activity, rheological properties, and good compatibility with skin, Carbomer 1342 has been used as a primary emulsifier in some emulsions. These types of excipients function as thickeners to form a structural matrix, improve the cosmetic feel of the products, and facilitate ease of spreading on the skin. Interestingly, the increasing use of oily excipients such as petrolatum and of gelling agents such as Carbopol in emulsions has made it difficult to tell the difference between a cream and an ointment or a gel.

3.3 ADVANCES IN USING MICROEMULSIONS AS TOPICAL AND TRANSDERMAL DRUG DELIVERY SYSTEMS

As stated at the beginning of the chapter, an emulsion is a fine dispersion of two immiscible phases, an oil phase and a water phase, in the presence of emulsifier(s). It is kinetically stable because of the reduced interfacial tension in the presence of the surface-active agents (surfactants and/or amphiphilic compounds). If a dispersion system contains sufficient levels of one or more surface-active agents to cover the entire interface of the two immiscible phases, but not enough to make the oil and water phases completely miscible, the interfacial tension between the water and oil droplets can be reduced to near zero. This system is defined as a microemulsion. Table 3.4 shows the similarities and differences between emulsion and microemulsion systems.

Generally, microemulsion systems consist of high levels of emulsifiers (10–80%) and often a cosurfactant such as an amphiphilic short-chain alcohol or its ester. As a structurally complementary component, a short-chain alcohol together with a relatively long-chain emulsifier occupies the oil–water interface to form a more integrated layer. Though it depends somewhat upon the type and concentration of emulsifier/cosurfactant and the ratio of oil and

Table 3.4 Comparison of emulsions and microemulsions

Property	Emulsion	Microemulsion
Basic composition	Oil phase/water phase Emulsifier(s) <5%	Oil phase/water phase Emulsifier(s) 10–80% Cosurfactant >5%
Stability	Thermodynamically unstable	Thermodynamically stable
Appearance	Non-optically isotropic Turbid to milky, opaque	Optically isotropic Transparent to translucent
Particle size (μm)	0.1–100	<0.2
HLB of emulsifier and type of formulation	HLB 4–6, W/O HLB 8–18, O/W	HLB 3–8, W/O HLB 8–18, O/W
Preparation	Require external energy, agitation or homogenization	Spontaneous, mild mixing
Drug solubility	Lower	Higher
Drug deliverability	Depends	Depends

water components, the resulting reduction of surface energy leads to smaller droplet sizes. Unlike conventional emulsion systems, microemulsions are thermodynamically stable. The preparation of microemulsions requires only minimal external mixing to form a uniform dispersion. One advantage (and probably the biggest advantage) in choosing microemulsions as a dosage form is that no additional efforts are required to meet shelf-life requirements – physical stability of the dispersion results in a long shelf-life. Moreover, the solubility of lipophilic drugs can be substantially increased due to high levels of surfactants/cosurfactants. Also, the microstructure of the dispersions contributes to increased solubility. A number of publications have reported improved bioavailability of active ingredients from microemulsions. Several reviews have discussed the potential use of microemulsions for topical and transdermal delivery (Tenjaria, 1999; Pappinen and Urtti, 2006). However, the mechanism of topical and transdermal delivery using microemulsion systems is not yet fully understood, making it challenging to design a desirable formulation or improve the delivery of a microemulsion. Because relatively few excipients have been approved as safe for human use, there are a limited number of emulsifiers and cosurfactants available for the development of microemulsion products. Among them, polysorbate 80 and lecithin are the most widely used emulsifiers, with IPM, isopropanol, and ethanol as the most common choice for the cosurfactants. It is well known that some of these excipients potentially enhance drug permeation through the skin without forming microemulsions. Therefore, enhanced drug delivery observed from microemulsions is possibly just a result of the excipient effect. Similar to the discussion in the previous section, it is critical for formulation scientists reviewing the literature to find out whether the thermodynamic activity of the active compounds in the microemulsions is comparable with that in the control formulations, and whether the dose applied in the study is comparable to clinical applications.

When developing a topical or transdermal delivery system, it is preferable to avoid using a high concentration (>5%) of surfactants and cosurfactants such as IPM. In particular, the combination of a short-chain alcohol and IPM with emulsifiers may lead to elevated skin irritation. Most reported studies of microemulsions do not include *in vivo* skin irritation tests. It is general practice that safety studies for pharmaceutical products require much higher levels of skin exposure to the formulations when compared to those for the projected clinical doses. For instance, the dosing amount for skin tolerance or systemic exposure studies could be more than five to 10 times higher than clinical doses. Therefore, the intended excipients and vehicles must be safe at these elevated levels.

A few studies have been published comparing microemulsions with emulsions, and to demonstrate the *in vitro/in vivo* correlation of microemulsion. As shown in Table 3.5, Rangarajan and Zatz (2003) investigated the skin delivery of 1% α-tocopherol from various formulations. A finite dose was applied to pig skin for 24 hours. When compared with an IPM solution, a microemulsion containing 10% IPM (ME-1) emerged as the best delivery system. A microemulsion containing benzyl alcohol and diethylene glycol monothyl ether (ME-2) exhibited properties similar to an emulsion, demonstrating the enhancement effect of IPM both in the microemulsion and the control solution. In a study evaluating *in vivo* anti-inflammatory effects, Dalmora et al. (2001) monitored the release of piroxicam from gels and microemulsion gels. The release of the compound was slower from the microemulsion system, and its anti-inflammatory effect in rats was weaker compared to control. However, the difference was substantially minimized when β-cyclodextrin was formulated in both the microemulsion and the control. In this case, the microemulsion functioned as a reservoir that reduced skin delivery of the agent. Chen et al. (2007) found that the skin flux

Table 3.5 Comparison of microemulsions with other topical formulations*

Method	Dosage form	Composition	Result
1% α-tocopherol (Rangarajan and Zatz, 2003), *in vitro* permeation using pig skin, finite dose for 24 h	Solution	IPM	9.74 ± 0.89% of applied dose permeated *in vitro*
	Emulsion	5.38% diisopropyl adipate, 5.38% mineral oil, 2% DEA-cetyl phosphate, 0.3% diazazodinyl urea, 0.3% carbomer, water	6.72 ± 1.4% of applied dose permeated *in vitro*
	ME-1	10% IPM, 16% polysorbate 80, 30% sorbitol, water	12.24 ± 0.98% of applied dose permeated *in vitro*
	ME-2	21.3% benzyl alcohol, 16.9% diethylene glycol monoethyl ether, 18.1% Tween 20, water	6.95 ± 1.28% of applied dose permeated *in vitro*
0.5% piroxicam (Dalmora et al., 2001), *in vitro* release in 300 min, *in vivo* anti-inflammatory effect in rats (% inhibition)	Gel-1	3% carbopol 934, phosphate buffer (pH 5.5)	Fast release, 42.7% of inhibited inflammation
	ME-1	3% carbopol 934, 30% HTAB, 30% ethanol, 6.6% IPM, phosphate buffer (pH 5.5)	Slow release, 32.9% of inhibited inflammation
	Gel-2	3% carbopol 934, 0.5% β-CD, phosphate buffer (pH 5.5)	Slow release, 38.5% of inhibited inflammation
	ME-2	3% carbopol 934, 0.5% β-CD, 30% HTAB, 30% ethanol, 6.6% IPM, phosphate buffer (pH 5.5)	Slow release, 36.9% of inhibited inflammation

(*Continued*)

Table 3.5 Comparison of microemulsions with other topical formulations*—Cont'd

Method	Dosage form	Composition	Result
0.003% triptolide (Chen et al., 2007), *in vitro* permeation using mice skin, 0.5 g/cm^2 dose for 8 h	Gel	0.75% carbomer, 1% TEA, 20% propylene glycol, water	Flux: 0.018 ± 0.003 μg/cm^2 h
	ME gel	3% IPM, 40% Tween 80, 20% propylene glycol, 0.75% carbomer, 1% TEA, water	Flux: 0.043 ± 0.001 μg/cm^2 h
0.1% hydrocortisone acetate (Moldovan et al., 2006), *in vitro* release for 13 h, *in vivo* tape stripping and skin whitening in human subject (n = 6, dose 150 mg/cm^2)	Emulsion	30% paraffin oil, 3% cetyl alcohol, 3% polysorbate 80, water	Slow release, low concentration in stratum corneum, 10.3 ± 0.637% of bleach effect
	ME	25% IPM, 58.33% polysorbate 80–propylene glycol (9:1), water	Fast release (2.5-fold), high concentration in stratum corneum (2.5-fold), 13.6 ± 2.675% of bleach effect

* Abbreviations: ME-microemulsion, IPM-isopropyl myristate, DEA-diethanolamine, HTAB-hexadecyltrimethylammonium bromide, β-CD, β-cyclodextrin; TEA-triethanolamine.

of triptolide from a microemulsion system was faster than that from a control formulation. It should be mentioned that an infinite dose (approximately 0.5 g formulation per cm^2) was applied as a donor phase in this study. At an extremely low drug loading (0.003%), the active agent in both formulations appeared to be much below the solubility level. In the presence of 3% IPM and 10% Tween 80, the solubility of triptolide in the microemulsion system should be increased significantly, resulting in a lower driving force for drug delivery in the microemulsion compared to that in the control solution. However, the permeation rate of triptolide across mice skin was higher from the microemulsion, indicating that the enhancement effect of IPM, and possibly Tween 80, dominated the process. A very interesting study reported recently was conducted by Moldovan et al. (2006). The investigators prepared a microemulsion and a conventional emulsion using 0.1% hydrocortisone acetate as an active ingredient for a head-to-head comparison. A 13-hour drug release test was performed using a modified Franz diffusion cell and a semipermeable membrane. Also, at 24 hours after applying 150 mg/cm^2 formulation to the skin of six volunteers, drug concentrations in the stratum corneum were measured using a tape-stripping technique. The efficacy was observed through scoring the degree of whiteness in the treated sites. As shown in Table 3.5, the microemulsion achieved a faster release, higher drug levels in the stratum corneum, and better efficacy compared with the emulsion. Although the study involved a small group of subjects and the driving force from these formulations was unknown, a clear correlation between *in vitro* release and *in vivo* delivery, as well as efficacy, was demonstrated when comparing an emulsion with a microemulsion dosage form.

Similar to the liposome formulations, microemulsions provide a potentially novel drug dosage form for topical and transdermal drug delivery. Further investigations are necessary to evaluate whether the advantages outweigh the limitations in developing microemulsion-type formulations for pharmaceutical and cosmetic products. In the authors' view, microemulsions may fit a small niche of topical and transdermal formulations. Unless more convincing results can be demonstrated, the better drug solubility and physical stability of microemulsions achieved at the expense of using higher levels of surfactants/cosurfactants may not be good enough reasons to replace emulsions for most dermatological and cosmetic products.

REFERENCES

Ambrosone, L., Murgia, S., Cinelli, G., Monduzzi, M., Ceglie, A., 2004. Size polydispersity determination in emulsion systems by free diffusion measurements via PFG-NMR. J. Phys. Chem. B 108, 18472–18478.

Arts, T.J.C., Laven, J., van Voorst Vader, F., Kwaaitaal, Th., 1994. Zeta potentials of tristearoylglycerol crystals in olive oil. Colloid Surfaces A: Physiochem. Eng. Aspects 85, 149–158.

Barry, B.W., 1983. Dermatological Formulations. Marcel Dekker, New York, pp. 313–322.

Callaghan, P.T., Jolley, K.W., Humphrey, R.S., 1983. Diffusion of fat and water in cheese as studied by pulse field gradient nuclear magnetic resonance. J. Colloid Interface Sci. 93, 521–529.

Chen, H.B., Mou, D.S., Du, D.R., Chang, X.L., Zhu, D.D., Liu, J., Xu, H.B., Yang, X.L., 2007. Hydrogel-thickened microemulsion for topical administration of drug molecule at an extremely low concentration. Int. J. Pharm. 341, 78–84.

Dalmora, M.E., Dalmora, S.L., Oliveira, A.G., 2001. Inclusion complex of piroxicam with β-cyclodextrin and incorporation in cationic microemulsion. In vitro drug release and in vivo topical anti-inflammatory effect. Int. J. Pharm. 222, 45–55.

Davis, A.F., Gyurik, R.J., Hadgraft, J., Pellett, M.A., Walter, K.A., 2002. Formulation strategies for modulating skin permeation. In: Walter, K.A. (Ed.), Dermatological and Transdermal Formulations. Marcel Dekker, New York, pp. 271–318.

DeMerlis, C.C., Goldring, J.M., Schoneker, D.R., 2008. New excipient evaluation procedure. Pharm. Technol. 32 (1), 104–105.

Eccleston, G.M., 2007. Emulsion and microemulsion. In: Swarbrick, J. (Ed.), Encyclopedia of Pharmaceutical Technology. Informa Healthcare, New York, pp. 1548–1565.

Eros, I., Abu-Eida, E.Y., Csóka, I., Sánta, Z., Cserne, A., Köver, T., 2003. Optimization of drug release from dermatological semisolid preparations. Drug Dev. Res. 59, 316–325.

Flynn, G.L., Weiner, N.D., 1993. Topical and transdermal delivery – provinces of realism. In: Gurny, R. (Ed.), Dermal and Transdermal Drug Delivery. APV, Stuttgart, pp. 33–65.

Flynn, G.L., Shah, V.P., Tenjarla, S.N., Corbo, M., DeMagistris, D., Feldman, T.G., Franz, T.J., Miran, D.R., Pearce, D.M., Sequeira, J.A., Swarbrick, J., Wang, J.C.T., Yacobi, A., Zatz, J.L., 1999. Assessment of value and applications of in vitro testing of topical dermatological drug products. Pharm. Res. 16 (9), 1325–1330.

Forgiarini, A., Esquena, J., González, C., Solans, C., 2001. Formation of nano-emulsions by low-energy emulsification methods at constant temperature. Langmuir 17, 2076–2083.

Groning, R., Braun, F.J., 1996. Three dimensional solubility parameters and their use in characterizing the permeation of drugs through the skin. Pharmazie 51, 337–341.

Hadgraft, J., Williams, D.G., Allan, G., 1993. Azone: mechanisms of action and clinical effect. In: Walter, K.A., Hadgraft, J. (Eds), Pharmaceutical Skin Penetration Enhancement. Informa Health Care, New York, pp. 175–197.

Hakansson, B., Pons, R., Soederman, O., 1999. Structure determination of a highly concentrated W/O emulsion using pulsed-field-gradient spin-echo nuclear magnetic resonance "diffusion diffractograms". Langmuir 15, 988–991.

Hedin, N., Furó, I., 2001. Ostwald ripening of an emulsion monitored by PGSE NMR. Langmuir 17, 4746–4752.

Higuchi, T., 1960. Physical chemical analysis of percutaneous absorption process from cream and ointments. J. Soc. Cosmet. Chem. 11, 85–97.

Higuchi, T., 1963. Mechanism of sustained-action medication. Theoretical analysis of rate of release of solid drugs dispersed in solid matrices. J. Pharm. Sci. 52, 1145–1149.

Hindmarsh, J.P., Su, J.H., Flanagan, J., Singh, H., 2005. PFG-NMR analysis of intercompartment exchange and inner droplet size distribution of W/O/W emulsions. Langmuir 31, 9076–9084.

Inactive Ingredient Search for Approved Drug Products. <http://www.accessdata.fda.gov/cder/iig>

Li, X., Cox, J.C., Flumerfelt, R.W., 1992. Determination of emulsion size distribution by NMR restricted diffusion measurement. AIChE J. 38, 1671–1674.

Loennqvist, I., Khan, A., Soederman, O., 1991. Characterization of emulsions by NMR methods. J. Colloid Interface Sci. 144, 401–411.

Losso, J.N., Khachatryan, A., Ogawa, M., Godber, J.S., Shih, F., 2005. Random centroid optimization of phosphatidylglycerol stabilized lutein-enriched oil-in-water emulsions at acidic pH. Food Chem. 92, 737–744.
Martin, A., Newberger, J., Adjei, A., 1980. Extended Hildebrand solubility of theophylline in binary polar solvents. J. Pharm. Sci. 69, 487–491.
Martin, P.L., Yen, S.S.C., Bumier, A.M., Hermann, H., 1979. Systemic absorption and sustained effects of vaginal estrogen creams. JAMA 242 (24), 2699–2702.
Melzer, E., Kreuter, J., Daniels, R., 2003. Ethylcellulose: a new type of emulsion stabilizer. Eur. J. Pharm. Biopharm. 56, 23–27.
Moldovan, M.A., Leucuta, S.E., Bakri, A., 2006. Increase of the cutaneous bioavailability of hydrocortisone acetate using a microemulsion as vehicle. Farmacia 5, 35–44.
Nagarajan, R., Ruckenstein, E., 2000. Molecular theory of microemulsions. Langmuir 16, 6400–6415.
Pappinen, S., Urtti, A., 2006. Microemulsions in topical drug delivery. In: Smith, E.W., Maibach, H.I. (Eds), Percutaneous Penetration Enhancers, 2nd ed. CRC Press, Boca Raton, FL, pp. 109–116.
PDR, 2007. PDR® Electronic Library™. Thomson.
Pillai, R., Shah, V., Abriola, L., Caetano, P., Flynn, G.L., 2001. Release of hydrocortisone from a cream matrix: dependency of release on suspension concentration and measurement of solubility and diffusivity. Pharm. Dev. Technol. 6 (3), 373–384.
Pongcharoenkiat, N., Narsimhan, G., Lyons, R.T., Hem, S.L., 2002. The effect of surface charge and partition coefficient on the chemical stability of solutes in O/W emulsion. J. Pharm. Sci. 91, 559–570.
Potts, R.O., Guy, R.H., 1992. Predicting skin permeability. Pharm. Res. 9, 663–669.
Price, W.S., 1997. Pulsed-field gradient nuclear magnetic resonance as a tool for studying translational diffusion: Part 1. Basic theory. Concepts Magn. Reson. 9, 299–336.
Rangarajan, M.A., Zatz, J.L., 2003. Effect of formulation on the topical delivery of α-tocopherol. J. Cosmet. Sci. 54, 161–174.
Rigg, L.A., Hermann, H., Yen, S.S.C., 1978. Absorption of estrogens from vaginal creams. N. Engl. J. Med. 298 (4), 195–197.
Roberts, M.S., Cross, S.E., Pellett, M.A., 2002. Skin transport. In: Walter, K.A. (Ed.), Dermatological and Transdermal Formulations. Marcel Dekker, New York, pp. 89–195.
Roland, I., Piel, G., Delattre, L., Evrard, B., 2003. Systematic characterization of oil-in-water emulsions for formulation design. Int. J. Pharm. 263, 85–94.
Scheuplein, R.J., Blank, I.H., 1971. Permeability of the skin. Physiol. Rev. 51 (4), 702–747.
Schiff, I., Tuichinsky, D., Ryan, K.J., 1977. Vaginal absorption of estrone and 17β-estradiol. Fertil. Steril. 28 (10), 1063–1066.
Seth, B.L., 1992. Comparative pharmacokinetics and bioavailability study of percutaneous absorption of diclofenac from two topical formulations containing drug as a solution gel or as an emulsion gel. Arzneimittel-Forschung 42 (2), 120–122.
Seth, P.L., 1993. Percutaneous absorption of ibuprofen from different formulations: comparative study with gel, hydrophilic ointment and emulsion cream. Arzneimittel-Forschung 43 (8), 919–921.
Shah, V.P., Elkins, J.S., Williams, R.L., 1999. Evaluation of the test system used for in vitro release of drugs for topical dermatological drug products. Pharm. Dev. Technol. 4 (3), 377–385.
Sloan, K.B., Koch, S.A.M., Siver, K.G., Flowers, F.P., 1986. Use of solubility parameters of drug and vehicle to predict fluxes through skin. J. Invest. Dermatol. 87, 244–252.
Spangler, M., Harris, D., 1999. A novel percolation method for determining solubilities of pharmaceutical agents in semisolid vehicles. Pharm. Dev. Technol. 4 (4), 545–552.
Stoughton, R.B., McClure, W.O., 1983. Azone: a new non-toxic enhancer of cutaneous penetration. Drug Dev. Ind. Pharm. 9 (4), 725–744.

Tanaka, S., Takashima, Y., Murayama, H., Tsuchiya, S., 1985. Solubility and distribution of dexamethasone acetate in oil-in-water creams and its release from the creams. Chem. Pharm. Bull. 33 (9), 3929–3934.

Tenjaria, S., 1999. Microemulsions: an overview and pharmaceutical applications. Crit. Rev. Ther. Drug Carrier Systems 16 (5), 461–521.

Van den Enden, J.C., Waddington, D., Van Aalst, H., Van Kralingen, C.G., Packer, K.J., 1990. Rapid determination of water droplet size distributions by PFG-NMR. J. Colloid Interface Sci. 140, 105–113.

Walter, K.A., Brain, K.R., 2002. Dermatological formulation and transdermal systems. In: Walter, K.A. (Ed.), Dermatological and Transdermal Formulations. Marcel Dekker, New York, pp. 319–400.

Welch, C.F., Rose, G.D., Malotky, D., Eckersley, S.T., 2006. Rheology of high internal phase emulsions. Langmuir 22, 1544–1550.

Wiechers, J.W., Kelly, C.L., Blease, T.G., Dederen, J.C., 2004. Formulation for efficacy. Int. J. Cosmet. Sci. 26, 173–182.

Wilkes, G.L., Brown, I.A., Wildnauer, R.H., 1973. The biomechanical properties of skin. CRC Crit. Rev. Bioeng. 1 (4), 453–495.

CHAPTER 4

Iontophoretic Transdermal Drug Delivery

Longsheng Hu, Priya Batheja, Victor Meidan *and* **Bozena B. Michniak-Kohn***

Ernest Mario School of Pharmacy, Rutgers – The State University of New Jersey, Piscataway, NJ 08854, USA

Contents

4.1 Introduction 95
4.2 History and development of iontophoresis 97
4.3 Advantages and limitations of iontophoresis 97
4.4 Iontophoretic delivery principles and system components 98
4.5 Drug transport in iontophoresis 100
4.6 Pathways for iontophoretic transport and factors affecting the same 102
4.7 Properties of the drug/ion: concentration, competing ions, and molecular size 103
4.8 Properties of the iontophoretic system: pH, current 103
4.9 Combination strategies to improve transdermal iontophoretic drug delivery 105
  4.9.1 Iontophoresis with electroporation 106
  4.9.2 Iontophoresis with chemical enhancers 106
  4.9.3 Iontophoresis with microneedles 107
  4.9.4 Iontophoresis with sonophoresis 108
  4.9.5 Ion-exchange materials 109
  4.9.6 Other combinations 110
4.10 Reverse iontophoresis 110
4.11 Applications of iontophoresis 111
4.12 Current and future development 113
References 114

4.1 INTRODUCTION

Skin, our integumentary system, is the largest organ of the body and has been successfully utilized as a means to achieve local pharmacological action (topical delivery) or for systemic effects (transdermal drug delivery). Transdermal delivery of drugs serves to circumvent the degradation of drugs in the

* Author to whom correspondence should be addressed.

Handbook of Non-Invasive Drug Delivery Systems ISBN 9780815520252

gastrointestinal (GI) tract and their subsequent metabolism due to hepatic first-pass effects. It also allows controlled drug delivery to the target tissues, thereby reducing potential adverse effects. The skin is a multilayered organ, and the outermost layer (the stratum corneum) is the rate-limiting step to drug permeation as it hosts dead protein-filled corneocytes intercalated by a complex lipid structure that together form a strong permeation barrier. This permeation barrier provided by the stratum corneum not only effectively regulates transepidermal water loss and maintains homeostasis, but greatly limits the range and the type of molecules that can be therapeutically delivered by the otherwise accessible and large surface area of the skin. Commercial transdermal drug delivery systems (TDDS) constitute patches that deliver low-molecular-weight molecules such as fentanyl, clonidine, nicotine, scopolamine, and various hormones (testosterone, estradiol, and the birth control combinations of norelgestromin/ethinyl estradiol). Various penetration enhancement techniques have been investigated over the years in order to extend the application of transdermal delivery to agents that target other therapeutic areas. These techniques, broadly classified as chemical and physical enhancement techniques, are reversible methods that alter the skin barrier in order to increase the permeation of compounds through the skin and achieve therapeutically effective concentrations in the underlying tissues. Chemical enhancement methods have been widely studied and they work to increase the drug flux by one or a combination of the following mechanisms: (i) disruption of the lipid structure in the stratum corneum; (ii) opening up the dense protein structure in the corneocytes; or (iii) altering the chemical environment in the stratum corneum, thereby increasing the partitioning of a second active agent into the skin (Barry, 2001). However, the type and amount of chemical enhancers that can be used in transdermal applications are restricted by their skin irritation potential (Karande et al., 2005). Physical enhancement methods constitute electrically assisted methods (iontophoresis, electroporation), mechanical methods (microneedles), velocity-based techniques (such as jet propulsion), or other methods such as ultrasound, laser, and photomechanical waves (Cross and Roberts, 2004; Brown et al., 2006). Of the various enhancement methods, transdermal iontophoresis, which uses low-level electric current (~0.5 mA/cm^2) to enhance the transdermal delivery of charged substances through the skin, has been extensively studied, and devices for drug delivery and for non-invasive blood monitoring have been developed and successfully commercialized. IONSYS™ (fentanyl iontophoretic transdermal system, Ortho-McNeil, Inc., Raritan, NJ) (Herndon, 2007), Hybresis™ Iontophoresis Drug Delivery System (Iomed, Inc., Salt Lake City, UT) (Parkinson et al., 2007), and GlucoWatch® G2 Biographer for monitoring of glucose (Potts et al., 2002) are some iontophoretic devices that are commercially available or nearing commercialization. Iontophoresis has also demonstrated great value for the delivery of macromolecules (proteins and peptides) as it enhances their transdermal permeation while protecting them from the harsh proteolytic environments encountered with oral delivery (Kanikkannan, 2002; Pillai et al., 2003; Raiman et al., 2004). In addition to transdermal delivery, iontophoresis has shown great potential in other fields and is being increasingly researched for its application in the treatment of hyperhidrosis (excessive sweating) (Kavanagh et al., 2004; Kreyden, 2004), ocular drug delivery (Eljarrat-Binstock and Domb, 2006), and for delivery through the nail for the treatment of fungal nail infections (Narasimha Murthy et al., 2007). This chapter details the concepts, principles, and applications of iontophoresis and explores its current standing and potential in the increasingly growing transdermal drug delivery segment.

4.2 HISTORY AND DEVELOPMENT OF IONTOPHORESIS

The history of iontophoresis has been detailed by Banga and Chien (1988) and the concept of the application of an electric current to increase skin penetration was described as early as the 1700s. One of the earliest and most well-documented experiments in this field, however, was conducted by Ludec at the beginning of the twentieth century, where he demonstrated the importance of polarity when using direct current to administer strychnine and cyanide to two rabbits (Chien and Banga, 1989). Though iontophoresis was explored significantly in the 1930s and 1940s, its use has declined in subsequent years, with more recent investigations focusing on applications such as sweat testing for the diagnosis of cystic fibrosis (pilocarpine), hyperhidrosis (tap water, atropine), local anesthesia (lidocaine with epinephrine), and for ulcers (histamine, zinc oxide) (Rai and Srinivas, 2005). Additionally, over the last few decades, iontophoresis has come to be used as a technique researched primarily for its use in the delivery of charged ions and macromolecules through the skin for systemic circulation. A wide range of drugs, such as propranolol hydrochloride (Chesnoy et al., 1999), apomorphine (van der Geest et al., 1997a, b), buspirone hydrochloride (Meidan et al, 2003a, b), insulin (Zakzewski et al., 1998; Pillai et al., 2003), thyrotropin-releasing hormone (Burnette and Marrero, 1986), nafarelin and luteinizing hormone-releasing hormone (LHRH) (Raiman et al., 2004), have been investigated for their transdermal delivery potential by iontophoresis alone, or in conjunction with other physical and chemical enhancement techniques (Tiwari and Udupa, 2003).

4.3 ADVANTAGES AND LIMITATIONS OF IONTOPHORESIS

Iontophoresis provides a non-invasive method for the delivery of a wide variety of agents at therapeutically significant levels due to its enhancement effects on skin permeation. When used transdermally, it shares the benefits associated with transdermal drug administration, primarily the circumvention of the hepatic first-pass effects and the harsh environment present in the gastrointestinal tract. These benefits are especially significant for the administration of proteins and peptides, which do not achieve therapeutically significant concentrations by the oral delivery route. Iontophoresis allows the effective delivery of drugs with short biological half-lives, as it shortens the time frame from administration to delivery to the target tissues. Patient compliance is significantly improved due to infrequent dosing, and the treatment can be terminated when desired. This technique also enables a great degree of programmability as therapy can be monitored and tailored at pre-programmed rates that may be customized to the patient's condition. This programmability, in conjunction with the direct dependence of drug delivery on the applied current, results in a significant reduction in the variability of the drug levels achieved.

Advantages of the iontophoresis technique are accompanied by limitations of the procedure, the most obvious being the potential for irritation and pain on application of the electric current. Although the technique is considered safe, increasing the intensity of the applied current can lead to skin irritation, burning, blisters, necrosis, and erythema (Ledger, 1992; Burton, 1995). In addition, iontophoresis is usually restricted to compounds that can be formulated in their ionic form. For ionic compounds, the pH of the skin is not the same in the deeper layers as it is at the surface, and changes (from 4 to 7.3) are encountered during transdermal transport. This may lead to ineffective iontophoresis for some molecules due to the effect of pH on ionization. This phenomenon may be especially relevant for proteins and peptides where charge is an important factor in their delivery and stability. Effective iontophoretic transport of charged species may also be affected by an ion-competition factor due to the presence of other charged moieties. Also, degradation or adsorption of the drug under the electrode has been a major concern. Lastly, the engineering of a device that is inexpensive, safe, convenient, and ensures therapeutically relevant concentrations may be a challenge.

4.4 IONTOPHORETIC DELIVERY PRINCIPLES AND SYSTEM COMPONENTS

Iontophoresis is based on the principle that "like charges repel", while "opposite charges attract". A set of electrodes is used to create these repulsive and attractive forces that provide the external energy to drive the ions through the skin, leading to enhancement of drug permeation. These principles have led to the development of two types of primary system, anodal and cathodal iontophoresis, both consisting of a basic iontophoretic setup with a set of electrodes (anode and cathode), drug and salt reservoirs, and a current source. Positively charged ions can be delivered by placing them under the positive electrode (anode) due to repulsion, and at the same time creating an attractive force by placing the negative electrode at a distal site on the skin. At the same time, negatively charged ions from the body (primarily Cl^-) move toward the anode. This system is called anodal iontophoresis, and a typical setup is illustrated in Figure 4.1. For cathodal iontophoresis, the electrodes are reversed so that a negatively charged drug is delivered through the skin. Figure 4.1 illustrates a system that comprises silver–silver chloride (Ag/AgCl) electrodes, the most commonly used electrodes for iontophoretic systems (Raynauld and Laviolette, 1987; Phipps et al., 1989). The Ag^+ and Cl^- react at the anode–solution interface to form insoluble AgCl, which is deposited at the electrode. In order to maintain electroneutrality, either a cation moves out of this compartment into the skin, or an anion from the skin is taken up into this chamber. Similarly, in the cathodal compartment, reduction of the AgCl to Ag^+ at the electrode leads

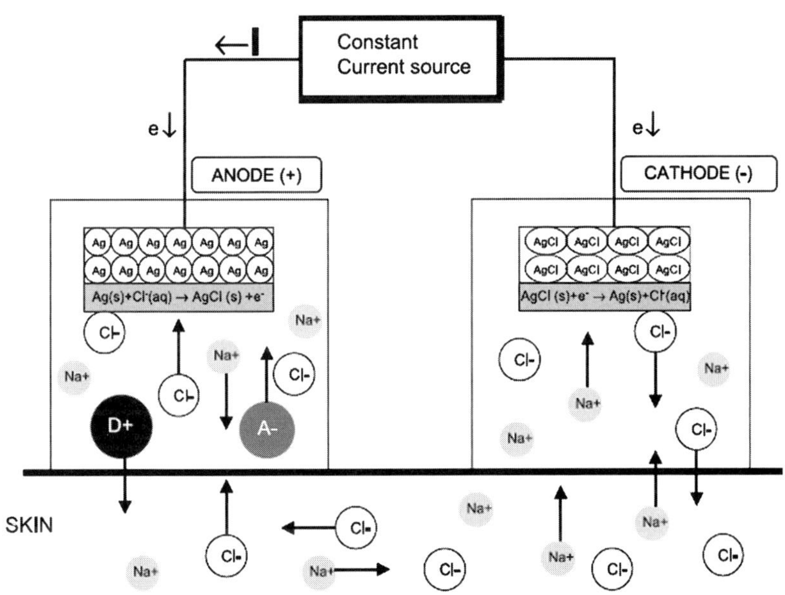

Figure 4.1 Basic setup of an iontophoretic system using a silver–silver chloride electrode. On application of an electric current, the ionizable drug (D^+) is transported from the anodal compartment into the skin, while endogenous ions such as Cl^- move out of the skin. In the cathodal compartment, the AgCl electrode releases Cl^- ions into the compartment, which may be transported into the skin, or a cation may be transported into the chamber from the skin. With permission from Kalia et al. (2004).

to release of a Cl^- ion into solution, which necessitates the uptake of a cation from the skin or loss of an anion from the chamber (Kalia et al., 2004).

Several reasons make the Ag/AgCl system the most preferred electrode system. First, this pair of electrodes resists changes in pH, one of the primary requirements for efficient iontophoresis. Since the Ag/AgCl pair is a reversible electrode, the electrochemical reactions that occur do so at voltages that are lower than those required for electrolysis of water, preventing pH changes and increasing the drug delivery efficiency. A reusable iontophoresis electrode device was invented by Higo et al. (2001) to avoid the release of metal from the electrode into the body. It also claims no change in pH or fluctuation in voltage/current during operation. This electrode component can also include depolarizers (potassium iodide, sodium sulfite, thioglycolic acid, ascorbic acid, etc.), pH regulators (calcium oxide, zeolite, hydroxyapatite, etc.), and water (as an electrolyte).

4.5 DRUG TRANSPORT IN IONTOPHORESIS

A large portion of iontophoretic permeation enhancement occurs due to the effect of the applied electric current on the migration of charged molecules across the skin, a phenomenon called electromigration or electrorepulsion. The applied electric current also gives rise to a water transport phenomenon, called electro-osmosis, which also contributes to the iontophoretic transport of molecules. While electrorepulsion occurs due to the repulsion of similarly charged molecules, electro-osmosis is a facet of the cation permselectivity that occurs as a result of the net negative charge on the skin at physiological pH (7.4) (Burton, 1995). This cation permselectivity can facilitate cation transport, present an electrostatic barrier to anion transport, and enhance the transport of neutral or polar solutes. This results in a net volume flow, called electro-osmosis, in the direction of movement of the positive ions. Consequently, neutral molecules such as water can be transported by this process. Electro-osmostic flow is favored in the anode to cathode direction, while cathodic delivery is retarded.

Iontophoretic transport is primarily driven by electrorepulsion and electro-osmosis, and to a smaller extent by passive diffusion. Thus, the total iontophoretic flux, J_{total} (Guy et al., 2000, 2001), can be described as:

$$J_{total} = J_e + J_c + J_p \tag{4.1}$$

where J_e = flux contribution due to electrorepulsion, J_c = flux contribution due to electro-osmosis, and J_p = flux contribution due to passive diffusion.

It has been considered that in all practical iontophoretic applications, J_p will form a negligible fraction of J_{total}, with the contribution becoming even smaller with increasing molecular weight of the agents.

$$J_e = \left(\frac{1}{Z_X F}\right)\frac{z_X u_X c_X}{\sum_{n=0}^{i} z_i u_i c_i} I_D \tag{4.2}$$

where z_X, u_X, and c_X refer respectively to the charge, mobility, and concentration of the drug in the membrane, F is Faraday's constant, and I_D is the applied current density. The denominator is a summation of the products of the charge, mobility, and concentration of each ion in the system:

$$J_c = vC_X \tag{4.3}$$

where C_X is the concentration of the compound and v is the electro-osmotic solvent velocity.

While the quantification of the contributions of each of these processes is difficult, it has been suggested that electrorepulsion enhances the transport of small cations such as Na^+. Electrorepulsion falls off sharply with increasing molecular weight of the species and electro-osmosis becomes the more dominant mechanism, especially with bulkier cations of higher molecular weight ($\geq$1000 daltons) (Guy et al., 2001; Higo et al., 2001). Similarly, for anions, there will be a point where the increase in molecular weight will lead to the effects of electrorepulsion and electro-osmosis canceling out and the cathodal delivery will cease to be efficient (Green et al., 1991; Guy et al., 2000). The contribution that each component makes to the total flux depends on several factors, such as the physiochemical properties of the permeant including the degree and type of charge, the properties of the solution, and the properties of the membrane. Merino et al. (1999a) measured the *in vitro* iontophoretic delivery of a small negatively chargeable compound, 5-fluorouracil (5-FU; pK_a ~ 8), through excised porcine skin as a function of the pH of the solutions in contact with both sides of the skin. They found that at a pH of 8.5, when 5-FU was predominantly negatively charged, the electrorepulsive transport (cathode to anode) overwhelmed the electro-osmotic convective flux in the opposite direction. At a pH of 3, when 5-FU was almost completely un-ionized and the skin carried a net positive charge, the transport was primarily electro-osmostic and was higher in the cathode to anode direction. At pH 4, the anodal and cathodal fluxes were equal due to the neutralization of the skin at this pH (skin pI ~ 4). However, at pH 7.4, about 25% of 5-FU was ionized and the cathodal electrorepulsive forces carrying the ionized compound were about equal to the electro-osmostic flow that carried the uncharged species in the opposite anodal direction.

Just as they influence the mechanism of iontophoretic transport, the properties of the drug and the vehicle also influence the route of penetration in iontophoretic drug delivery (Panchagnula et al., 2000); the effect of these factors on the iontophoretic flux and the permeation routes will be discussed in the following sections.

4.6 PATHWAYS FOR IONTOPHORETIC TRANSPORT AND FACTORS AFFECTING THE SAME

The three primary routes of passive skin permeation are transcellular, paracellular, and transappendageal (Figure 4.2). Iontophoretic transport is said to occur primarily through the transappeandageal (shunt pathway) and paracellular routes, with the appendageal pores (such as the hair follicles and the sweat glands) being the major transport route. The dominance of this transappendageal iontophoretic transport has been determined by several techniques such as vibrating probe electrodes (Cullander and Guy, 1991), and visualization techniques such as transmission electron microscopy using mercuric chloride (Monteiro-Riviere et al., 1994) and laser confocal microscopy (Regnier and Preat, 1998; Turner and Guy, 1998). Iontophoresis can also enhance skin delivery by potential-dependent pore formation in the stratum corneum that is attributed to a "flip-flop" gating mechanism that occurs due to the restructuring of the polypeptide helices on the application of an electric current.

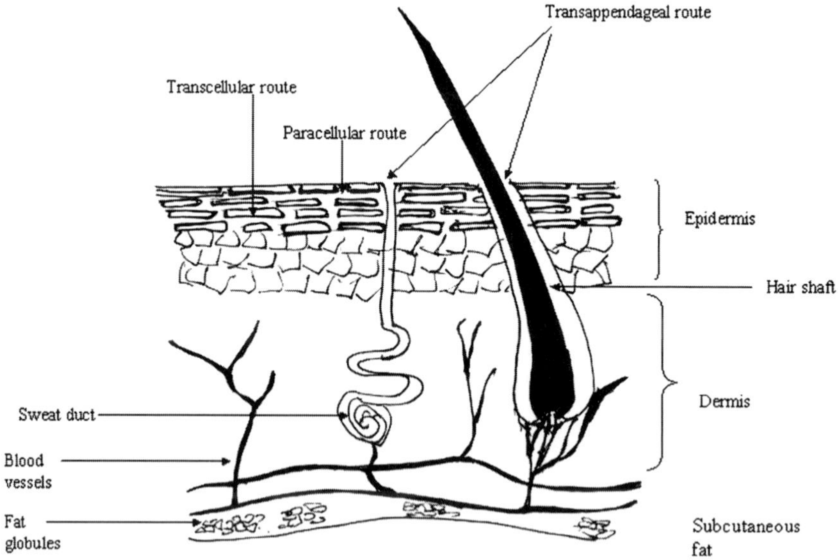

Figure 4.2 Pathways of drug transport through skin. With permission from Batheja et al. (2006).

Though iontophoretic transport is capable of producing 100-fold enhancement relative to passive diffusion (for a cation having a relative molecular size equal to unity) (Guy et al., 2000), the efficiency of iontophoretic transport is frequently below 100% due to several variables that influence the transport efficiency, predominantly the properties of the permeant, the vehicle and other formulation factors, the membrane and components of the system.

4.7 PROPERTIES OF THE DRUG/ION: CONCENTRATION, COMPETING IONS, AND MOLECULAR SIZE

Various studies (Del Terzo et al., 1989; Singh et al., 1999) have shown that an increase in the drug concentration results in an increase in the iontophoretic flux. However, the concentration–flux profile often showed a curvilinear pattern that plateaued off as the concentration was further increased. Van der Geest et al. (1997a) demonstrated a plateau in the flux profile during the iontophoretic delivery of apomorphine through human stratum corneum. This was attributed to attaining the maximum solubility concentration of apomorphine in the donor, and a postulated skin resistance caused by increasing concentrations of apomorphine. The effect of drug concentration on the flux is also negatively influenced by the presence of competing ions, either co-ions (same charge) or counter-ions (opposite charge) from the formulation itself, the presence of a buffer added to maintain pH, or from endogenous ions. This competition from other ions results in a reduction in the transport number of the primary drug ion, especially if the competing ions are more mobile, such as those present in buffers (Singh and Maibach, 1994). Another important contributor to the iontophoretic flux is the molecular size of the permeant. A decrease in permeability coefficient with increasing molecular size has been established regardless of whether the solutes are positively charged, negatively charged, or uncharged (Yoshida and Roberts, 1993). This phenomenon is further complicated by the contribution of electro-osmosis to the iontophoretic flux with increasing size of the molecules.

4.8 PROPERTIES OF THE IONTOPHORETIC SYSTEM: pH, CURRENT

pH plays a very critical role in iontophoresis by keeping a large proportion of the drug in its ionized form. The importance of pH in the efficiency of iontophoretic transport has been clearly demonstrated in several studies. Siddiqui et al. (1985) showed that the passive diffusion flux of lignocaine was high and iontophoretic flux was low when it was in the un-ionized form.

However, at pH values of 5.2 and 3.4, where the drug is increasingly ionized, the flux increased by 4 and 8.5 times respectively. pH also plays an important role in the iontophoretic delivery of proteins and peptides, as it can control the net charge on these molecules based on their isoelectric points. In addition, the pH of the membrane (skin) also regulates the flux as it determines the electromigration and electro-osmotic contributions to the flux.

Iontophoretic flux has shown a direct dependence on the strength of the applied current, with the flux values increasing linearly with increases in current density (Figure 4.3) for several compounds such as lithium (Phipps et al, 1989), verapamil (Wearley and Chien, 1990), apomorphine (van der Geest et al., 1997a), methylphenidate hydrochloride (Singh et al., 1999), and thyrotropin-releasing hormone (TRH) (Burnette and Marrero, 1986). The application of higher current densities is, however, limited by a saturation phenomenon observed with higher currents, and also by the concern for causing skin discomfort and irritation (Kasting and Keister, 1989). Cazares-Delgadillo et al. (2007) used constant-current iontophoresis for transdermal delivery of cytochrome C across porcine skin. While studying the impact of increasing the current density from 0.15 to 0.5 mA/cm^2, they found that although the flux increased linearly with the applied current at lower current densities, the flux values of cytochrome C at 0.3 and 0.5 mA/cm^2 were not statistically different, probably due to a saturation phenomenon observed after a limiting transport number is achieved.

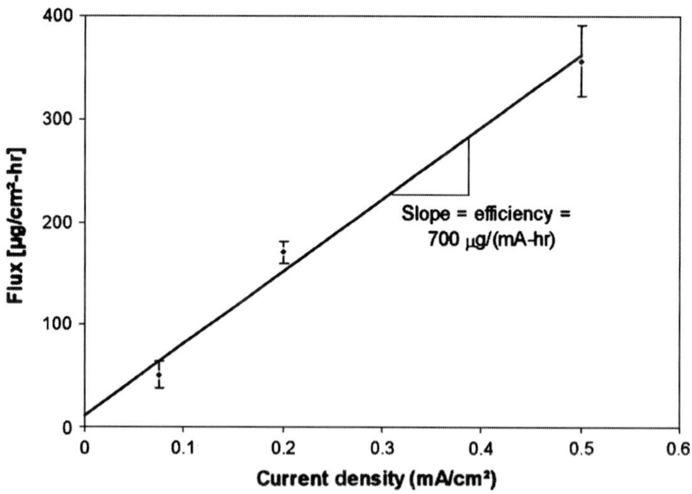

Figure 4.3 Relationship between applied current density and methylphenidate iontophoretic flux at a methylphenidate concentration of 0.5 M (pH = 3.5, $n = 4$). With permission from Singh and Maibach (1994).

Iontophoretic devices employ either constant- or pulsed-current systems, and both systems have been used with success in both *in vitro* and *in vivo* transdermal studies (Pacini et al., 2006; Cazares-Delgadillo et al., 2007). Though comparisons have been made, the advantage of one type of current profile over the other has not been confirmed. A pulsed direct current profile was shown to be more efficient in transporting LHRH and nafarelin across the human epidermis as compared to constant direct current or alternating current (Raiman et al., 2004). On the other hand, Tiwari and Udupa (2003) found constant direct current to be more effective than pulsed current for the transdermal permeation of ketorolac through rat skin. It has been widely discussed by various groups that the use of an appropriate time interval between the pulses in a pulsed-current setup may allow the skin to depolarize and return to its original state, thereby leading to more efficient transport (less buildup of charge) and safer delivery (minimal accumulation of charges).

In order to account for the different factors that influence iontophoretic transport, models developed for iontophoretic delivery have strived to study iontophoretic transport as a function of various factors that influence the delivery. Some models have characterized limited variables such as the relationship of the solute structure and size to the iontophoretic transport (Stokes–Einstein and free volume models) (Yoshida and Roberts, 1993), while other models such as the ionic mobility–pore model are more comprehensive. This model (Roberts et al., 1998; Lai and Roberts, 1999) incorporates a range of important factors such as the size, mobility and concentration of the solute, applied current, presence of extraneous ions, the conductivity of the solvent, interaction of the solute molecules with the pore walls and the subsequent restriction of the pore size, the epidermal permselectivity, and electro-osmosis.

4.9 COMBINATION STRATEGIES TO IMPROVE TRANSDERMAL IONTOPHORETIC DRUG DELIVERY

Over the last two decades, a tremendous amount of work has been directed towards developing various enhancement techniques to improve transdermal iontophoretic drug delivery (Batheja et al., 2006; Benson and Namjoshi, 2008). The major enhancement techniques include electroporation (Vanbever et al., 1996; Sharma et al., 2000; Crawford and Brown, 2003), microneedles (Matriano et al., 2002; Prausnitz, 2004), chemical enhancers (Fang et al., 2002; Femenía-Font et al., 2005), sonophoresis (Bommannan et al., 1992; Mitragotri et al., 2000; Jaskari et al., 2001; Merino et al., 2003a, b), and ion-exchange materials (Kankkunen et al., 2002; Hanninen et al., 2003; Vuorio et al., 2003). Some of these techniques have been used in combination with iontophotesis and are detailed below.

4.9.1 Iontophoresis with electroporation

Electroporation involves application of high voltage (>50 V) to create transient pores in the stratum corneum, leading to a large increase in drug transport, delivery of macromolecules, a quick drug effect onset, and insignificant or only minor skin damage (Golden et al., 1987; Regnier and Preat, 1998; Hofmann, 1999). A combination of iontophoresis and electroporation could possibly further enhance drug transport, allowing rapid delivery of a bolus dose and precise control of drug delivery. The synergistic effects of iontophoresis (0.4 mA/cm^2) and electroporation (150 or 300 V, 10 ms, 10 pulses) on the *in vivo* percutaneous absorption of human insulin in rats were investigated by Tokumoto et al. (2006). Blood glucose levels were significantly changed by the combined use of electroporation and iontophoresis, reflecting a plasma insulin concentration increase by approximately 23% in the 150-V group and 53% in the 300-V group 2 hours after administration against the initial level. Conversely, in some cases, lower combined effects than the effects with each individual treatment have been reported. Yang et al. (2002) studied the effects of iontophoresis and electroporation on the transdermal delivery of buprenorphine from solutions and hydrogels. The combination of electroporation and iontophoresis (500-V pulses followed by 0.3 mA/cm^2) did not enhance the flux as compared to iontophoresis alone. However, the permeation lag time was significantly reduced, indicating that a more rapid onset was achieved by combining iontophoresis and electroporation. Heller et al. (1999) published a review of the use of electroporation for enhanced chemotherapy of cutaneous tumors and seemed to indicate that the technique was reasonably well tolerated by subjects. However, some subjects experienced strong muscle contractions occurring with each electrical pulse, and around 25% suffered mild muscle fatigue after treatment. The practical application of combining electroporation with iontophoresis is still in its initial development stage and has yet to gain wide acceptance in the transdermal drug delivery field.

4.9.2 Iontophoresis with chemical enhancers

Chemical enhancers are known to increase skin permeability to therapeutic drugs. Enhancers offer several to more than a hundred times higher drug penetration in the flux, depending on the properties and concentrations of the penetrants and enhancers, as well as the other ingredients in the formulations. It is believed that some of the chemical enhancers can increase permeability of the stratum corneum by acting as solvents to dissolve the skin lipids or to denature skin proteins. In other cases, enhancers can modify drug solubility parameters in the vehicle or in the skin to increase drug penetration. In addition, these compounds affect the partitioning of the drug from the applied formulations by altering the lipid structure. Chemical enhancers, however, sometimes offer limited enhancements to skin permeability. The combination of enhancers and iontophoresis has demonstrated synergistic effects in enhancing drug penetration, as well as reducing side-effects such as irritation caused by high concentration of enhancers or stronger electric forces. The most commonly studied enhancers in combination with iontophoresis are fatty acids, terpenes, surfactants, Azone®, alcohols, glycols, glycerides, and sulfoxides/similar compounds. Femenía-Font et al. (2006) studied the effects of Azone pretreatment followed by 0.50 mA/cm^2 iontophoresis on the transdermal absorption of sumatriptan through human skin. Using the combination provided synergistic effects, increasing the drug flux by 5-fold compared with Azone alone, or 2-fold compared with iontophoresis alone. Similarly, fatty acids and iontophoresis were found to provide synergistic effects in promoting the delivery of midodrine hydrochloride through human skin (Wang et al., 2003). Using 5% oleic acid as the enhancer in combination with iontophoresis resulted in more than a 15-fold drug flux increase when compared to 5% oleic acid alone. The authors concluded that, using the combination of 5% oleic acid pretreatment followed with the electrical current treatment at 0.1 mA/cm^2, a daily delivery of midodrine hydrochloride can provide an adequate clinical application. The effect of sodium dodecyl sulfate, an anionic surfactant, on the iontophoretic transport of a neutral drug, hydrocortisone, across hairless mouse skin was found to be highly concentration dependent and driving mode dependent. Below its critical micelle concentration (CMC), increased surfactant concentration resulted in an increase in drug flux. Above the CMC, the drug transport reached a plateau and leveled off, suggesting that the drug transport is retarded by the iontophoresis, possibly due to electrostatic attraction (Wang et al., 2000).

4.9.3 Iontophoresis with microneedles

Microneedles have significant potential in devices for diagnostics, healthcare monitoring, and drug delivery by mechanically perforating the outer skin layer and allowing for transdermal drug delivery or fluid sampling. The processing techniques incorporate one or more technologies that enable the precise machining, extrusion, casting, and/or forming from one to an array or grid of microneedles. Evolving microneedle systems will be well positioned to address a significant segment of the large-molecule biological drugs expected to emerge from the convergence of automated discovery and genome mapping. However, before microneedles find widespread use, researchers still need to find ways of perfecting the techniques for optimally inserting them into the skin, and complete the integration of microneedles into a full diagnostic, monitoring, or drug delivery system. 3M and Alza Corp are two pioneering companies in designing microneedle systems to deliver vaccines, proteins, and peptide-based drugs. Studies at 3M have achieved successful delivery of water-soluble, polar, ionic, and large molecules (19 500 Da). The combination of iontophoresis and microneedles was used to enhance trandermal drug delivery for certain compounds, including peptides. Lin et al. (2001) designed a Macroflux® and iontophoresis combined transdermal delivery system for the delivery of an antisense oligonucleotide system. Lin et al. (2007) found that a combination of microneedle pretreatment and subsequent iontophoresis significantly enhanced the transdermal delivery of high-molecular-weight compounds such as fluorescein isothiocyanate (FITC)–dextran flux (FD-4, FD-10, FD-40, FD-70, and FD-2000; average molecular weights of 3.8, 10.1, 39.0, 71.2, and 200.0 kDa) compared with microneedle pretreatment alone or iontophoresis alone. Vemulapalli et al. (2008) investigated the transdermal iontophoretic delivery of methotrexate, alone and in combination, with microneedles *in vitro* and *in vivo* in the hairless rat. It was found that a synergistic 25-fold enhancement of delivery was observed *in vivo* with the combined techniques compared with either technique when used alone.

4.9.4 Iontophoresis with sonophoresis

Sonophoresis is the movement of drug molecules through the skin under the influence of ultrasound. This technique typically uses a low-frequency pressure wave of less than 100 kHz. The application of ultrasound to the skin can disrupt the stratum corneum lipid bilayer. As a result, drug molecules are allowed to permeate through the skin more easily. The combination of sonophoresis and iontophoresis significantly enhanced the transdermal delivery of sodium nonivamide acetate over either technique alone (Wang et al., 2002). Le et al. (2000) studied the combined effect of low-frequency ultrasound and iontophoresis for transdermal heparin delivery. The combined treatment enhanced heparin flux about 56-fold, which was significantly higher than the sum obtained using ultrasound alone (3-fold) and iontophoresis alone (15-fold). Thus, the effect of ultrasound and iontophoresis on transdermal heparin transport was truly synergistic. Yukio et al. (2006) conducted an *in vivo* study of the percutaneous penetration of ascorbic acid by sonophoresis and iontophoresis in human subjects. Their results demonstrated that the combined treatment of iontophoresis and sonophoresis promoted the permeation of ascorbic acid and collagen synthesis when compared to either technique alone.

4.9.5 Ion-exchange materials

The ion-exchange process is a stoichiometric and reversible process wherein an ion from the solution is replaced with a similarly charged ion attached to an immobile solid phase to fulfill the electroneutrality requirement. One of the novel applications of ion-exchange materials is to enhance transdermal drug delivery.

Conaghey et al. (1998) studied the *in vitro* iontophoretic transdermal delivery of nicotine by ion-exchange resins in agar hydrogel. Both strong and weak resins were used and the heterogeneous vehicles were shown to have advantages over comparable simple hydrogel vehicles in their versatility, in their capacities to store the drug, and to control the delivery rate during iontophoresis. Kankkunen et al. (2000, 2002, 2004) studied controlled transdermal delivery of the zwitterionic levodopa by iontophoresis and ion-exchange fiber. Ion-exchange kinetics and transdermal permeation of a cationic (presumably more stable) model drug, metaraminol, were compared to the corresponding data for levodopa. Levodopa was rapidly oxidized in the presence of water, especially at basic pH. At acidic pH, stability was improved significantly. Ion-exchange groups and pH had a clear effect on the release of both the levodopa and metaraminol from the ion-exchange fiber. Iontophoretic enhancement of drug permeation across the skin was more significant with the positively charged metaraminol than with the zwitterionic levodopa. Ion-exchange fibers provide a promising alternative to control drug delivery, and can store drugs that are readily degraded. The same group of workers also studied the *in vivo* delivery of tacrine, an anti-Alzheimer's disease agent, and other drugs such as sodium salicylate, propranolol, and nadolol. These studies further demonstrated that the transdermal delivery of drugs using iontophoresis with cation- and anion-exchange fibers can be used as controlled drug delivery vehicles. Drug release from the cation-exchange fiber into physiological saline was dependent on the lipophilicity of the drug. The release rates of lipophilic tacrine and propranolol were significantly slower than that of hydrophilic nadolol. Permeation of tacrine across the skin was directly related to the iontophoretic current density and drug concentration used. Anion-exchange fiber, Smopex®-108 (Amidoxime) with ion-exchange capacity of 3.4 mmol/g, was tested with anionic sodium salicylate. The iontophoretic flux enhancement of sodium salicylate from the fiber was substantial. Overall, the ion-exchange fibers act as a drug reservoir, controlling the release and iontophoretic transdermal delivery of the drug.

4.9.6 Other combinations

The enhancement of transdermal drug delivery by combinations of other techniques has been investigated. An electrotransport device comprising blades was invented to deliver large molecules into the skin (Sun et al., 2003, 2004). The blades were approximately 800 μm in length and were able to disrupt the stratum corneum without causing damage to the underlying dermis. This invention delivered substantial insulin levels in pigs and a significant reduction of glucose levels using 800-μm blades and iontophoresis (0.33 mA/cm^2, 30 minutes). Glucose levels were maintained below 140 mg/dL for more than 8 hours. This device can be used in combination with shock waves, heat, chemical enhancers, and ultrasound for transdermal drug delivery. A combination of laser and iontophoresis was found to be more effective than the combination of iontophoresis and electroporation to enhance 5-fluorouracil permeation through skin (Fang et al., 2004).

4.10 REVERSE IONTOPHORESIS

Reverse iontophoresis is a technique involving the application of a small electrical current, usally less than 0.5 mA/cm^2, to the skin to extract analytes (charged and uncharged). The mechanism of extraction involves either electromigration of charged molecules to the electrode of opposite polarity, or electro-osmosis of neutral molecules to the cathode or, for cations potentially, a combination of the two (Merino et al., 1997; Roberts et al., 1997). The technique has the advantage of being truly non-invasive, and therefore warrants consideration as an alternative to blood sampling for drug monitoring. Reverse iontophoresis has been used successfully to develop the GlucoWatch Biographer (Cygnus, Inc., USA), a device for the continuous monitoring of blood glucose levels in diabetics (Tamada et al., 1999; Pitzer et al., 2001). This Food and Drug Administration (FDA)-approved device allows patients to obtain a reading of their glucose levels every 20 minutes over a 12-h test period using only one finger stick blood test to calibrate the system.

Other applications have been envisaged for both drug monitoring (Glikfeld et al., 1989; Sekkat et al., 2002) and diagnostic purposes (Merino et al., 1999b; Wascotte et al., 2007), and additional efforts to improve the extraction efficiency have been undertaken extensively (Wascotte et al., 2005). However, the limitations of these non-invasive biological sampling techniques would be in their ability to measure reliably and accurately low levels of analytes. To date, only one device of this kind (GlucoWatch Biographer) has been commercialized.

Sekkat et al. (2002) assessed the feasibility of using reverse iontophoresis to monitor two model drugs, caffeine and theophylline, which are often administered to premature neonates. Their experimental results suggested that the reverse iontophoretic extraction of the two drugs across the skin to both cathode and anode increased linearly with time. Extraction to the cathode for both drugs was significantly greater (t-test, $P < 0.05$) than that to the anode. The passive diffusion of the two drugs across intact skin was too small to be detectable. It was concluded that they were primarily electrotransported by electro-osmosis, i.e. a convective solvent flow in the direction of counter-ion movement from the anode towards the cathode. This resulted in a significant increase of the percutaneous movement of relatively poorly permeable neutral theophylline and caffeine.

Wascotte et al. (2007) studied the potential of the technique to measure urea and potassium levels non-invasively, and to track their concentrations during hemodialysis. These *in vitro* experiments demonstrated that urea and potassium can be quantitatively and proportionately extracted by reverse iontophoresis, even when the subdermal concentrations of the analytes varied with time. These results suggested that the non-invasive monitoring of urea and potassium to diagnose renal failure during hemodialysis is feasible, and that *in vivo* measurements are warranted. Wascotte et al. (2005) also investigated the use of a thermoreversible gel as a collector vehicle in reverse iontophoresis applications. Their *in vitro* iontophoresis experiments investigated the simultaneous extraction of lithium (analyte of interest) and sodium (an internal standard) into either a control buffer or a gelled receptor. They found that both the lithium extraction fluxes and the lithium-to-sodium ratio of the extraction fluxes were linearly related to the subdermal lithium concentration. The results showed that thermoreversible polymer solutions can be used as a collector vehicle to offer a simple and convenient way of handling samples in reverse iontophoresis applications.

4.11 APPLICATIONS OF IONTOPHORESIS

Transdermal iontophoretic systems have the potential to enhance the delivery of drugs at levels that are therapeutically significant. They also enable the programming or control of the drug kinetics through the device, and therefore enhance patients' compliance. These properties make the system suitable for the delivery of a wide range of molecules, including some macromolecules with limited passive permeation. Enhancement of transdermal permeation by iontophoresis and its combination with other techniques has been demonstrated for several small molecules including apomorphine, sumatriptan, 5-fluorouracil, rotigotine, buspirone hydrochloride, sodium salicylate, hydrocortisone, lidocaine, and buprenorphine (Barry, 2001).

Iontophoretic drug delivery systems have also generated interest for enhancing the delivery of macromolecules such as leuprolide acetate, insulin, vasopressin, parathyroid hormone, and peptides. These types of molecules are often polar, carry a charge, and cannot be administered via the oral route due to poor absorption and stability (Rastogi et al., 2003). Insulin has been a popular candidate for iontophoretic delivery and has been investigated extensively, mostly in combination with chemical enhancer pretreatment or other enhancement techniques. Most of these studies have led to a significant flux enhancement of insulin and, in many cases, delivery of the therapeutic dose. Recently, Akimoto et al. (2006) reported similar results and found that the iontophoretically enhanced skin permeation of insulin increased linearly with the density of the applied pulsed direct current. Several other macromolecules such as LHRH and nafarelin have also been studied for iontophoretic transport. The effectiveness of transdermal iontophoresis of peptides, like smaller molecules, is greatly influenced by several factors such as the charge, molecular weight, lipophilicity and, above all, the physical and chemical stability of the peptides in the delivery system and the proteolytic activity of the skin. Iontophoresis is also finding value in drug delivery via other drug administration routes such as transcorneal and trans-scleral routes (Eljarrat-Binstock et al., 2005). Results from the transcorneal iontophoresis of dexamethasone in the rabbit showed that after a short iontophoretic treatment of 1 mA for only 1 minute, a 30-fold higher concentration was recorded in the cornea than after the common treatment of frequent drop instillation (every 5 min for 1 h).

A method of loading tubular sections of cortical bone has also been investigated (Khoo et al., 2006). The bone was held vertically in an antibiotic bath with a cylindrical outer electrode and a wire electrode down the center of the bone. An electrical potential of approximately 90 V was applied to drive antibiotics into the bone to prevent infection during allograft implantation. This *in vivo* study indicated that iontophoresis is a safe and inexpensive method that can be executed in the operating theater. Iontophoresed bone delivers a high local dose of antibiotic, which may prevent early biofilm formation initiated during allograft handling and exposure to theater air. With no early infections and no re-infections, it is appealing to continue further feasibility assessment of this technique.

Most recently, transungual iontophoretic transport of the neutral permeants mannitol (MA) and urea (UR), and the positively charged permeant tetraethylammonium ion (TEA) across fully hydrated human nail plates was investigated *in vitro* (Hao et al., 2008). Enhanced transport of the neutral permeants across fully hydrated human nail plates was achieved at a higher current of 0.3 mA, and transport of positively charged TEA was significantly enhanced at a lower current of 0.1 mA. These results indicated that it is possible to increase the flux of a drug compound across the nail plate to treat nail infections using the iontophoretic drug delivery system.

4.12 CURRENT AND FUTURE DEVELOPMENT

Several factors – including an aging patient population, biological drug therapies for chronic conditions, and an emphasis on patient self-monitoring and self-care – are driving innovations in the field of drug delivery. There is no doubt that the skin will become one of the major routes of drug administration in the next few decades. Iontophoretic systems using an electrically driven penetration enhancement method have succeeded in overcoming the formidable barrier presented by the stratum corneum and have been shown to be a promising technique for delivering various agents, including macromolecules. To date, iontophoretic drug delivery systems make up the majority of the relatively few physically enhanced transdermal delivery systems that have been successfully developed and commercialized. The key factors in commercialization are the net efficiency of the device, its portability, and convenience of administration. Some of the most well-known devices (see Table 4.1) are the GlucoWatch® Biographer for monitoring glucose (Potts et al., 2002), LidoSite® patch (Vyteris Inc.; http://www.vyteris.com), the E-TRANS® (fentanyl hydrochloride; http://www.alza.com/alza/etrans), and IOMED's new Hybresis™ Iontophoresis Drug Delivery System (Parkinson et al., 2007). Vyteris Inc. has received US FDA approval for a first-of-a-kind system, called LidoSite® (lidocaine hydrochloride and epinephrine bitartrate). This system consists of a LidoSite Patch, a LidoSite Controller, and a portable microprocessor-controlled, battery-powered direct current source, and is indicated for local analgesia during superficial dermatological procedures. Alza Corp's E-TRANS fentanyl hydrochloride system is an FDA-approved, patient-controlled alternative for acute pain management in a medically supervised setting. The advantages of this device are the ability to program it to deliver fixed amounts and an "on-demand" button, enabling the patient to administer

Table 4.1 FDA-approved iontophoresis devices

Company	Device	Molecule	Application
Alza Corp.	E-TRANS®	Fentanyl hydrochloride	Pain management
Vyteris Inc.	LidoSite®	Lidocaine hydrochloride	Pain management
Ortho-McNeil Inc.	IONSYS™	Fentanyl	Pain management
Animas Corp.	GlucoWatch® G2 Biographer	Glucose	Monitoring glucose

the drug when needed. The Hybresis™ System by IOMED is an integrated mini-controller and patch system that uses no lead wires and can deliver in-clinic treatment in 3 minutes. The 3-minute Skin Conductivity Enhancement feature provides a rapid decrease in the resistance of the skin and normalizes or reduces its variability, thereby making the patch-only wear time shorter and more predictable.

Combination strategies with other physical enhancement techniques such as electroporation, chemical enhancers, microneedles, ultrasound, ion-exchange materials, lasers, blades, or heat have led to the observation of significantly higher flux levels as compared to passive transdermal delivery and any one of the techniques alone. Also, electrically assisted delivery systems provide the advantage of controlled drug delivery with customized drug input rates and an option of ceasing drug transport when desired. However, the skin irritation resulting from combined strategies may be a cause for concern. To date, most clinical studies have been limited to smaller molecules such as lidocaine, metoclopramide plus hydrocortisone, cortisone, vincristine, and fentanyl. The transdermal iontophoretic delivery of therapeutic agents such as protein drugs will continue to be an area of high growth, especially in the areas of vaccine, chronic CNS disorders (Parkinson's, depression, and attention deficit hyperactivity disorder) and pain management.

REFERENCES

Akimoto, M., et al., 2006. Progress in Electromagnetics Research Symposium, Cambridge, MA, 26–29 March 2006, p. 157.

Banga, A.K., Chien, Y.W., 1988. Iontophoretic delivery of drugs: fundamentals, developments and biomedical applications. J. Control. Release 7, 1–24.

Barry, B.W., 2001. Novel mechanisms and devices to enable successful transdermal drug delivery. Eur. J. Pharm. Sci. 14 (2), 101–114.

Batheja, P., Thakur, R., Michniak, B., 2006. Transdermal iontophoresis. Expert Opin. Drug Deliv. 3 (1), 127–138.
Benson, H.A.E., Namjoshi, S., 2008. Proteins and peptides: strategies for delivery to and across the skin. J. Pharm. Sci. Published online, 15 Jan.
Bommannan, D., et al., 1992. Sonophoresis. II. Examination of the mechanism(s) of ultrasound-enhanced transdermal drug delivery. Pharm. Res. 9, 1043–1047.
Brown, M.B., et al., 2006. Dermal and transdermal drug delivery systems: current and future prospects. Drug Deliv. 13 (3), 175–187.
Burnette, R.R., Marrero, D., 1986. Comparison between the iontophoretic and passive transport of thyrotropin releasing hormone across excised nude mouse skin. J. Pharm. Sci. 75 (8), 738–743.
Burton H.S., Jr., 1995. Iontophoresis. In: Smith, E.W., Maibach, H.I. (Eds), Percutaneous Penetration Enhancers. CRC Press, Boca Raton, FL, pp. 351–368.
Cazares-Delgadillo, J., et al., 2007. Transdermal delivery of cytochrome C – a 12.4 kDa protein – across intact skin by constant-current iontophoresis. Pharm. Res. 24 (7), 1360–1368.
Chesnoy, S., et al., 1999. Structural parameters involved in the permeation of propranolol HCl by iontophoresis and enhancers. J. Control. Release 58 (2), 163–175.
Chien, Y.W., Banga, A.K., 1989. Iontophoretic (transdermal) delivery of drugs: overview of historical development. J. Pharm. Sci. 78 (5), 353–354.
Conaghey, O.M., et al., 1998. Iontophoretically assisted in vitro membrane transport of nicotine from a hydrogel containing ion exchange resins. Int. J. Pharm. 170, 225–237.
Crawford, N., Brown, C.L. III, 2003. US patent 6662044.
Cross, S.E., Roberts, M.S., 2004. Physical enhancement of transdermal drug application: is delivery technology keeping up with pharmaceutical development? Curr. Drug Deliv. 1 (1), 81–92.
Cullander, C., Guy, R.H., 1991. Sites of iontophoretic current flow into the skin: identification and characterization with the vibrating probe electrode. J. Invest. Dermatol. 97 (1), 55–64.
Del Terzo, S., Behl, C.R., Nash, R.A., 1989. Iontophoretic transport of a homologous series of ionized and nonionized model compounds: influence of hydrophobicity and mechanistic interpretation. Pharm. Res. 6 (1), 85–90.
Eljarrat-Binstock, E., Domb, A.J., 2006. Iontophoresis a non-invasive ocular drug delivery. J. Control. Release 110 (3), 479–489.
Eljarrat-Binstock, E., et al., 2005. Transcorneal and transscleral iontophoresis of dexamethasone phosphate using drug loaded hydrogel. J. Control. Release 106 (3), 386–390.
Fang, I., et al., 2004. Transdermal iontophoresis of 5-fluorouracil combined with electroporation and laser treatment. Int. J. Pharm. 270, 241–249.
Fang, J.Y., et al., 2002. Transdermal iontophoresis of sodium nonivamide acetate. V. Combined effect of physical enhancement methods. Int. J. Pharm. 235, 95–105.
Femenía-Font, A., et al., 2005. Effect of chemical enhancers on the in vitro percutaneous absorption of sumatriptan succinate. Eur. J. Pharm. Biopharm 51, 50–55.
Femenía-Font, A., et al., 2006. Combination strategies for enhancing transdermal absorption of sumatriptan through skin. Int. J. Pharm. 323, 125–130.
Glikfeld, P., Hinz, R.S., Guy, R.H., 1989. Noninvasive sampling of biological fluids by iontophoresis. Pharm. Res. 6, 988–990.
Golden, G.M., et al., 1987. Role of stratum corneum lipid fluidity in transdermal drug flux. J. Pharm. Sci. 76, 25–28.
Green, P.G., et al., 1991. Iontophoretic delivery of amino acids and amino acid derivatives across the skin in vitro. Pharm. Res. 8 (9), 1113–1120.
Guy, R.H., et al., 2000. Iontophoresis: electrorepulsion and electroosmosis. J. Control. Release 64 (1–3), 129–132.
Guy, R.H., Delgado-Charro, M.B., Kalia, Y.N., 2001. Iontophoretic transport across the skin. Skin Pharmacol. Appl. Skin. Physiol. 14 (Suppl. 1), 35–40.

Hanninen, K., et al., 2003. Rate and extent of ion-exchange process: the effect of physico-chemical characteristics of salicylate anions. J. Control. Release 91, 449–463.
Hao, J., et al., 2008. Transungual iontophoretic transport of polar neutral and positively charged model permeants: effects of electrophoresis and electroosmosis. J. Pharm. Sci. 97, 881–893.
Heller, R., et al., 1999. Clinical applications of electrochemotherapy. Adv. Drug Deliv. Rev. 35, 119–129.
Herndon, C.M., 2007. Iontophoretic drug delivery system: focus on fentanyl. Pharmacotherapy 27 (5), 745–754.
Higo, N., et al., 2001. US patent 6330471.
Hofmann, G.A., 1999. US patent 6009345.
Jaskari, T., et al., 2001. Ion-exchange fibers and drugs: an equilibrium study. J. Control. Release 70, 219–229.
Kalia, Y.N., et al., 2004. Iontophoretic drug delivery. Adv. Drug Deliv. Rev. 56 (5), 619–658.
Kanikkannan, N., 2002. Iontophoresis-based transdermal delivery systems. BioDrugs 16 (5), 339–347.
Kankkunen, T., et al., 2000. Controlled transdermal iontophoresis by ion-exchange fiber. J. Control. Release 67, 179–190.
Kankkunen, T., et al., 2002. Improved stability and release control of levodopa and metaraminol using ion-exchange fibers and transdermal iontophoresis. Eur. J. Pharm. Sci. 16, 273–280.
Kankkunen, T., et al., 2004. Transdermal iontophoresis of tacrine in vivo. Pharm. Res. 19, 705–708.
Karande, P., et al., 2005. Design principles of chemical penetration enhancers for transdermal drug delivery. Proc. Natl. Acad. Sci. USA 102 (13), 4688–4693.
Kasting, G.B., Keister, J.C., 1989. Application of electrodiffusion theory for a homogeneous membrane to iontophoretic transport through skin. J. Control. Release 8 (3), 195–210.
Kavanagh, G.M., Oh, C., Shams, K., 2004. BOTOX delivery by iontophoresis. Br. J. Dermatol. 151 (5), 1093–1095.
Khoo, P.P., et al., 2006. Iontophoresis of antibiotics into segmental allografts. J. Bone Joint Surg. Br. 88 (9), 11149–11157.
Kreyden, O.P., 2004. Iontophoresis for palmoplantar hyperhidrosis. J. Cosmet. Dermatol. 3 (4), 211–214.
Lai, P.M., Roberts, M.S., 1999. An analysis of solute structure – human epidermal transport relationships in epidermal iontophoresis using the ionic mobility: pore model. J. Control. Release 58 (3), 323–333.
Le, L., et al., 2000. Combined effect of low-frequency ultrasound and iontophoresis: applications for transdermal heparin delivery. Pharm. Res. 17, 1151–1154.
Ledger, P., 1992. Skin biological issues in electrically enhanced transdermal delivery. Adv. Drug Deliv. Rev. 9 (2–3), 289–307.
Lin, W., et al., 2001. Transermal delivery of antisense oligonucleotides with microprojection patch (Macroflux) technology. Pharm. Res. 18, 1789–1793.
Lin, W., et al., 2007. Enhancement of skin permeation of high molecular compounds by a combination of microneedle pretreatment and iontophoresis. J. Control. Release 118, 189–195.
Matriano, J.A., et al., 2002. Macroflux microprojection array patch technology: a new and efficient approach for intracutaneous immunixation. Pharm. Res. 19, 63–70.
Meidan, V.M., Al-Khalili, M., Michniak, B.B., 2003a. Enhanced iontophoretic delivery of buspirone hydrochloride across human skin using chemical enhancers. Int. J. Pharm. 264 (1–2), 73–83.
Meidan, V.M., Al-Khalili, M., Michniak, B.B., 2003b. Iontophoretic delivery of buspirone hydrochloride in hairless mouse skin. AAPS Pharm. Sci. 5 (2), 134–144.
Merino, G., et al., 2003a. Ultrasound-enhanced transermal transport. J. Pharm. Sci. 92, 1125–1137.

Merino, G., et al., 2003b. Frequency and thermal effects on the enhancement of transdermal transport by sonophoresis. J. Control. Release 88, 85–94.
Merino, V., Kalia, Y.N., Guy, R.H., 1997. Transdermal therapy and diagnosis by iontophoresis. Trends Biotechnol. 15, 288–290.
Merino, V., et al., 1999a. Electrorepulsion versus electroosmosis: effect of pH on the iontophoretic flux of 5-fluorouracil. Pharm. Res. 16 (5), 758–761.
Merino, V., Lopez, A., Hochstrasser, D., Guy, R.H., 1999b. Noninvasive sampling of phenylalanine by reverse iontophoresis. J. Control. Release 61, 65–69.
Mitragotri, S., et al., 2000. Determination of threshold energy dose for ultrasound-induced transdermal drug transport. J. Control. Release 63, 41–52.
Monteiro-Riviere, N.A., Inman, A.O., Riviere, J.E., 1994. Identification of the pathway of iontophoretic drug delivery: light and ultrastructural studies using mercuric chloride in pigs. Pharm. Res. 11 (2), 251–256.
Narasimha Murthy, S., Wiskirchen, D.E., Bowers, C.P., 2007. Iontophoretic drug delivery across human nail. J. Pharm. Sci. 96 (2), 305–311.
Pacini, S., et al., 2006. Transdermal delivery of heparin using pulsed current iontophoresis. Pharm. Res. 23 (1), 114–120.
Panchagnula, R., et al., 2000. Transdermal iontophoresis revisited. Curr. Opin. Chem. Biol. 4 (4), 468–473.
Parkinson, T.M., Szlek, M.A., Isaacson, J.D., 2007. Hybresis™ the hybridization of traditional with low-voltage iontophoresis. Drug Deliv. Technol. Vol. 7, No. 4, 54.
Phipps, J.B., Padmanabhan, R.V., Lattin, G.A., 1989. Iontophoretic delivery of model inorganic and drug ions. J. Pharm. Sci. 78 (5), 365–369.
Pillai, O., et al., 2003. Transdermal iontophoresis of insulin. II. Physicochemical considerations. Int. J. Pharm. 254 (2), 271–280.
Pitzer, K.R., Desai, S., Dunn, T., Edelman, S., Jayalakshmi, Y., Kennedy, J., Tamada, J.A., Potts, R.O., 2001. Detection of hypoglycemia with the GlucoWatch biographer. Diabetes Care 24, 881–885.
Potts, R.O., Tamada, J.A., Tierney, M.J., 2002. Glucose monitoring by reverse iontophoresis. Diabetes Metab. Res. Rev. 18 (Suppl. 1), S49–S53.
Prausnitz, M.R., 2004. Microneedles for transdermal drug delivery. Adv. Drug Deliv. Rev. 56, 581–587.
Rai, R., Srinivas, C.R., 2005. Iontophoresis in dermatology. Indian J. Dermatol. Venereol. Leprol. 71 (4), 236–241.
Raiman, J., et al., 2004. Delivery and stability of LHRH and Nafarelin in human skin: the effect of constant/pulsed iontophoresis. Eur. J. Pharm. Sci. 21 (2–3), 371–377.
Rastogi, S.K., et al., 2003. Passive and iontophoretic transport enhancement of insulin through porcine epidermis by depilatories: permeability and Fourier transform infrared spectroscopy studies. AAPS Pharm. Sci. Tech. 4 (3), 1–9.
Raynauld, J.P., Laviolette, J.R., 1987. The silver–silver chloride electrode: a possible generator of offset voltages and currents. J. Neurosci. Meth. 19 (3), 249–255.
Regnier, V., Preat, V., 1998. Localization of a FITC-labeled phosphorothioate oligodeoxynucleotide in the skin after topical delivery by iontophoresis and electroporation. Pharm. Res. 15 (10), 1596–1602.
Roberts, M.S., Lai, P.M., Cross, S.E., Yoshida, N.H., 1997. In: Mechanisms of Transdermal Drug Delivery. Marcel Dekker, New York, pp. 291–349.
Roberts, M.S., Lai, P.M., Anissimov, Y.G., 1998. Epidermal iontophoresis: I. Development of the ionic mobility–pore model. Pharm. Res. 15 (10), 1569–1578.
Sekkat, N., et al., 2002. Reverse iontophoretic monitoring in premature neonates: feasibility and potential. J. Control. Release 81 (1–2), 83–89.
Sharma, A., et al., 2000. Transdermal drug delivery using electroporation. I. Factors influencing in vitro delivery of terazosin hydrochloride in hairless rats. J. Pharm. Sci. 89, 528–535.

Siddiqui, O., Roberts, M.S., Polack, A.E., 1985. The effect of iontophoresis and vehicle pH on the in-vitro permeation of lignocaine through human stratum corneum. J. Pharm. Pharmacol. 37 (10), 732–735.
Singh, P., Maibach, H.I., 1994. Iontophoresis in drug delivery: basic principles and applications. Crit. Rev. Ther. Drug Carrier Syst. 11 (2–3), 161–213.
Singh, P., et al., 1999. Transdermal iontophoretic delivery of methylphenidate HCl in vitro. Int. J. Pharm. 178 (1), 121–128.
Sun, Y., et al., 2003. US patent 6532386.
Sun, Y., et al., 2004. US patent 6678554.
Tamada, J.A., Garg, S., Jovanovic, L., Pitzer, K.R., Fermi, S., Potts, R.O., 1999. Noninvasive glucose monitoring: comprehensive clinical results. J. Am. Med. Assoc. 282, 1839–1844.
Tiwari, S.B., Udupa, N., 2003. Investigation into the potential of iontophoresis facilitated delivery of ketorolac. Int. J. Pharm. 260 (1), 93–103.
Tokumoto, S., et al., 2006. Effect of electroporation and pH on the iontophoretic transdermal delivery of human insulin. Int. J. Pharm. 326 (1–2), 13–19.
Turner, N.G., Guy, R.H., 1998. Visualization and quantitation of iontophoretic pathways using confocal microscopy. J. Invest. Dermatol. Symp. Proc. 3 (2), 136–142.
Vanbever, R., et al., 1996. Transdermal delivery of fentanyl by electroporation. I. Influence of electrical factors. Pharm. Res. 13, 559–565.
van der Geest, R., Danhof, M., Bodde, H.E., 1997a. Iontophoretic delivery of apomorphine. I: In vitro optimization and validation. Pharm. Res. 14(12), 1798–1803.
van der Geest, R., et al., 1997b. Iontophoretic delivery of apomorphine. II: An in vivo study in patients with Parkinson's disease. Pharm. Res. 14(12), 1804–1810.
Vemulapalli, V., et al., 2008. Synergistic effect of iontophoresis and soluble microneedles for transdermal delivery of methotrexate. J. Pharm. Pharmacol. 60, 27–33.
Vuorio, M., et al., 2003. Ion-exchange fibers and drugs: a transient study. J. Control. Release 91, 439–448.
Wang, J.Y., et al., 2002. Transdermal iontophoresis of sodium nonivamide acetate. V. Combined effect of physical enhancement methods. Int. J. Pharm. 235, 95–105.
Wang, Y., et al., 2000. The effect of sodium dodecyl sulfate on the iontophoretic transport of hydrocortisone across hairless mouse skin. Pharm. Dev. Technol. 5, 533–542.
Wang, Y., Michniak, B., et al., 2003. Effects of fatty acids and iontophoresis on the delivery of midodrine hydrochloride and the structure of human skin. Pharm. Res. 20 (10), 1612–1618.
Wascotte, V., et al., 2005. Reverse iontophoresis of lithium: electrode formulation using a thermoreversible polymer. Eur. J. Pharm. Biopharm. 59 (1), 237–240.
Wascotte, V., et al., 2007. Monitoring of urea and potassium by reverse iontophoresis in vitro. Pharm. Res. 24 (6), 1131–1137.
Wearley, L.L., Chien, Y.W., 1990. Iontophoretic transdermal permeation of verapamil. III: Effect of binding and concentration gradient on reversibility of skin permeation rate. Int. J. Pharm. 59 (2), 87–94.
Yang, J.Y., et al., 2002. The effects of iontophoresis and electroporation on transdermal delivery of buprenorphine from solutions and hydrogels. J. Pharm. Pharmacol. 54, 1329–1337.
Yoshida, N.H., Roberts, M.S., 1993. Solute molecular size and transdermal iontophoresis across excised human skin. J. Control. Release 25 (3), 177–195.
Yukio, M., et al., 2006. Percutaneous penetration of ascorbic acid by sonophoresis and iontophoresis. Aesth. Dermatol. 16, 102–108.
Zakzewski, C.A., et al., 1998. Transdermal delivery of regular insulin to chronic diabetic rats: effect of skin preparation and electrical enhancement. J. Control. Release 50 (1–3), 267–272.

CHAPTER 5

Ultrasound-based Technology for Skin Barrier Permeabilization

Victor Meidan [1*] *and* **Bozena B. Michniak-Kohn** [2]

[1] Strathclyde Institute of Pharmacy and BioMedical Sciences, Strathclyde University, 27 Taylor Street, Glasgow G4 0NR, Scotland, UK; email: victor.meidan@strath.ac.uk

[2] Ernest Mario School of Pharmacy, Rutgers – the State University of New Jersey, Piscataway, NJ 08854, USA

Contents

5.1 Introduction 119
5.2 What is ultrasound? 120
5.3 Ultrasound bioeffects 121
  5.3.1 Cavitation 121
  5.3.2 Heating 122
  5.3.3 Microstreaming 122
5.4 Historical background 122
5.5 Sonophoresis for enhanced drug delivery 123
  5.5.1 Mechanisms 125
5.6 Ongoing developments 127
  5.6.1 Effective delivery of macromolecules 127
  5.6.2 "Reverse sonophoresis" for transdermal diagnostics 128
  5.6.3 Advances in transducer fabrication 129
5.7 Current challenges and trends 130
References 131

5.1 INTRODUCTION

Phonophoresis, or sonophoresis, is defined as the use of ultrasonic energy in order to enhance the topical or transdermal delivery of drugs. This chapter will introduce and discuss the nature of ultrasonic energy, its mode of propagation, and physical properties. This is followed by a description of the key biophysical phenomena that are observed when ultrasound interacts with biological tissues. The historical development of sonophoretic treatment is then discussed with emphasis placed on the evolution and identification of low-frequency sonophoresis as a novel technology for skin

* Author to whom correspondence should be addressed.

Handbook of Non-Invasive Drug Delivery Systems
ISBN 9780815520252

barrier permeabilization. This modality is then described with respect to its mechanistic aspects, efficacy, and applications. Finally, the chapter summarizes the current challenges, critical questions, and future potential pertaining to this technology.

5.2 WHAT IS ULTRASOUND?

Ultrasound is an acoustic vibration characterized by longitudinal compression waves that propagate at frequencies beyond the human auditory threshold, i.e. 0.02 MHz. Thus, ultrasound transmission induces compression and expansion of the medium through which it is passing, leading to associated pressure variations. The intensity parameter, measured in $\mathrm{W/cm^2}$, defines the quantity of energy conveyed by an ultrasonic wave as it passes through any fixed point and this parameter is used to establish exposure.

Ultrasound is normally generated by a transducer that converts electricity to sound by utilizing the piezoelectric effect. Piezoelectricity in this context refers to the behavior of specific ceramic materials, which expand or contract when a voltage is applied across them. For continuous wave ultrasound production, an alternating voltage is applied to the transducer, resulting in the continuous emission of ultrasound of the same frequency. To generate pulsed ultrasound, short bursts of alternating voltage are repeatedly applied to the transducer. The duration of each "voltage-on" time relative to the duration of "voltage-off" time determines the duty cycle of the emitted ultrasound. Under the intensities and conditions used in sonophoresis, generated ultrasound tends to retain its frequency and waveform regardless of the medium that the beam is passing through.

As ultrasound propagates through a material, that medium's resistance to sound wave propagation depends upon the acoustic impedance (Z), which is related to the mass density of the medium (ρ) and the speed of ultrasound propagation (C) according to the equation:

$$Z = \rho \times C \tag{5.1}$$

The specific acoustic impedances for skin, bone, and air are 1.6×10^6, 6.3×10^6, and $400.0\ \mathrm{kg/(m^2\,s)}$ respectively (Williams, 1983). One consequence of the high acoustic impedance of air is that ultrasound will not propagate well between the transducer source and skin surface. Hence, investigations into sonophoresis require the use of a suitable coupling medium such as an aqueous solution, gel, oil, or emulsion. This facilitates an efficient transfer of ultrasound between the transducer and the skin surface.

Once ultrasound is propagating through body tissue, biological effects may be triggered if the tissues absorb the incident energy. An absorption coefficient (a) can be defined, which is used to establish absorption in each tissue. In the ideal case of longitudinal ultrasonic waves that are perpendicularly incident on a homogeneous tissue, the following equation applies:

$$I(x) = I_0 \times e^{-ax} \tag{5.2}$$

where $I(x)$ is the intensity at depth x, I_0 is the intensity at the surface, and a is the absorption coefficient.

When ultrasound propagates through an interface between media exhibiting a mismatch in acoustic impedance, it can undergo reflection, thus interfering with its own incident component to form a standing wave. Such a wave exhibits a regular repeating pattern of nodes and antinodes. Nodes are characterized by zero displacement while antinodes are characterized by displacement varying from positive to negative at the same frequency as the incoming wave, but at twice its amplitude. A partial standing wave develops if the reflector is less than 100% efficient. *In vivo*, a standing wave field can form at a bone–soft tissue interface. Standing waves can also arise *in vitro*, within diffusion cells used to measure cutaneous drug flux. For example, it was determined that the steel mesh supporting a skin sample in a Franz cell reflected back at least 75% of energy to form a standing wave within the skin and coupling gel (Meidan et al., 1998b).

Another important effect that can develop at a bone–soft tissue interface is mode conversion. This is characterized by partial transformation of longitudinal ultrasonic waves into transverse waves. Any transverse wave component will not readily propagate, so energy is rapidly released as heat at the interface.

5.3 ULTRASOUND BIOEFFECTS

Ultrasonic waves can interact with biological tissues in a variety of ways. These can be classified into cavitational effects, heating, and microstreaming. Although these are discussed separately below, it should be noted that in practice these phenomena are interrelated as, for example, cavitation can cause heating as well as microstreaming.

5.3.1 Cavitation

Cavitation describes a very wide spectrum of acoustically induced bubble phenomena (Humphrey, 2007) that are primarily caused by ultrasound-induced pressure variations in the sonicated medium. Stable cavitation is the relatively gentle volumetric pulsations of gas-filled bodies, while transient cavitation is the destructive and free radical-generating creation and implosion of vapor-filled voids. All cavitation is powered by the energy from the incident ultrasonic beam, about 10% of which may be re-radiated from the bubble as an outgoing spherical wave. The remainder may be lost as heat, or may induce shock waves that can perturb surrounding tissues. Induction of cavitation varies inversely with ultrasound frequency and directly with ultrasound intensity. The frequency effect is due to the fact that, at higher frequencies, there is insufficient time for gas molecules to diffuse into growing cavities during the brief rarefaction phase of the acoustic cycle.

5.3.2 Heating

As ultrasonic waves propagate through biological tissue, any absorption of the energy will increase the temperature of the exposed medium. Biological tissues that possess a high ultrasound absorption coefficient, such as bone, undergo extensive heating while those that possess a low absorption coefficient, such as muscle tissue, undergo only mild heating (Suslick, 1988). Frequency is a key parameter, as a tissue's absorption coefficient increases directly with ultrasonic frequency. Furthermore, at any fixed frequency, temperature elevation varies directly with acoustic intensity and exposure time.

5.3.3 Microstreaming

This phenomenon describes the acoustically triggered unidirectional current flow that is produced by ultrasonic reflections and other distortions occurring in liquid or semiliquid media during wave propagation (Clarke et al., 2004). Although the microcurrents typically have low velocities, the phenomenon is associated with very high rates of velocity change with distance. Hence, large hydrodynamic shear stresses may be induced and these can affect exposed biological tissue. Microstreaming can also develop secondarily to cavitation. Acoustic streaming may be significant when the biological tissue contains liquid or semiliquid elements, when there is a mismatch in acoustic impedance between adjacent sonicated tissues, or when a continuous wave application is employed.

5.4 HISTORICAL BACKGROUND

Since the 1960s, physiotherapists have used the combination of ultrasound plus a steroid or analgesic cream or gel in order to treat a broad range of muscular and arthritic conditions (Meidan et al., 1995). The idea was that the ultrasound enhances topical drug absorption. Unfortunately, most of this therapy was carried out in a rather subjective and ad hoc manner. Many of the physiotherapy clinic-based studies of the 1960s and 1970s were characterized by a lack of suitable control groups, incomplete descriptions of the methodologies used, and nonvalidation of ultrasound dosimetry. The final point is particularly pertinent since some of the older physiotherapy-type ultrasonic generators suffered from poor calibration and the quantity of emitted energy was not truly reflected by the control dial on the device (Williams, 1983). Consequently, as reviewed by Meidan et al. (1995), much of the data derived from these early studies were scientifically inadequate.

Starting in the late 1980s, a series of more refined, carefully controlled initiatives were undertaken by both pharmaceutical and chemical engineering teams. Research now involved calibration of the acoustic generators by application of radiation pressure force measurements (Ueda et al., 1996; Meidan et al., 1998b, 1999), calorimetric techniques (Hikima et al., 1998), or hydrophones (Johnson et al., 1996; Mitragotri et al., 1996). In terms of frequency, intensity, mode, and treatment time, individual groups employed different parameters. Generally, the frequency varied between 0.5 and 3 MHz while intensity was usually below 3 W/cm^2. Treatment time varied tremendously between a few minutes and several hours. Some researchers used pulsed beams, of various duty cycles, in order to reduce ultrasonic heating. In some of these studies it was determined that ultrasound was ineffective. For instance, negative results were reported regarding attempts to ultrasonically enhance the absorption of: benzydamine (Benson et al., 1989), trolamine salicylate (Oziomek et al., 1991), zidovudine (Pelucio-Lopes et al., 1993), digoxin (Machet et al., 1996), and hydrocortisone (Machet et al., 1998), as well as various anesthetics (Williams, 1990) through human skin. In contrast, other researchers reported moderate success in achieving accelerated cutaneous delivery. For example, Murphy and Hadgraft (1990) documented some enhancement of methyl and ethyl nicotinates across human skin *in vivo*. Thus, use of ultrasonic frequencies greater than 0.5 MHz is either ineffective or only marginally effective in promoting transdermal drug delivery.

5.5 SONOPHORESIS FOR ENHANCED DRUG DELIVERY

In a landmark initiative, Tachibana and Tachibana (1991) reported that ultrasound at a very low frequency could enhance transcutaneous insulin penetration in hairless rats *in vivo*. The group developed a system whereby the animals were immersed in a beaker filled with insulin solution (20 U/mL) and exposed to 48 kHz ultrasound. It was found that blood glucose levels decreased by 50% in 4 hours. Further research established that low-frequency ultrasound could also promote transdermal insulin absorption in rabbits (Tachibana, 1992). In sequential work, this team demonstrated that exposure to 48 kHz ultrasound could increase the duration of anesthesia produced by transdermal lidocaine treatment in hairless rats (Tachibana and Tachibana, 1993).

An array of subsequent studies confirmed that low-frequency (20–500 kHz) ultrasound was indeed highly efficacious in enhancing the transdermal absorption of a diverse range of penetrants. Frequencies lower than 100 kHz were reported to be particularly effective. Mitragotri et al. (1996) compared a frequency of 20 kHz with a frequency of 1 MHz in terms of the permeabilization of human cadaver skin *in vitro*. Enhancement ratios, defined as the ratio of the sonophoretic to passive permeability, were employed to quantify permeabilization for butanol, corticosterone, salicylic acid, and sucrose. They reported that the enhancement induced by 20 kHz ultrasound was up to 1000-fold greater than that induced by 1 MHz ultrasound. Low-frequency ultrasound has also been used successfully to accelerate the *in vitro* transdermal flux of other small molecules such as mannitol (Merino et al., 2003), morphine (Monti et al., 2001), and caffeine (Boucaud et al., 2001). The extent of reported transport enhancement has typically been of up to one order of magnitude but even greater improvements have been observed, depending upon the model system employed. Furthermore, the number of compounds that can be efficaciously delivered with this technology is increasing all the time (Park et al., 2007).

Some research has been conducted with human volunteers (Katz et al., 2004). In this double-blind trial of 42 subjects, skin was sonicated at 55 kHz for 10 seconds and a local anesthetic or placebo cream was then applied to the treatment site for various fixed durations. The skin was pricked with a needle. Pain scores and patient preferences for each treatment were quantified at successive intervals. It was found that sonication followed by 5, 10, or 15 minutes of anesthetic cream application yielded pain and preference rates that were statistically indistinguishable from anesthetic cream application for 1 hour without sonication. There were no significant adverse effects. Hence, low-frequency ultrasound pretreatment appears to be effective in accelerating local anesthesia onset in this model, with results occurring as early as 5 minutes.

Importantly, low-frequency ultrasound does not usually appear to cause gross morphological changes in human skin. For instance, one team assessed the histological effects of 20 kHz ultrasound on human skin samples *in vitro*

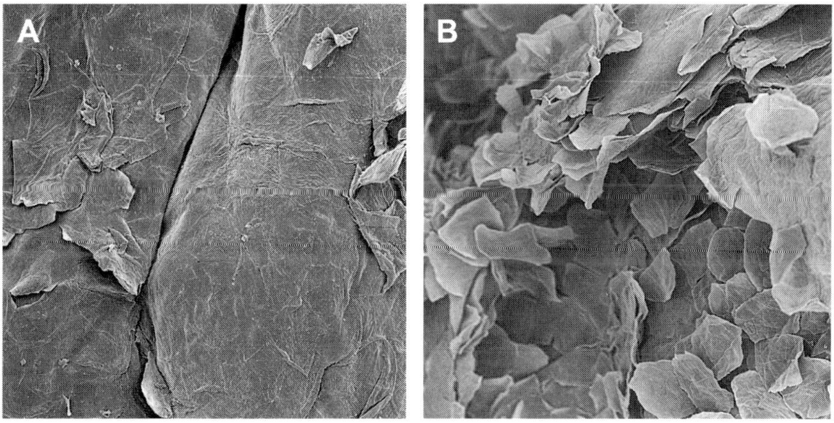

Figure 5.1 Surface images of porcine skin derived via scanning electron microscopy (original magnification ×500); 20 kHz ultrasound was applied for 5 minutes using a pulsed (1 s on, 9 s off) beam of SATA intensity 0.37 W/cm^2 (V. Meidan, unpublished data). (A) Before sonication. (B) One hour after sonication.

(Boucaud et al., 2001). At intensities below 2.5 W/cm^2, no structural modifications were evident. In other *in vitro* studies (Mitragotri et al., 1996), no detrimental changes were observed in human skin following sonication at 20 kHz. *In vivo* studies with dogs also produced evidence that low-frequency ultrasound at low intensities and without any drug was apparently safe for sonophoresis, inducing only minor urticarial reactions (Singer et al., 1998). Brief applications of 20 kHz ultrasound were reportedly well tolerated in human volunteers (Kost et al., 2000). However, some researchers have observed ultrasound-induced tissue changes. Yamashita et al. (1997) detected some minor detachment of keratinocytes around human hair follicles following *in vitro* sonication for 5 minutes at 48 kHz. Meidan (unpublished data) showed that a 5-minute *in vitro* application of a pulsed 20-kHz beam caused previously flat, intact, and confluent porcine corneocytes to undergo noticeable disordering, detachment, and uplifting (see Figure 5.1).

5.5.1 Mechanisms

Over the last 15 years or so, appreciable research has been directed towards understanding the mechanisms responsible for low-frequency sonophoresis. It is now established that permeabilization is caused by cavitational activity in the stratum corneum and/or adjoining coupling medium. Acoustic spectral analysis has indicated that transient cavitation is associated with shock waves produced by symmetrical bubble collapse, as well as high-velocity microjets caused by asymmetrical implosions (Tezel et al., 2002b).

Both these phenomena are associated with tremendously high pressures and temperatures that develop for extremely brief moments of time. The results of recent mathematical modeling suggest that acoustic cavitation causes the growth of bubbles within the stratum corneum (Lavon et al., 2007). During this process, adjacent bubbles merge to form larger bubbles that continue to enlarge, creating channels in the stratum corneum. These channels grow with time so as to traverse the entire thickness of the stratum corneum, leading essentially to the formation of new internal transport routes.

Whatever the precise nature of cavitation, its effects on skin microstructure have been studied. Guy's group (Alvarez-Roman et al., 2003) determined that a 2-hour exposure to pulsed 20-kHz ultrasound resulted in the release of about 30% of intercellular stratum corneum lipids from porcine skin. In human skin, exposure to 168-kHz ultrasound was found to result in the formation of air-filled voids, 20 μm in diameter, within the stratum corneum lipid bilayers (Wu et al., 1998). The size and distribution of these voids were reported to be consistent with the phenomenon being induced by acoustic cavitation. In other work, quantum dots of 20 nm diameter were used as fluorescent tracers to reveal the perturbations occurring in porcine epidermis following ultrasonication at 20 kHz (Paliwal et al., 2006). This approach allowed visualization of stratum corneum alterations occurring at the submicroscopic level. It was found that sonication promoted the creation of new lacunar spaces and voids in the stratum corneum and/or enlarged those already in existence. These changes were essentially leading to the generation of a three-dimensional porous network, thus facilitating enhanced transdermal permeation of drugs and macromolecules. Good mathematical models, describing transdermal solute permeation through such a network, have already been published (Tezel et al., 2002a).

Given the fact that cavitation is the principal mechanism of low-frequency sonophoresis, it has been possible to determine how different variables affect transport enhancement. The magnitude of enhancement depends upon four principal acoustic parameters – frequency, intensity, duty cycle, and duration. In general, enhancement increases with decreasing ultrasound frequency; this is due to the fact that ultrasound induces proportionally more cavitational activity at lower frequencies (Williams, 1983). Sonophoretic permeabilization also increases with increasing intensity although two thresholds, both a lower cutoff intensity and upper decoupling intensity, have been identified (Mitragotri and Kost, 2004). Lengthening the duration of application or using a greater duty cycle in the case of pulsed ultrasound (i.e. increasing the "pulse-on" time) will similarly

Table 5.1 Ultrasound parameters and the resulting effects on sonophoretic enhancement

Change in acoustic parameter	Expected effect on enhancement	Comments
Increase in frequency	Decrease	Auditory sensitivity to high-pitched sounds poses a lower threshold on frequency
Increase in intensity	Increase	There is a lower threshold intensity and upper decoupling intensity
Increase in duration	Increase	Prolonged durations may cause excess tissue heating
Increase in duty cycle	Increase	High duty cycles may cause excess tissue heating

tend to increase enhancement. Table 5.1 summarizes the relationship between these parameters and their expected effects.

Although cavitation is the principal mechanism, ultrasonic heating may also contribute to skin permeabilization. Merino et al. (2003) deployed a 20-kHz ultrasound source and tracked mannitol permeation through full-thickness porcine ear skins mounted in Franz cells. The group devised a special control application that replicated the thermal effects of the beam but without the acoustic–mechanical effects. It was possible to show that ~25% of the transport enhancement was induced by heating and that ~75% of the enhancement was caused by other nonthermal components, probably cavitation.

5.6 ONGOING DEVELOPMENTS

5.6.1 Effective delivery of macromolecules

One of the most promising potential applications has been the idea of using low-frequency ultrasound to permeabilize the skin for the delivery of therapeutic macromolecules. In particular, the concept of non-invasive and needle-free insulin delivery has attracted considerable interest. As previously mentioned, positive findings were published in the early 1990s for insulin delivery in rats (Tachiban and Tachibana, 1991) and rabbits (Tachibana, 1992). Relevant research was performed by Mitragotri et al. (1995), who exposed human cadaver skin *in vitro* to pulses of 20-kHz ultrasound at 125 W/cm^2. It was found that ultrasound enhanced the transdermal flux of insulin (as well as γ-interferon and erythropoietin). In other experiments (Mitragotri et al., 1995), diabetic hairless rats were treated with insulin while undergoing simultaneous sonication with a 20-kHz pulsed beam of intensity 0.225 W/cm^2. These applications decreased the animals' blood glucose level from approximately 400 to 200 mg/dL in 30 minutes while concomitant changes in plasma insulin levels occurred. It has been shown that low-frequency sonophoresis of insulin was also possible in hairless rats (Boucaud et al., 2002). Most recently, the feasibility of transdermal insulin delivery was demonstrated in an *in vivo* pig model (Park et al., 2007). A 1-hour application of very-low-intensity ultrasound at 20 kHz resulted in a dramatic and significant drop in the animals' blood glucose concentrations.

Important data have also been acquired with respect to the sonophoretic delivery of heparin and low-molecular-weight heparin (Mitragotri and Kost, 2001). Low-frequency ultrasound facilitated sustained macromolecule transport, unlike injections, which only allowed transient delivery of these agents.

5.6.2 "Reverse sonophoresis" for transdermal diagnostics

Sonophoretic permeabilization of the skin can be deployed in reverse, with the aim of developing a non-invasive transdermal monitoring system that can be used to quantify clinically relevant blood analytes. Although the concept of such a diagnostic device is highly desirable, collecting a sufficient amount of analyte for accurate monitoring has historically proved problematic. Research has indicated that "reverse sonophoresis" works best when the skin is first permeabilized by low-frequency ultrasound and then the site is subjected to a vacuum. This facilitates a sufficient amount of analyte to be drawn upwards to the skin surface. A suitable sensor then assays the analyte (Kost et al., 2000; Lavon and Kost, 2004).

Such an approach was assessed in a clinical trial involving type 1 diabetic patients (Kost et al., 2000). The aim was to determine whether a brief (<2 min) application of 20-kHz pulsed ultrasound could extract enough glucose from human skin over the ensuing several hours and also to determine whether glucose flux was proportional to plasma glucose levels. Crucially, it was determined that, following sonication, skin permeability remained high for approximately 15 h but decreased to normal levels by 24 h. Moreover, following skin sonication, the magnitude of extracted glucose flux correlated well with venous blood glucose concentrations. Site-to-site variations of glucose permeability were measured within the same volunteer as well as between volunteers. Interestingly, variability due to site-associated differences was about the same as that due to inter-volunteer differences. This reveals the necessity of initially performing a one-point calibration between transdermal glucose flux and a single blood sample. This can then be applied to predict subsequent blood glucose values with a relatively low error margin. A similar clinical trial was performed by Chuang et al. (2004), who monitored, for over 8 h, the glucose levels of 10 diabetic volunteers through ultrasonically permeabilized skin. The investigators observed a good correlation between sensor output readings and plasma glucose concentrations. Several areas were identified where refinements could be undertaken in order to enhance the system's sensitivity and accuracy.

The safety of the ultrasound-assisted diagnostic approach still requires further investigation. Although pilot studies indicated that ultrasound did not induce detrimental effects on the skin with no damage or irritation being visually apparent (Kost et al., 2000; Chuang et al., 2004), further evaluations, particularly within the context of multiple extractions, are clearly warranted.

5.6.3 Advances in transducer fabrication

A major hurdle to the therapeutic utilization of low-frequency sonophoresis has been the large size and weight of pre-existing ultrasound generators. Most commercial generators are heavy, bulky, and immobile tabletop devices that have to be connected to a standard outlet. Moreover, the transducer probes are often too large. These generators were originally developed for creating emulsions, cleaning, lysing cells, or sonochemical purposes. For practical sonophoretic applications, a smaller lightweight, portable, and inexpensive device is needed.

Perhaps the most established purpose-designed device for sonophoresis is the SonoPrep® system. This Food and Drug Administration (FDA)-approved device, developed by Echo Therapeutics (Franklin, MA) operates by emitting 55-kHz ultrasound from a probe with a surface area of 0.8 cm^2. The equipment is portable, lightweight, and battery operated. A key attribute of the SonoPrep system is an integrated real-time skin impedance feedback mechanism. An incorporated sensor measures changes in skin impedance during sonication and ceases output when a desired impedance value has been attained. Since impedance is related to skin permeability, this feature prevents cutaneous overexposure to ultrasound. The device is not suitable for use on compromised skin.

Another interesting system is the cymbal transducer and multi-element array, which is essentially a novel flextensional transducer capable of emitting low frequencies such as 20 kHz (Park et al., 2007). The resulting fabricated device is lightweight and highly compact. It was recently established that the cymbal array was suitable for efficaciously delivering insulin in an *in vivo* pig model.

5.7 CURRENT CHALLENGES AND TRENDS

Skin permeabilization by ultrasound still faces several challenges and opportunities on several diverse fronts, as Figure 5.2 illustrates. One interesting route involves combining low-frequency ultrasound with another enhancer in order to develop synergistic enhancement strategies. Sodium lauryl sulfate has already been identified as an effective chemical enhancer that acts in this manner when combined with ultrasound (Lavon et al., 2005). It may be useful to carry on searching for other such synergistic compounds. Other types of physical enhancers such as electroporation and iontophoresis are available (Meidan and Michniak, 2004). These may also provide synergistic enhancement with ultrasound, but it is probably premature to combine such complex technologies at this time.

Low-frequency sonophoresis may also have considerable potential for facilitating needle-free vaccinations and immunizations. Transcutaneous immunization provides contact with the skin's immune system via densely concentrated antigen-presenting Langerhans cells. These cells induce T-cell-mediated immune reactions against numerous antigens. In order for this technique to be practical, the vaccine, which is usually a large molecule,

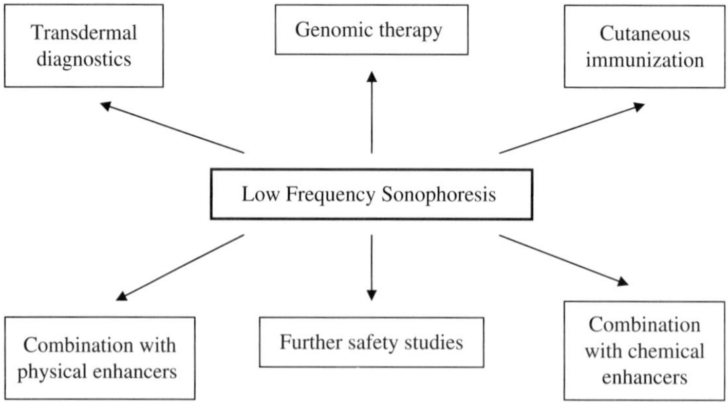

Figure 5.2 Flow chart illustrating the current challenges and potential directions of ultrasound-assisted skin permeabilization technology.

must penetrate the stratum corneum. Low-frequency ultrasound could be used to permeabilize the stratum corneum to a vaccine as well as a complementary adjuvant. Promising work in this context has already been published (Tezel at al., 2005).

Another exciting potential application lies in the realm of topical genomic therapy. The concept is that sonophoretic treatment would allow either a gene–vector complex or oligonucleotides (Tezel et al., 2004) to penetrate into the stratum corneum. The target could be either cutaneous cells or, alternatively, systemic circulation. Highly debilitating monogenic cutaneous disorders, such as epidermolysis bullosa, might one day be treated in this way.

One current knowledge gap relates to the influence of low-frequency ultrasound on formulations. There is little known about the effect of this energy on the stability of topical emulsions, creams, suspensions, and gels that are used as vehicles for active compounds. Active agents typically constitute a small fraction of the entire topical dosage form. More research needs to be conducted in this area.

To date, ultrasound applications for both diagnostic and therapeutic purposes are continually increasing (ter Haar, 2007). Consequently, there is mounting concern with respect to ultrasonic bioeffects as they relate to potential tissue damage and safety issues (Kimmel, 2006; Blomley et al., 2007; Humphrey, 2007; Miller, 2007). In particular, the bioeffects of low-frequency sonophoresis are currently poorly understood. Induced skin changes seem to be predominantly cavitational in nature. Yet there are still questions relating to the exact mechanisms and reversibility of the perturbations. Hence, further research in this area, both *in vitro* and *in vivo*, is clearly warranted so that this exciting technology can be utilized further.

REFERENCES

Alvarez-Roman, R., Merino, G., Kalia, Y.N., Naik, A., Guy, R.H., 2003. Skin permeability enhancement by low frequency sonophoresis; lipid extraction and transport pathways. J. Pharm. Sci. 92, 1138–1146.

Benson, H.A.E., McElnay, J.C., Harland, R., 1989. Use of ultrasound to enhance percutaneous absorption of benzydamine. Phys. Ther. 69 (2), 113–118.

Blomley, M., Claudon, M., Cosgrove, D., 2007. WFUMB Safety Symposium on Ultrasound Contrast Agents: clinical applications and safety concerns. Ultrasound Med. Biol. 33, 180–186.

Boucaud, A., Machet, L., Arbeille, B., Machet, M.C., Sournac, M., Mavon, A., Patat, F., Vaillant, L., 2001. In vitro study of low frequency ultrasound-enhanced transdermal transport of fentanyl and caffeine across human and hairless rat skin. Int. J. Pharm. 228, 69–77.

Boucaud, A., Garrigue, M.A., Machet, L., Vaillant, L., Patat, F., 2002. Effect of sonication parameters on transdermal delivery of insulin to hairless rats. J. Pharm. Sci. 91, 113–119.

Chuang, H., Taylor, E., Davidson, T.W., 2004. Clinical evaluation of a continuous minimally invasive glucose flux sensor placed over ultrasonically permeated skin. Diabetes Technol. Ther. 6, 21–30.

Clarke, L., Edwards, A., Graham, E., 2004. Acoustic streaming: an *in vitro* study. Ultrasound Med. Biol. 30, 559–562.

ter Haar, G.R, 2007. Therapeutic applications of ultrasound. Prog. Biophys. Mol. Biol. 93, 111–123.

Hikima, T., Hirai, Y., Tojo, K., 1998. Effect of ultrasound application on skin metabolism of prednisolone 21-acetate. Pharm. Res. 15, 1680–1683.

Humphrey, V.F., 2007. Ultrasound and matter physical interactions. Prog. Biophys. Mol. Biol. 93, 195–211.

Johnson, M.E., Mitragotri, S., Patel, A., Blankschtein, D., Langer, R., 1996. Synergistic effects of chemical enhancers and therapeutic ultrasound on transdermal drug delivery. J. Pharm. Sci. 85, 670–679.

Katz, N.P., Shapiro, D.E., Herrmann, T.E., Kost, J., Custer, L.M., 2004. Rapid onset of cutaneous anesthesia with EMLA cream after pretreatment with a new ultrasound-emitting device. Anesth. Analg. 98, 371–376.

Kimmel, E., 2006. Cavitation bioeffects. Crit. Rev. Biomed. Eng. 34, 105–161.

Kost, J., Mitragotri, S., Gabbay, R.A., Pishko, M., Langer, R., 2000. Transdermal monitoring of glucose and other analytes using ultrasound. Nature Med. 6, 347–350.

Lavon, I., Kost, J., 2004. Ultrasound and transdermal drug delivery. Drug Discov. Today 9, 670–676.

Lavon, I., Grossman, N., Kost, J., 2005. The nature of ultrasound-SLS synergism during enhanced transdermal transport. J. Control. Release 107, 484–494.

Lavon, I., Grossman, N., Kost, J., Kimmel, E., Enden, G., 2007. Bubble growth within the skin by rectified diffusion might play a significant role in sonophoresis. J. Control. Release 117, 246–255.

Machet, L., Pinton, J., Patat, F., Arbeille, B., Pourcelot, L., Vaillant, L., 1996. In vitro phonophoresis of digoxin across hairless mice and human skin: thermal effects of ultrasound. Int. J. Pharm. 133, 39–45.

Machet, L., Cochelin, N., Patat, F., Arbeille, B., Machet, M.C., Lorette, G., Vaillant, L., 1998. In vitro phonophoresis of mannitol, oestradiol and hydrocortisone across human and hairless mouse skin. Int. J. Pharm. 165, 169–174.

Meidan, V.M., Michniak, B.B., 2004. Emerging technologies in transdermal therapeutics. Am. J. Ther. 11, 312–316.

Meidan, V.M., Walmsley, A.D., Irwin, W.J., 1995. Phonophoresis – is it a reality? Int. J. Pharm. 118, 129–145.

Meidan, V.M., Docker, M.F., Walmsley, A.D., Irwin, W.J., 1998b. Phonophoresis of hydrocortisone with enhancers: an acoustically defined model. Int. J. Pharm. 170, 157–168.

Meidan, V.M., Walmsley, A.D., Docker, M., Irwin, W.J., 1999. Ultrasound-enhanced diffusion into coupling gel during phonophoresis of 5-fluorouracil. Int. J. Pharm. 185, 205–213.

Merino, G., Kalia, Y.N., Delgado-Charro, M.B., Potts, R.O., Guy, R.H., 2003. Frequency and thermal effects on the enhancement of transdermal transport by sonophoresis. J. Control. Release 88, 85–94.

Miller, D.L., 2007. Overview of experimental studies of biological effects of medical ultrasound caused by gas body activation and inertial cavitation. Prog. Biophys. Mol. Biol. 93, 314–330.

Mitragotri, S., Kost, J., 2001. Transdermal delivery of heparin and low-molecular-weight heparin using low frequency ultrasound. Pharm. Res. 18, 1151–1156.

Mitragotri, S., Kost, J., 2004. Low frequency sonophoresis: a review. Adv. Drug Deliv. Rev. 56, 589–601.
Mitragotri, S., Blankschtein, D., Langer, R., 1995. Ultrasound-mediated transdermal protein delivery. Science 269, 850–853.
Mitragotri, S., Blankschtein, D., Langer, R., 1996. Transdermal drug delivery using low-frequency sonophoresis. Pharm. Res. 13, 411–420.
Monti, D., Gianelli, R., Chetoni, P., Burglasi, P., 2001. Comparison of the effect of ultrasound and of chemical enhancers on transdermal permeation of caffeine and morphine through hairless mouse skin in vitro. Int. J. Pharm. 229, 131–137.
Murphy, T.M., Hadgraft, J., 1990. A physico-chemical interpretation of phonophoresis in skin penetration enhancement. In: Scott, R.C., Guy, R.H., Hadgraft, J. (Eds), Prediction of Percutaneous Penetration, Methods, Measurements and Modelling. IBC Technical Services, London, pp. 333–336.
Oziomek, R.S., Perrin, D.H., Herold, D.A., Craig, C.R., 1991. Effect of phonophoresis on serum salicylate levels. Med. Sci. Sports Med. 23, 397–401.
Paliwal, S., Menon, G.K., Mitragotri, S., 2006. Low frequency sonophoresis: ultrastructural basis for stratum corneum permeability assessed using quantum dots. J. Invest. Dermatol. 126, 1095–1101.
Park, E.J., Werner, J., Smith, N.B., 2007. Ultrasound mediated transdermal insulin delivery in pigs using a lightweight transducer. Pharm. Res. 24, 1396–1401.
Pelucio-Lopes, C., Machet, L., Vaillant, L., Patat, F., Lethieq, M., Furet, Y., Pourcelot, L., Lorette, G., 1993. Phonophoresis of azidothymidine (AZT). Int. J. Pharm. 96, 249–252.
Singer, A.J., Homan, C.S., Church, A.L., McCain, S.A., 1998. Low frequency sonophoresis: pathologic and thermal effects in dogs. Acad. Emerg. Med. 5, 35–40.
Suslick, K.S. (Ed), 1988. Ultrasound: Its Chemical, Physical and Biological Effects. VCH, New York.
Tachibana, K., 1992. Transdermal delivery of insulin to alloxan-diabetic rabbits by ultrasound exposure. Pharm. Res. 9, 952–954.
Tachibana, K., Tachibana, S., 1991. Transdermal delivery of insulin by ultrasonic vibration. J. Pharm. Pharmacol. 43, 270–271.
Tachibana, K., Tachibana, S., 1993. Use of ultrasound to enhance the local anesthetic effect of topically applied aqueous lidocaine. Anesthesiology. 78, 1091–1096.
Tezel, A., Sens, A., Mitragotri, S., 2002a. Incorporation of lipophilic pathways into the porous pathway model for describing skin permeabilization during low frequency sonophoresis. J. Control. Release 83, 183–188.
Tezel, A., Sens, A., Mitragotri, S., 2002b. Investigations of the role of cavitation in low frequency sonophoresis using acoustic spectroscopy. J. Pharm. Sci. 91, 444–453.
Tezel, A., Dokka, S., Kelly, S., Hardee, G.E., Mitragotri, S., 2004. Topical delivery of anti-sense oligonucleotides using low-frequency sonophoresis. Pharm. Res. 21, 2219–2225.
Tezel, A., Paliwal, S., Shen, Z., Mitragotri, S., 2005. Low-frequency ultrasound as a transcutaneous immunization adjuvant. Vaccine 23, 3800–3807.
Ueda, H., Ogihara, M., Sugibayashi, K., Morimoto, Y., 1996. Change in the electrochemical properties of skin and the lipid packing in stratum corneum by ultrasonic irradiation. Int. J. Pharm. 85, 217–224.
Williams, A.R. (Ed), 1983. Ultrasound: Biological Effects and Potential Hazards. Academic Press, London.
Williams, A.R., 1990. Phonophoresis: an in vivo evaluation using three topical anaesthetic preparations. Ultrasonics 28, 137–140.
Wu, J., Chappelow, J., Yang, J., Weimann, L., 1998. Defects generated in human stratum corneum specimens by ultrasound. Ultrasound Med. Biol. 24, 705–710.
Yamashita, N., Tachibana, K., Ogawa, K., Tsujita, N., Tomita, A., 1997. Scanning electron microscopic evaluation of the skin surface after ultrasound exposure. Anat. Rec. 247, 455–461.

CHAPTER 6

Microneedles – Minimally Invasive Transdermal Delivery Technology

Diksha Kaushik, Brian Kilfoyle, Rashmi Thakur *and* **Bozena B. Michniak-Kohn**
Ernest Mario School of Pharmacy, Rutgers – the State University of New Jersey, Piscataway, NJ 08854, USA

Contents

6.1 Introduction 135
6.2 Microneedle geometry 136
  6.2.1 Solid microneedles 137
  6.2.2 Hollow microneedles 139
  6.2.3 Biodegradable microneedles 140
6.3 Types of approach 140
  6.3.1 Poke with patch technique 142
  6.3.2 Coat and poke technique 142
  6.3.3 Dip and scrape technique 142
6.4 Factors influencing transdermal permeation 143
  6.4.1 Depth of permeation 143
  6.4.2 Coating strategies 143
  6.4.3 Microneedle geometry and force of penetration 148
6.5 Microneedle technology in drug delivery 151
6.6 Commercialization of microneedle technology 157
6.7 Conclusion 160
References 161

6.1 INTRODUCTION

The transdermal route of drug delivery overcomes many of the limitations associated with the more common oral and parenteral routes. It offers advantages including ease of use, controlled drug delivery, avoidance of first-pass metabolism, and lack of pain (Sinko, 2005). In spite of these advantages, transdermal drug delivery is limited by the barrier function of the outermost nonviable layer of epidermis, the stratum corneum. Several approaches to overcome this barrier function of the skin have been proposed, including chemical modification, iontophoresis, ultrasound, and electroporation.

Handbook of Non-Invasive Drug Delivery Systems ISBN 9780815520252

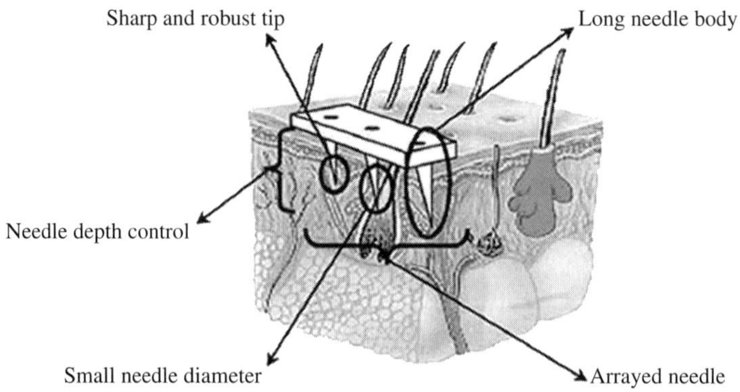

Figure 6.1 Schematic view of microneedle array features and requirements for transdermal drug delivery: sharp and robust tip, needle depth control, and small needle diameter. Adapted from Moon et al. (2005).

These methods have met with limited degrees of success, especially in delivery of hydrophilic or large-molecular-weight active substances.

Henry et al. (1998) published a paper describing a novel approach of using microfabricated microneedles to enhance drug delivery across the skin. The microneedles were approximately 150 μm in length, long enough to cross the stratum corneum (opening up small channels to allow drug delivery) but small enough not to stimulate the nerves in the dermal layer (Figure 6.1). The microneedles could be inserted into human skin with a force of approximately 10 N. Although the idea was conceptualized and patented by Alza Corporation in the 1970s (Gerstel and Place, 1976), it was not demonstrated until microfabrication tools for making such small needles became available (Prausnitz, 2004). This promising transdermal drug delivery system (TDDS) has been referred to as "pain free" because the sensation caused by microneedle insertion is statistically indistinguishable from the sensation caused by a smooth surface (Kaushik et al., 2001).

This chapter provides an in-depth review of this promising technique, including the most recent developments in the field.

6.2 MICRONEEDLE GEOMETRY

Microneedles range from 100 to 2000 μm in length and have the potential to make microchannels in the skin that, when coupled with drug reservoirs, act as delivery channels for the applied drug. Different materials have been used in the fabrication of microneedles, including silicon (wafers), stainless steel, titanium, glass, and numerous polymers. Each material requires a slightly different method of fabrication and has different physical characteristics. Additionally, microneedles have been fabricated in-plane, where the microneedle is parallel to the substrate surface, and out-of-plane, where the microneedle is perpendicular to the substrate surface (Reed and Lye, 2004). This section will discuss some of the more common methods of microneedle fabrication found in the literature.

6.2.1 Solid microneedles

Microfabrication technologies, originally developed to produce microelectronic devices, have increasingly found their way into machining devices for application in biological systems. These devices are called microelectromechanical systems (MEMS) (Tao and Desai, 2003; Ji et al., 2006). One of the earliest and most common methods of microneedle fabrication utilized the LIGA technique (from the German "Lithographie" (lithography), "Galvanik" (electroplating), and "Abformung" (molding)) (Khumpuang et al., 2006). This technique utilized the transfer of a microlithographic pattern on to a pure silicon wafer (Figure 6.2) and made it possible to shrink the needle (McAllister et al., 2000). This fabrication process was used by Hashmi et al. (1995) to design and fabricate the first microneedles. Later, Henry et al. (1998) demonstrated the application of this device regarding its ability to enhance drug delivery across the skin. These

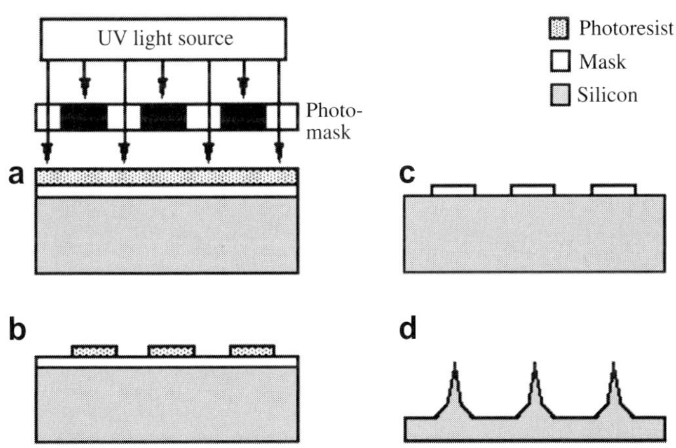

Figure 6.2 Microneedle fabrication sequence. This process consists of (a) photo-defining the photoresist by selective exposure to UV light through a photomask, (b) developing the photoresist, (c) etching the mask layer and removing the photoresist, and (d) performing an anisotropic etch to undercut the masks, leaving behind arrays of sharp microprobes or solid microneedles. From McAllister et al. (2000).

microneedles were fabricated by a process involving deep reactive ion etching in which a masking agent was deposited on to the silicon wafers. The wafers were then placed in a reactive ion etcher where the areas protected by the mask remained and formed microneedles. Since these techniques showed initial success, similar methods have been used to both selectively etch away parts of the silicon wafer and to add new structural layers to form the array of microneedles.

Microneedle geometry and height have been varied to further strengthen and sharpen the tip, reducing both the force required for microneedle insertion and the probability of microneedle breakage (Khumpuang et al., 2006). Hollow microneedles made from silicon have been fabricated by similar techniques (Teo et al., 2005; Nordquist et al., 2007).

Other methods that use lithographic processes similar to that used by Henry et al. (1998) have been developed for the fabrication of microneedles. Wilke et al. (2005) demonstrated a wet etching technology to fabricate silicon microneedles that utilized corner undercutting with potassium hydroxide to cut into the silicon wafer. Ji et al. (2006) fabricated silicon microneedle arrays by using a combination of isotropic etching with deep etching and wet etching. Moon and Lee (2003) fabricated polymethylmethacrylate (PMMA) substrate microneedle arrays using an X-ray lithographic (LIGA) process similar to MEMS. This X-ray lithographic method involved vertical deep X-ray exposure followed by an inclined deep X-ray exposure with a deep X-ray mask whose pattern possessed a hollow triangular array. By altering factors such as the mask design, the gap between the mask and the PMMA, and the angle of the X-ray exposure, microneedles of desired dimension, length, and sharpness can be obtained. Such microneedles not only penetrated the skin without pain but also involved no microneedle breakage. Microneedles of dimensions ranging from 750 to 1000 μm with base areas of 190–300 μm and tapered angles of 15–20° have been fabricated using this method (Moon and Lee, 2003).

Microneedles have also been fabricated from stainless steel, which is not only biocompatible and inexpensive but also has good mechanical strength (Verbaan et al., 2007). One of the more common fabrication techniques utilizes an infrared laser to cut microneedles in-plane from sheets of stainless steel that are approximately 75 μm thick. Engineering computer programs can be used to control the exact microneedle geometries by controlling the path of the laser used in the fabrication process. The needles are then manually bent in such a manner that they are perpendicular to the stainless steel sheet. They are then electropolished to remove excess debris and to reduce them to their final dimensions (Martanto et al., 2004). Laser cutting of stainless steel sheets allows for complex geometries, including contoured surfaces in the form of barbs and serrated edges (Gill and Prausnitz, 2007a).

6.2.2 Hollow microneedles

The hollow needle was invented in 1844 and the intravenous route of drug administration has been utilized since (McAllister et al., 2000). With access to microfabrication techniques in the 1990s the logical step was taken to reduce the size of the hypodermic needle. By reducing the size to the microneedle scale, pain and tissue trauma associated with hypodermic needle injection was reduced greatly or removed entirely. It has been hypothesized that microneedle insertion causes significant compaction of the skin decreasing water content, porosity, and pore size, thereby reducing the flow conductivity of the skin below the needle tip (Martanto et al., 2006a).

Single hollow glass microneedles are another class of microneedles that are often used at laboratory scale for small proof-of-concept studies. The use of hollow glass microneedles is encouraged due to their ease of small-scale manufacture and their ability to provide visualization of fluid flow. These hollow glass microneedles are made by pulling fire-polished borosilicate glass pipettes using a micropipette puller (Martanto et al., 2006b). Hollow microneedles are typically designed to allow constant infusion of drug solution through the hollowed out center of the device. Studies have shown that there is a large resistance to flow potentially caused by dermal tissue compression. Martanto et al. (2006b) showed that this flow resistance can be overcome by partially retracting the glass hollow microneedle after insertion into the skin. It was determined that microneedle insertion resulted primarily in skin indentation and only a fraction of the microneedle actually went into the skin causing significant dermal tissue compression (Martanto et al., 2006a).

Davis et al. (2005) have designed and tested hollow microneedles for continuous delivery of insulin. They fabricated the microneedle arrays with a modified LIGA process. The process utilized a UV laser to micromachine molds. The molds were then coated with nickel by electrodeposition. The nickel microneedle arrays were released from the molds by etching the polymer mold. Arrays contained microneedles measuring 500 μm in length with a 75 μm tip, and testing showed that they were strong enough to pierce living skin without breaking.

6.2.3 Biodegradable microneedles

Owing to fragile and nonbiocompatible properties of silicon in silicon microneedles, biodegradable microneedles have been introduced. Park et al. (2006) reported the use of polylactic acid (PLA), polyglycolic acid (PGA), and their copolymers, poly(lactic-co-glycolic acid) (PLGA), to fabricate microneedles. Ultraviolet lithography, wet silicon etching, and reactive ion etching with a novel lens-based technique have been used to create microneedle master structures. These master structures were then used to form polydimethylsiloxane (PDMS) molds. Once the master structures were removed from the molds, pellets of biocompatible material were placed in the molds, heated slightly above the melting point of the polymer, and placed under vacuum to create microneedles made of the biocompatible polymer (Figure 6.3). Park et al. (2006) demonstrated that microneedles made of biodegradable polymers with beveled tips, chisel tips, and tapered cones were all mechanically suitable to pierce and increase the permeability of the skin.

Similar mold-based fabrication methods have been used to create maltose-based microneedles from 200 to 2000 μm. These microneedles are designed to dissolve instantly if they break off in the skin, removing potential complications associated with metal and silicon microneedles. Maltose microneedles have been characterized and have shown the ability to create microchannels in the skin and to therefore enhance transdermal permeability (Kolli and Banga, 2008).

Ito et al. (2006a) reported the design and fabrication of self-dissolving microneedles where the active substance was loaded into the polymer solution that was used to create the microneedle. Polymer bases such as dextrin, chondroitin sulfate, and albumin were used in the fabrication of these microneedles. The fabrication procedure involved addition of drug to the polymer base followed by spreading of polymer–drug solution using polyethylene/polypropylene tips. The tips were then pulled perpendicularly to the mixture to obtain thread-like needles. The threads were subsequently dried in a desiccator and microneedles were obtained by cutting from the tips.

6.3 TYPES OF APPROACH

Even though the concept of drug delivery via microneedles did not occur until 1998, there has been rapid and considerable progress in the field. Many different approaches utilizing microneedles have been investigated. These

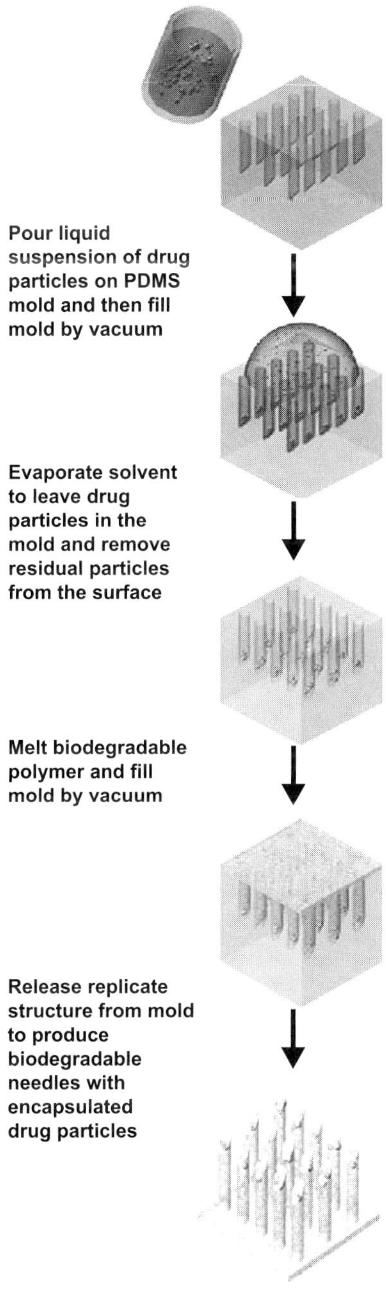

Figure 6.3 Method to fabricate polymer microneedles that encapsulate drug for controlled release. First, a suspension of drug particles is filled into a microneedle mold. Evaporation of the solvent leaves solid drug particles partially filling the mold. Pellets of biodegradable polymer are then melted into the mold under vacuum. Cooling and solidification of the polymer yields biodegradable polymer microneedles with encapsulated drug particles. From Park et al. (2006).

approaches include the use of hollow microneedles, the poke with patch technique, the coat and poke technique, and the dip and scrape technique (Prausnitz, 2004).

6.3.1 Poke with patch technique

The poke with patch technique utilizes an array of microneedles to pierce the stratum corneum. Once the channels exist in the stratum corneum, a reservoir of drug is placed on the skin and allowed to pass through the channels created by the microneedles. Studies have been performed comparing removal of the microneedle array before applying the drug reservoir and leaving the microneedle array in the skin for varying lengths of time while applying the drug reservoir (Martanto et al., 2004). The results indicate that shorter microneedle insertion time results in a greater pharmacological effect.

Recently, a novel microneedle patch for active insulin delivery was demonstrated by Nordquist et al. (2007). This patch utilized the poke with patch approach in the form of a patch–needle hybrid to deliver insulin. The microneedles used were hollow silicon microneedles fabricated by utilizing MEMS technology placed in a 5×5 array. The device was designed in such a way that it included a drug dispenser that stored and dispensed drug at a predetermined flow rate. The dispenser utilized a thermally expandable silicone material that expanded into the drug reservoir, causing the drug to move. Such studies depict the potential for microneedle technology to revolutionize the drug delivery field.

6.3.2 Coat and poke technique

The coat and poke technique involves coating the microneedle arrays with drug and then inserting them into the area of administration. In this approach, there is no drug reservoir; the drug solution is placed directly on the microneedle before insertion. The coating solution usually contains drug, viscosity agent, surfactant, and solvent. In this technique, withdrawal speed, drying time, and temperature need to be controlled to ensure homogeneity across microneedles (Gill and Prausnitz, 2007b). Jiang et al. (2007) also reported coated stainless microneedles for ocular drug delivery.

6.3.3 Dip and scrape technique

The dip and scrape technique involves dipping the microneedles into a solution of drug and then scraping it along the skin to form micro-abrasions. These abrasions provide pathways for the drug to permeate the skin in much higher concentrations than would be seen by topical administration (Prausnitz, 2004).

6.4 FACTORS INFLUENCING TRANSDERMAL PERMEATION

6.4.1 Depth of permeation

The depth of microneedles that permeate the skin is a most crucial factor in ensuring optimum systemic delivery of the active substance via the microneedles applied on the skin. *In vitro* microscopic studies performed by Thakur and Michniak (2007) in order to assess the depth of microneedles that penetrate the skin showed that microneedles of length 300 and 600 μm permeated skin up to depths of 97 and 262 μm respectively. This observation was explained by the fact that when microneedles of shorter length are pressed on the skin surface with a certain degree of force, a portion of microneedle length and effort is involved in skin tissue compression, thereby leaving little microneedle length available for skin penetration. In the case of long microneedles, even after skin compression, a significant length of microneedle is available to penetrate the skin. Experiments pertaining to length of microneedles and their depth of permeation have also revealed that the viscoelasticity characteristics of the skin result in initial skin indentation up to a certain length of microneedles followed by actual piercing of the skin when microneedles are pressed on the skin.

In vitro transdermal studies performed by Thakur and Michniak (2007) are further corroborated by Verbaan et al. (2007), who showed that microneedles of length less than 300 μm could not pierce the skin (skin thickness 330–440 μm) as they did not possess the critical length to overcome the bulk elastic tissue compression. Verbaan and coworkers also observed that the use of thinner and less flexible epidermal sheets facilitated successful piercing with microneedles. Experiments have also revealed dependency of skin thickness and the microneedle length required to penetrate the skin (Henry et al., 1998; McAllister et al., 2003; Chabri et al., 2004).

6.4.2 Coating strategies

Microneedles have the potential to be coated with a variety of active substances and have been used for transdermal delivery of various large molecules, including peptides (Cormier et al., 2004) and proteins (Matriano et al., 2002). Coated microneedles involve coating a drug formulation on to solid microneedles (nonpocketed) and inserting them into the skin for subsequent dissolution of the drug within the skin. For reliable and reproducible transdermal permeation of the active substance using coated microneedles, it is imperative to optimize the quality and quantity of the coating. Various parameters are involved in designing the coating formulation of the active agent to guarantee its desired spatial and temporal release, among which uniform coating on the microneedles using dip coating is the most critical. Uniform coating on microneedles can be achieved by controlling the surface tension and viscosity of the coating formulation. The properties of surface tension and viscosity of the coating formulation greatly affect the thermodynamics and hydrodynamics of the dip-coating procedure. Gill and Prausnitz (2007b) reported that microneedles made of stainless steel (with surface tension of 39.6 mJ/m^2), when dipped in aqueous solution (with surface tension of 72 mJ/m^2) of sulforhodamine, showed no drug coating owing to the large contact angle of 47.5 $\pm$ 3.0° formed between the stainless steel substrate and coating solution. The contact angle that was formed was due to the difference in the surface tension of the two substrates. But when Lutrol® F-68 NF, a surfactant, was added to the aqueous sulforhodamine dipping solution, the surface tension of the coating solution was reduced and a thin coating was obtained on the microneedles. In the same study, addition of 1% w/v aqueous solution of carboxymethylcellulose (CMC) (viscosity: 47 $\pm$ 0.5 nPa S) to the coating solution containing sulforhodamine led to increased drug loading on the microneedles due to the thicker drug coating. It was also observed that the combination of surfactant and viscosity enhancer modified the coating solution property. For instance, when coating solution consisted of drug and CMC, a de-wetting phenomenon was observed that concentrated coating in the center of the microneedles and not at the periphery. The occurrence of de-wetting was later removed by use of CMC–Lutrol F68 combination. Gill and Prausnitz noticed similar behavior with a Tween® 20 and sucrose combination.

The choice of excipients also influences the drug loading of the active substance in the coating formulation. Comparison of viscosity enhancers by Gill and Prausnitz revealed that 1% w/v CMC produced twice as much effect as 52% w/w sucrose, thus enabling higher drug content and lower excipient concentration in microneedle coatings. Likewise in the case of surfactants, Lutrol F-68 was preferred over Tween 20 as the former was a solid and resulted in a desirable hard coating on the microneedles as compared to the waxy coating produced by Tween 20-containing coating formulations. Therefore, an optimized microneedle coating formulation should involve the strategy of reduction in surface tension between the coating solution and the microneedle, and at the same time consider increments in viscosity by maintaining minimal drug–excipient and excipient interactions.

Instead of the use of surfactant to match the surface energies of the coating solution and microneedles, Gill and Prausnitz used another strategy that involved raising the surface energy of the stainless steel microneedles by making their surface hydrophilic. Surface hydrophilicity was maintained by precoating needles with a thin layer of silicon dioxide followed by subsequent coating with an aqueous solution of sulforhodamine. This excipient-free method ensured a uniform but thin coating on the microneedle. Another excipient-free approach involved precoating the stainless steel microneedles with PLGA, which forms a uniform and thin coating on the needles. PLGA provided a hydrodynamic drag that allows entrainment of liquid film and reduced surface tension-induced fluid contraction on the microneedle surface and therefore increased residence time for drying of a thin film on the PLGA surface. Also, incorporation of PLGA led to controlled release of the active substance from the coating. Even though excipient-free methods form thin uniform films on the microneedles, they are appropriate for coating sensitive protein solutions that are inherently viscous and act as self-viscosity enhancers to increase coating thickness.

For transdermal delivery of hydrophobic molecules through coated microneedles, the selection of excipients includes use of amphiphilic viscosity enhancers that are soluble in organic and aqueous solvents. Gill and Prausnitz reported coating formulations for delivery of curcumin using an ethanol-based solution of polyvinylpyrrolidone (PVP). The low surface tension of ethanol (21.8 mN/m at 25°C) eliminated the use of a surfactant in the coating formulation.

In order to simplify the coating formulation and minimize the use of excipients on the microneedles, molten drug coatings have been used that enable higher active mass loaded per microneedle. Gill and Prausnitz reported coating of molten lidocaine on microneedles using a dip-coating method. They also suggested delivery of hydrophobic active substances using molten polymers such as polyethylene glycol (PEG) and PVP.

In addition to solid microneedles, there are pocketed microneedles available that can be exploited to deliver liquid drug formulations, or drugs that are difficult to coat or have risk of being wiped off the skin during penetration. Also, molten drugs such as molten lidocaine can be delivered through pocketed microneedles. Gill and Prausnitz selectively filled pockets of microneedles with drug formulations that were devoid of surfactants but contained viscosity enhancers and excipients such as 25% sucrose to increase the solid content. While the removal of surfactant from the formulation inhibited surface wetting of the coating and increased the critical speed of liquid film formation, addition of viscosity enhancer and sucrose helped in retaining the coating solution within the pockets. Pocketed microneedles possess storage limitations due to evaporation of liquid used in coating formulation, thereby leading to variability in the dose incorporated. However, the storage problem can be overcome by packaging pocketed microneedles under pressure in a nitrogen atmosphere.

Gill and Prausnitz (2007b) explored coated microneedles to attain different release kinetics of the same or different drugs and proposed a method of avoiding drug–drug interactions in the case of simultaneous administration of multiple drugs. Gill and Prausnitz demonstrated sequential burst release of sulforhodamine that was separated by a delay, after a 1-minute dipping of coated microneedles in deionized water. Coating of microneedles was performed in sequential order with the innermost layer composed of water-soluble excipients and the active substance followed by a secondary layer of PLGA, and further followed by another layer of water-soluble excipient and the active substance.

For achieving reproducibility of transdermal permeation using coated microneedles, it is essential to quantify the mass of drug coated on the microneedles as a function of the coating formulation and microneedle parameters. Gill and Prausnitz investigated the effect of four factors on the mass of the drug, namely (Figure 6.4):

(a) The effect of drug (riboflavin) concentration in the coating solution
(b) The number of coating solution dips
(c) The number of microneedles in the array
(d) Drug (riboflavin) concentration during selective coating of microneedle pockets.

It was observed that the mass of riboflavin coated on the microneedles increased with an increase in concentration of the active substance in the coating solution and the number of dips. It was also noted that thicker coatings were limited by the loss of the drug during insertion into the skin. Interestingly, it was observed that an increase in array size from 5 to 50 microneedles did not increase the mass of the drug per microneedle but proportionally increased the mass of the drug on the array by 10-fold. Similarly, filling riboflavin selectively into pockets reduced the amount of drug per microneedle. Therefore, the above paradigms should be considered in order to achieve desired drug loading in the microneedles.

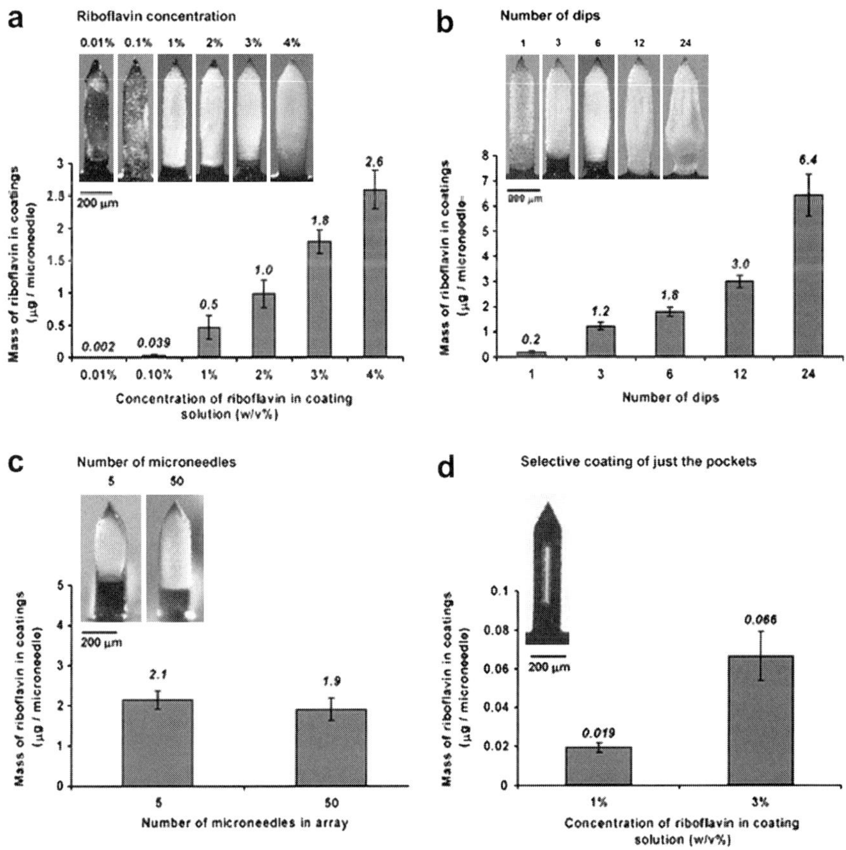

Figure 6.4 Effect of four factors on mass of riboflavin coated on microneedles: (a) effect of riboflavin concentration in coating solution; (b) number of coating solution dips; (c) number of microneedles in the array; (d) riboflavin concentration in microneedle coating. From Gill and Prausnitz (2007b).

Also, it is important to determine the optimized insertion time of coated needles into the skin for maximal and desired delivery of the active substance. Gill and Prausnitz observed that after a 5-minute insertion of riboflavin-coated microneedle rows with a five-needle shaft, 91% of the drug was delivered to the cadaver porcine skin, and 2% and 7% remained on the skin surface and microneedle surface respectively after removal from the skin.

Gill and Prausnitz noticed the effect of pocketed microneedle insertion speed on the delivery of the active substance in the liquid formulation across the skin. The insertion speed of solid microneedles that was applied in the study ranged from 0.5 to 2 mm/s. It was observed that at low insertion speeds (0.5–1 mm/s) there was wiping off of some liquid on the skin surface, which was prevented by applying higher insertion speeds of 1–2 cm/s. Higher insertion speed reduced the contact time between the liquid contained in the pockets and the skin surface, thus preventing liquid wipe-off on the skin surface.

6.4.3 Microneedle geometry and force of penetration

The microneedle geometry and force of penetration significantly affect transdermal permeation. Martanto et al. (2006a, b) investigated the effect on the flow rate of sulforhodamine through hollow microneedles inserted to a depth of 720–1080 μm on human cadaver skin, as a function of microneedle insertion and retraction distance, infusion pressure, microneedle tip geometry, presence of hyaluronidase, and time. On studying the effect of microneedle insertion and retraction (Figure 6.5), it was shown that decreasing the net insertion depth (initial insertion depth minus the retraction distance) resulted in larger flow rate. This phenomenon was explained by the fact that microneedles that are typically inserted deeply into the skin come into contact with compressed dermal tissue that blocks the flow from the microneedles. But when these microneedles are retracted slightly after insertion, compressive forces on the tissue are relieved and the tissue expands and increases the flow rate of the active substance.

On performing pressure studies, it was observed that an increase in infusion pressure of sulforhodamine (from 69 to 138 kPa) showed a proportional increase in flow rate (Figure 6.6). However, at an infusion pressure of 172 kPa, a disproportionate increase in flow rate was noticed relative to an increase in pressure due to the deformation of dermal tissue under high pressure, which imposed major resistance to the flow of the active substance. The comparison of bevel tip (hole is off to the side) microneedles and blunt tip (hole is at the tip) microneedles revealed that flow rate of sulforhodamine from the former is greater than from the latter as the former reduces the resistance by infusing the active substance to the side of the dense tissue formed (Figure 6.7). At the same time, Martanto et al. observed no effect of microneedle tip opening on flow rate of the drug as the compression of the skin tissue at the needle tip that provides resistance to flow is governed primarily by the depth of needle insertion and retraction (Figure 6.8). Martanto et al. reported that addition of hyaluronidase increased the infusion flow of the active substance through microneedles inserted on to skin by 7-fold due to breakdown of hyaluronan within skin collagen fibers,

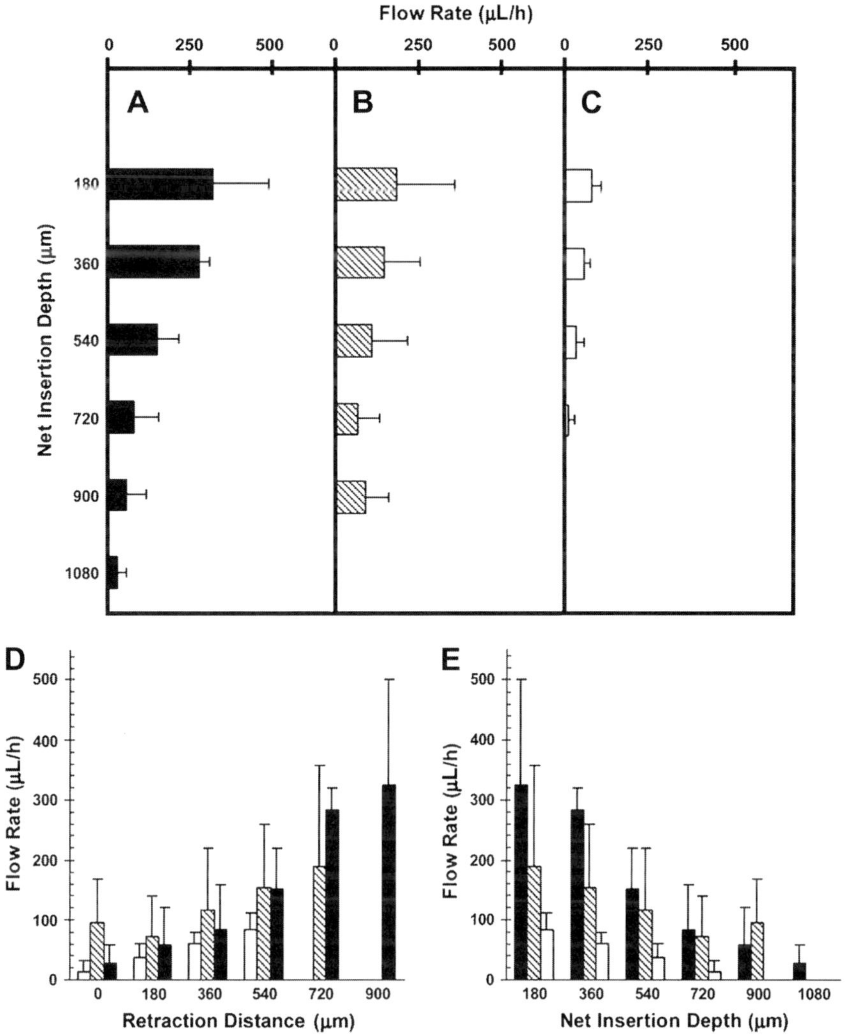

Figure 6.5 Effect of insertion depth and retraction distance on flow rate into human cadaver skin *in vitro*. Microneedles were initially inserted to a maximum insertion depth of 1080 μm (A), 900 μm (B), or 720 μm (C), and then retracted various distances back toward the skin surface to a final, net insertion depth. Pooling the data from parts A, B, and C, flow rate is shown as a function of retraction distance (D) and net insertion depth (E) for maximum insertion depths of 720 μm (white bars), 900 μm (striped bars), and 1080 μm (black bars). Microneedles had tip opening effective radii of 27–31 μm with bevel angles of 35–37°. Infusion was performed for 5 min at 138 kPa without needle retraction, after which microneedles were retracted by 180 μm every 5 min to a final insertion depth of 180 μm. From Martanto et al. (2006b).

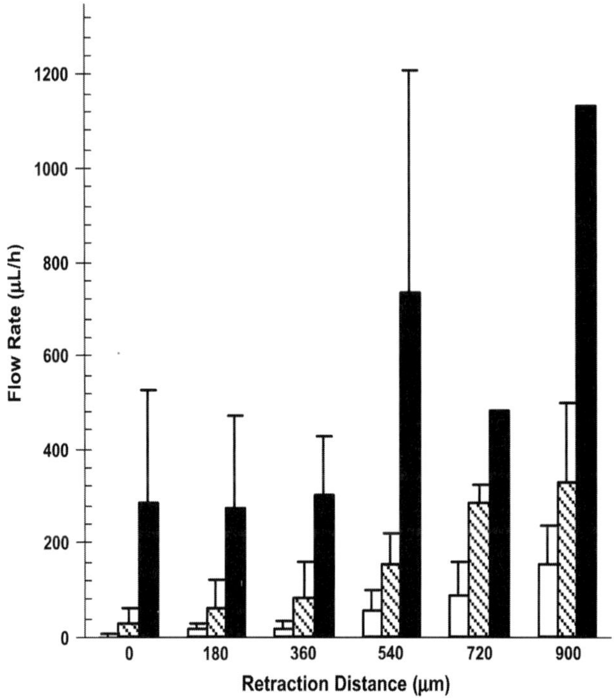

Figure 6.6 Effect of pressure on flow rate into human cadaver skin. Microinfusion flow rate was measured as a function of retraction distance at three pressures: 69 kPa (white bars), 138 kPa (striped bars), and 172 kPa (black bars). Microneedles having 35–37° beveled tips with 27- to 32-µm effective radius openings were inserted to a maximum depth of 1080 µm. From Martanto et al. (2006b).

thereby reducing flow resistance in the skin. This study not only supported the hypothesis that dermal tissue limits the flow through the microneedles, but also suggested the co-administration of hyaluronidase with microneedle infusion facilitated transdermal flow of the active substance (Figure 6.9). Martanto et al. also investigated the effect of time by measuring flow into the skin for 100 minutes for needles inserted and left in place and for needles inserted and retracted. In both cases, the cumulative volume of the active substance infused into the skin increased linearly with time. However, in the case of the insertion/retraction process, the drug flow rate was higher at the onset, but then decayed and achieved steady state. The higher flow rate at the onset was due to rapid filling of a small cavity that was formed in the skin tissue in the vicinity of the needle tip by the insertion/retraction process. But the needle tip also led to compression of the tissue that offered resistance to the flow rate and explained the attainment of the steady state.

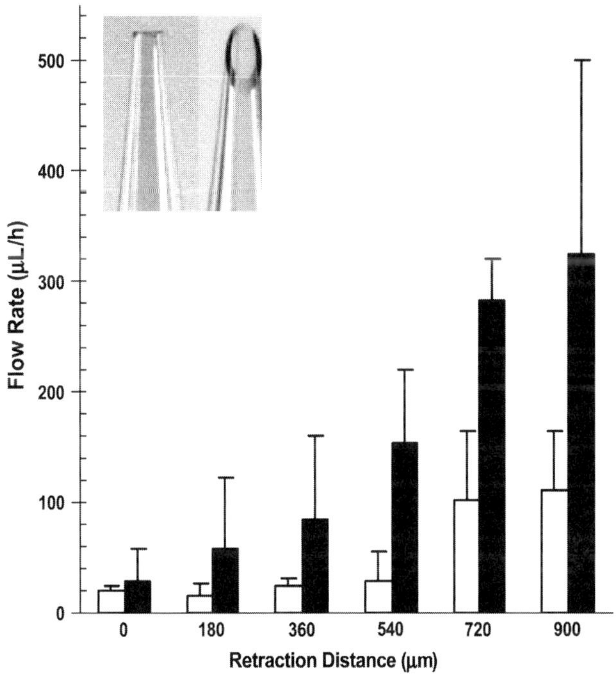

Figure 6.7 Effect of tip bevel on flow rate into human cadaver skin. Microinfusion flow rate was measured as a function of retraction distance during infusion using hollow microneedles with a blunt tip (white bars; left inset image) and a 35° beveled tip (black bars; right inset image). Microneedles with 27- to 32-μm effective radius openings were inserted to a maximum depth of 1080 μm, and infusion was carried out at 138 kPa. From Martanto et al. (2006b).

6.5 MICRONEEDLE TECHNOLOGY IN DRUG DELIVERY

For successful transdermal delivery of active substances, many approaches have been used that include use of chemical enhancers, electric fields, ultrasound, and thermal methods (Williams and Barry, 2004), but their success has been hindered by low permeability of active substance across the skin, especially in the case of high-molecular-weight compounds. Various scientific groups have demonstrated delivery of high-molecular-weight compounds such as insulin through various microneedle strategies. For instance, solid microneedles were used to deliver insulin to diabetic hairless rats *in vivo*. Microneedle arrays were inserted into the skin using a high-velocity injection followed by placement of a solution on the top of the microneedle array for 4 hours. A solid microneedle–insulin combination was successful in decreasing blood glucose level by 80% but was incapable of achieving insulin dose and flow rate adjustment (Martanto et al., 2004; Prausnitz, 2004).

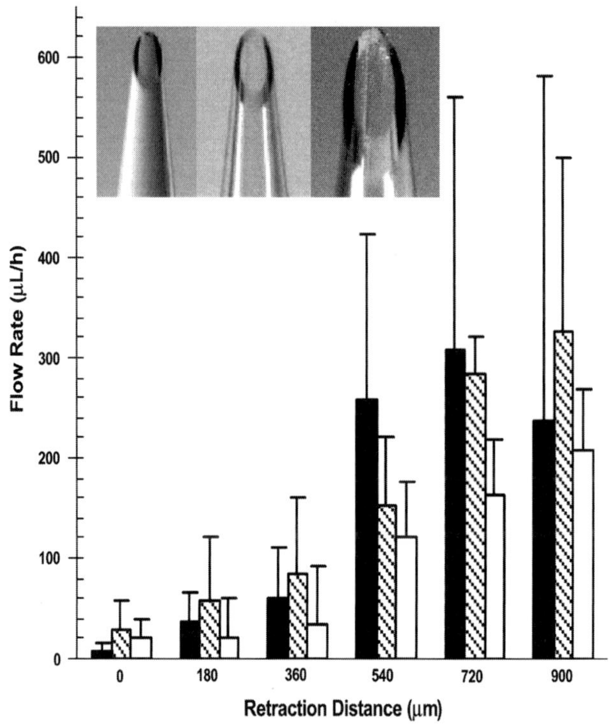

Figure 6.8 Effect of tip opening size on flow rate into human cadaver skin. Microinfusion flow rate was measured as a function of retraction distance using three different tip opening sizes: 22 μm (black bars; left inset image), 30 μm (striped bars; center inset image), and 48 μm (white bars; right inset image) effective radii. Microneedles having 35–37° beveled tips were inserted to a maximum depth of 1080 μm, and infusion was carried out at 138 kPa. From Martanto et al. (2006b).

The hollow microneedle technique has also been used for delivering insulin. McAllister et al. (2003) demonstrated a 70% drop in blood glucose level over a period of 5 hours after insertion of insulin-loaded single glass microneedles into diabetic hairless rat skin. In spite of the promising results, studies on hollow microneedles are fewer as hollow microneedles are inherently weaker than their solid counterparts, therefore making their design and fabrication difficult.

Recently, a novel approach has been proposed and demonstrated by Nordquist et al. (2007) that offers the possibility of insulin delivery that can not only be controlled but also exhibits patient compliance. The novel microneedle patch is made up of four components, namely (a) microneedle chip, (b) liquid reservoir, (c) expandable material, and (d) electric heater (Figure 6.10). The needles are organized on a 4 mm × 4 mm chip with 21

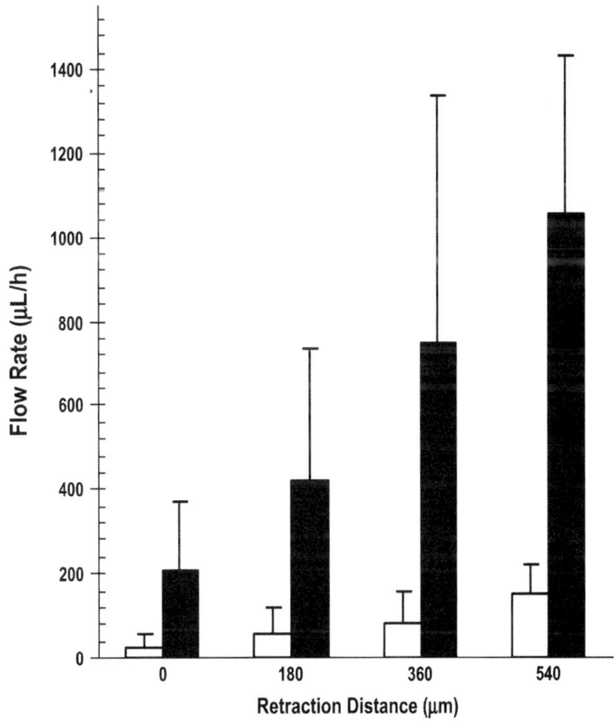

Figure 6.9 Effect of hyaluronidase on flow rate into human cadaver skin. Microinfusion flow rate was measured as a function of retraction distance during infusion in the absence (white bars) or presence (black bars) of hyaluronidase. Microneedles having 35–37° beveled tips with 27- to 32-μm effective radius openings were inserted to a maximum depth of 1080 μm, and infusion was carried out at 138 kPa. From Martanto et al. (2006b).

hollow needles (and four nonhollow needles owing to fabrication yield). The liquid reservoir stores and dispenses drug volume at a certain flow rate depending on the voltage applied to the heater. The working principle of the device is based on thermally expanding silicon material into a liquid reservoir that subsequently leads liquid containing drug to flow. Studies performed using this novel approach in adult male Sprague–Dawley rats showed that the microneedle patch had the advantage of showing less variability in insulin delivery, and faster onset of action, and offered controlled release of insulin.

Another approach of insulin delivery by microneedles was shown by Ito et al. (2006b), who designed self-dissolving insulin-loaded microneedles. The self-dissolving insulin-loaded microneedles were derived from dextrin and showed a dose-dependent hypoglycemic effect in mice with insulin physiological availabilities of 97.7%, 93.3 % and 91.3% at doses of 0.5, 1.0,

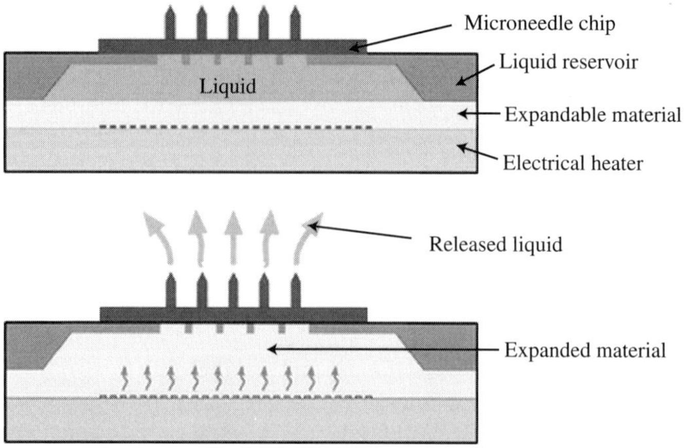

Figure 6.10 Diagrammatic representation of dispenser's working principle. When a voltage is supplied to the heater the expandable material expands into the liquid reservoir, consequently ejecting the liquid through the hollow microneedles. From Nordquist et al. (2007).

and 2.5 IU/kg respectively. Insulin present in the microneedle preparation showed stability for 1 month when stored at −80, 20, and 40°C.

In addition to insulin delivery by microneedles, Ito et al. (2006a) also evaluated percutaneous administration of erythropoietin (EPO) via self-dissolving EPO microneedles derived from three polymer bases – dextrin, albumin, and chondroitin sulfate. EPO is a 34 kDa compound that is administered to correct anemia in cancer patients. The bioavailability of EPO from microneedles prepared from various types of polymer base was not significantly different and the study showed the potential of EPO microneedles via transdermal administration.

Verbaan et al. (2007) and Ito et al. (2006a) investigated the transdermal permeation of different compounds with molecular weight ranging from 538 Da (Cascade blue), 10 kDa (dextran-Cascade blue) to 72 kDa (fluorescein isothiocyanate-coupled dextran). The studies divulged that pretreatment with microneedles enhanced the transport of compounds of molecular weight in the range of 538 Da–72 kDa across dermatomed skin and also showed that microneedles of length less than 300 μm could not pierce the skin *in vitro*.

Other large-molecular-weight compounds, such as desmopressin (Cormier et al., 2004), ovalbumin (Matriano et al., 2002), oligonucleotides (Lin et al., 2001), human growth hormone (Cormier and Daddona, 2003), proteins, and DNA (Choi et al., 2005), have also been delivered successfully by the microneedle approach.

Unlike Ito and coworkers, Park et al. (2006) did not confine the use of polymer to the microneedle base but fabricated polymer microneedles. Polymer microneedles are arrays of microneedles consisting of polylactide-co-glycolide polymer prepared using a mold-based method. Park and coworkers used polymer microneedle technology to encapsulate model drugs – calcein and bovine serum albumin (BSA) – either by a single-encapsulation method (encapsulation within the needle matrix) or a double-encapsulation method (by encapsulating the drug within carboxymethylcellulose or poly-L-lactide microneedles and then encapsulating drug-loaded microparticulates within polymer needles). Park et al. (2006) reported that polymeric microneedles could easily penetrate the skin and gave a controlled release profile of drug encapsulated and outlined numerous advantages of the polymer microneedle technique. Firstly, polymer microneedle technology offered the advantage of controlled release of drug that lends itself to be self-administered by patients as compared to surgical implants and hypodermic needle injection. Secondly, the intact polymer microneedle patch could be left in place on the skin for an extended period of time and later removed. Alternatively, a microneedle patch could be designed to remain anchored within the skin to prevent undesirable separation through the course of therapy and even after therapy has been achieved, thus enabling patient acceptance. Thirdly, large-scale production of polymer microneedle patches was cost-effective. Fourthly, polymer microneedle technology did not require an external power supply for operation, unlike other novel devices. Lastly, in the polymer microneedle technique, intersubject variability due to skin permeability was less important as drug delivery is governed by diffusive and degradation properties of the polymer. Despite numerous advantages, polymer microneedles have certain shortcomings. They are unsuitable for encapsulating thermolabile drugs as their fabrication involves subjecting the active substance to high temperature. Also, polymer microneedles are constrained by needle mechanical properties. Park et al. reported that with 2% drug loading polymer microneedles possessed sufficient mechanical strength, but when drug concentration was increased to 10% microneedles lost their mechanical strength. One of the most important drawbacks is that polymer microneedles offer very small dose sizes that can be loaded within them.

The potential of microneedle technology also extends to the arena of gene delivery. Chabri et al. (2004) demonstrated the utility of silicon microneedles in cutaneous gene delivery. Chabri and coworkers performed studies aimed at generating microconduits in the skin to facilitate access of lipid–polycation–pDNA (LPD) nonviral gene therapy vectors. The study involved the use of silicon-based microneedles that were fabricated using a novel isotope etching/Bosch reaction process. The results of the study revealed that the novel silicon-based microneedles managed to permeate the stratum corneum but could not deliver the LPD vector across the skin, leaving a significant amount of the LPD vector in the microconduits created by the needles between stratum corneum and epidermis. Even though cutaneous gene therapy via microneedles requires further exploration, it certainly has the potential for effective transfection of human keratinocytes *in vitro*.

Microneedle technology not only facilitates systemic delivery of active substances via the transdermal route, but also can be utilized to modify existing hypodermic needle design. Microhypodermic needles have emerged that can be fabricated from metal by surface micromachining techniques and have been investigated in two designs (Brazzle et al., 1998, 1999):

1. Fluid-coupled multiple-needle design
2. Single-needle design with multiple output ports.

Multiple-needle design is composed of structural supports and needle tips. Structural supports are hollow and the fluid-coupled multiple needle design assists in redirecting flow from clogged needles into adjacent unclogged needles. The advantage of multiple-needle design over a conventional hypodermic needle is that it limits penetration depth into the skin by the distance between its structural support and the needle tips, and therefore offers the advantage of a minimally invasive delivery system. Single-needle design is similar to its counterpart in diverting flow if one or more of its multiple ports are blocked, but it differs from multiple-needle design in the presence of a single needle in the model.

In addition to the systemic delivery of the active substances, microneedle technology can be used for targeted delivery of various drugs. A multichannel silicon-based neural probe fabricated by Chen et al. (1997) was reported to deliver both kainic acid (a neural stimulant) and γ-aminobutyric acid (a neural depressant) on insertion into guinea-pig brain *in vivo* with minimal tissue disruption. The study involved insertion of a neural microprobe into the superior colliculus and inferior colliculus of the guinea-pig brain followed by injection of a solution of kainic acid or γ-aminobutyric acid that subsequently resulted in an increased or decreased neuronal activity depending on the solution injected.

Microneedles also possess the potential to target drugs to arterial blood vessels. Studies performed by Reed et al. (1998) demonstrated insertion of microprobes of length 140 μm with insertion pressure of 100 mmHg in normal and atherosclerotic rabbit arteries *in vitro*. This approach can be exploited to target antirestenosis drugs into the arterial vessels to facilitate the widening of constricted arteries (McAllister et al., 2000).

Jiang et al. (2007) demonstrated ocular delivery of model drugs coated on microneedles via intrascleral and intracorneal routes by minimally invasive means. *In vitro* studies performed on human cadaveric sclera showed that microneedles were designed to insert into and not across the sclera and then sequentially deposited drug formulations of small model molecules, protein, and DNA in the sclera. *In vivo* studies in rabbits demonstrated that fluorescein, a model drug used in the study, is quickly delivered to the cornea and sustained drug release is achieved in the cornea by depot formation. Microneedle technology is therefore a promising candidate for minimally invasive treatment for diseases of the anterior and posterior segments of the eye.

Besides systemic and local delivery of the active substance at tissue and organ level, microneedles can also be used to deliver drug at the cellular level. Molecules to be delivered are either coated on to probes or the active substances are in the solution around the cells when microprobes are inserted. Microprobes have been reported to deliver genetic material into plant (Trimmer et al., 1995), nematode (Hashmi et al., 1995), and mammalian cells (Reed et al., 1998).

6.6 COMMERCIALIZATION OF MICRONEEDLE TECHNOLOGY

Due to wide application of microneedle technology, recent interest in such devices has emerged that has led to the commercialization of microneedles. Companies actively investing in microneedle research include the Alza Corporation (California) (Ameri et al., 2006), Apogee Technologies (Boston; http://www.apogeemems.com/medical/technology.html), BD technologies (North Carolina) (Alchas, 2002; Alarcon et al., 2007), Sandia National Laboratories (New Mexico; http://www.sandia.gov/mstc/technologies/microsensors/e_needles.html), BioValve Technologies (Massachusetts) (Ackley, 2006), Kumetrix (California) (Smart et al., 2007), NanoPass (Israel)

(Yeshurun et al., 2003), Sterling Medivations Inc. (Georgia) (Douglas and Hugo, 2003), and Therafuse (California) (Razzacki et al., 2004).

Alza/Johnson & Johnson has developed Macroflux® technology that involves use of a drug-coated titanium microprojection array of lengths less than 200 μm that are attached to a polymeric adhesive back. The array spreads to an area of 8 cm^2 with about 300 microprojections/cm^2. The maximal adhesive patch size is 10 cm^2. When the patch is applied to the skin, the drug-coated microprojections penetrate through the stratum corneum and release the drug into the microcapillaries for systemic distribution. The advantage of Macroflux technology in providing efficient drug delivery is mainly due to the high local drug concentration around the microprojections of the patch and also because of the large surface area of the patch array (Ameri et al., 2006).

In addition to large pharmaceutical companies such as Johnson & Johnson, many smaller firms are also involved in commercializing their in-house microneedle-based technologies. Apogee, a Massachusetts-based company, aims to develop technology called PyraDerm™ solution that comprises an array of silicon micro-pyramids surrounded with a proprietary coating containing therapeutic compounds that subsequently provide an easy-to-use patch applicator form. The novel Apogee technology is MEMS and nanotechnology based, and claims to possess not only the potentiality of encapsulating a range of therapeutic compounds but also achieving instantaneous and targeted release (http://www.apogeemems.com/medical/technology.html).

Becton Dickinson (BD) Technologies has developed a drug delivery device containing a needle assembly that is capable of facilitating intradermal injections using a variety of drug container types. The device is comprised of a hub and a limiter whereby the hub supports the needle while the limiter surrounds the needle. The limiter has a skin-engaging surface that is adapted to be received against the skin of an animal to be intradermally injected. The device is designed in such a way that the forward end of the needle extends beyond the skin engaging the surface up to a selected distance to limit the depth that the needle penetrates into the animal's skin. BD Technologies, using their in-house technique, reported preclinical studies that showed successful intradermal delivery of influenza vaccine that was at least as effective as conventional intramuscular injection (Alchas, 2002; Alarcon et al., 2007).

A novel way of detecting analytes in interstitial fluid (ISF) using microneedle technology has been introduced by Sandia National Laboratories, which has designed Electroneedles®. Electroneedles involve the use of microneedles to painlessly extract ISF and detect a wide range of analytes such as glucose, enzymes, proteins, antibodies, whole cells, and toxins that might be present in ISF (http://www.sandia.gov/mstc/technologies/microsensors/e_needles.html).

Likewise, Kumetrix has developed a unique medical diagnostics technology based on bioMEMS that involves a single-use disposable silicon microchip, with an integrated hollow microneedle that on puncturing the skin withdraws a submicroliter blood sample into the chip and performs a specific assay without subsequent blood transfer. The device is comprised of a chip integrated with the actuator of a handheld meter and a microprocessor. The microprocessor controls the puncturing of the skin and once intradermal penetration is achieved, *in situ* analysis of the analyte occurs in the chip, and the results are displayed and stored in the device's memory. This "lab-on-a-chip" device renders single-step blood testing that is painless, rapid, clean, and inexpensive, and eliminates the use of a skillful operator (Smart et al., 2007).

Another novel microneedle technology-based device has been designed by NanoPass that involves a single-use, fit-all device called *MicronJet*®, that can be used with any standard syringe or infusion device for painless intradermal delivery of any drug, protein, vaccine, etc. (Yeshurun et al., 2003).

Commercialization of microneedle-based technology has also been done by Sterling Medivations Inc., which has developed an infusion pump comprising numerous microprojections that are configured to enter the subcutaneous region of the patient's skin and provide a painless means of delivery across the stratum corneum. The design of the infusion device is such that it is sealed against leakage by the skin surrounding each projection and microprojections provide a flow path for either a basal or bolus injection of medication. The pump is composed of a drug reservoir containing a drug, a micro-actuator, and a housing having a foundation or lid that opens and closes so that the medication container can be inserted and supported in a delivery mode position. A micro-actuator is used to advance either a roller or a piston in communication with the medication container. Attached to the micro-actuator is a device that is mounted for movement along the medication container (Douglas and Hugo, 2003).

Another smart design combining microneedle technology with microfluidics has been demonstrated by TheraFuse, who have fabricated a next generation minimally invasive wearable drug infusion system that ensures accurate delivery. The system is composed of a disposable unit that is interfaced with a reusable unit. The disposable unit is comprised of a polymer containing a therapeutic compound and microneedles that are held to the skin by adhesives. The reusable unit contains the microfabricated logic, metering, and communication circuits. The TheraFuse device also includes a pressurized reservoir that is used to move the solution through a closed fluidic network, until injection occurs across the layers of the skin (Razzacki et al., 2004).

The needles are also attracting attention from nonpharmaceutical firms such as Hewlett-Packard Co. Recent reports by Hewlett-Packard Co. mention that researchers have developed a medical patch that uses their thermal inkjet technology found in printers to make such microneedles (Raju, 2006). This microneedle patch is equipped with basic electronics including a power source and spreads roughly to a 2.5 cm^2 area and is 3 mm thick. Also, the patch contains 400–1000 microneedles and can be programmed to precisely control the amount and timing of each dose that is delivered. The patch incorporating this feature has been licensed by Crospon Ltd. (Galway, Ireland), which plans to develop it into a commercial product (http://pcworld.about.com/od/futuretechnology/HP-s-thermal-inkjet-technology.htm). Another promising study has shown that microneedle-based intradermal delivery of the *Bacillus anthracis* recombinant protective antigen (Anthrax) vaccine elicits an immune response equivalent to that of intramuscular injection (IM) at a relatively lower dosage (Mikszta et al., 2006). Pain-free delivery, ease of self-administration by the patient, improved compliance, and shorter delivery times with lower dosages seem to be the winning features of this technology.

6.7 CONCLUSION

Microneedle array technology is a novel technology that has to be exploited but the emergence of this non-invasive technology opens a new arena of debate. The skin is the protection blanket of the body against external undesirable agents and creating percutaneous channels via microneedles opens up the security barrier of the body to foreign bodies. Even though preliminary studies suggest that the regenerative capacity of the skin can surmount the minor abrasions caused by microneedles, the kinetics and mechanisms of skin healing are still not clear. Also, the immunological repercussions of chronic administration of microneedles to the same site have not yet been defined (Birchall, 2006).

Another issue related to microneedle technology is the lack of reproducibility of the dose administered, which limits its application to drugs with a broad therapeutic window. Moreover, it is also required to generate more data relating to the fate of the drugs administered transdermally via microneedles and their kinetics in systemic circulation. This will not only broaden its application to a wide variety of active substances, but also to patients who can administer them in a safe and reproducible manner in the absence of medical supervision (Birchall, 2006).

The novel idea of cutaneous vaccination via microneedles is also of major concern as sufficient data are not available regarding the triggering of the immune system in response to microneedle structures, and also conclusive information is not available on the elimination of infection once the microneedle vaccine is discarded. Though microneedles (especially dry coated microneedle vaccines) offer more advantages as compared to liquid injectable vaccines, more research is required on improving the stability, refining the formulation, and the packaging of the coated microneedles (Birchall, 2006).

Since microneedles usage represents a minimally invasive technique for drug delivery, it is vulnerable to be abused for recreational drug administration, sedation for sexual exploitation, and for delivery of biologic and chemical toxins. Such factors are of crucial importance and require enthusiastic involvement of diverse groups, including dermatologists, toxicologists, immunologists, and social scientists, to resolve issues pertaining to the convenient, pain-free application of microneedles (Birchall, 2006).

In spite of a few limiting issues related to microneedles, microneedle technology offers immense potential in delivering all kinds of molecules, whether micromolecules or macromolecules, using any route from systemic to local, and also to all levels of life ranging from cellular to organ systems. Microneedle technology offers a beacon of hope that promises effortless, convenient, patient-compliant and minimally invasive forms of drug delivery, and has the potential to become the leading transdermal drug delivery system with improved technological advancement.

REFERENCES

Ackley, D.E., 2006. Microneedle array systems. US patent 7027478.

Alarcon, J.B., Hartley, A.W., Harvey, N.G., Mikszta, J.A., 2007. Preclinical evaluation of microneedle technology for intradermal delivery of influenza vaccines. Clin. Vaccine Immunol. 14 (4), 375–381.

Alchas, P., 2002. Intradermal delivery device including a needle assembly. US patent 6494865.

Ameri, M., Maa, C.M., Daddona, P., 2006. Microprojection array patch for transdermal delivery of vascular endothelial growth factors. World International Property Organization patent WO/2006/023684.

Birchall, J., 2006. Microneedle array technology: the time is right but is the science ready? Expert Rev. Med. Devices 3 (1), 1–4.

Brazzle, J., Papautsky, I., Frazier, A.B., 1998. Fluid-coupled hollow metallic micromachined needle arrays. Proc. SPIE Conf. Microfluidic Dev. Syst., Santa Clara, CA, pp. 116–124.

Brazzle, J., Mohanty, S., Frazier, A.B., 1999. Hollow metallic micromachined needles with multiple output ports. Proc. SPIE Conf. Microfluidic Dev. Syst., Santa Clara, CA, pp. 1765–1769.

Chabri, F., Bouris, K., Jones, T., Barrow, D., Hann, A., Allender, C., Brain, K., Birchall, J., 2004. Microfabricated silicon microneedles for nonviral cutaneous gene delivery. Br. J. Dermatol. 150 (5), 869–877.

Chen, J., Wise, K.D., Hetke, J.F., Bledsoe, S.C., Jr, 1997. A multichannel neural probe for selective chemical delivery at the cellular level. IEEE Trans. Biomed. Eng. 44 (8), 760–769.

Choi, S.-O., Park, J.-H., Kim, Y.-C., Gill, H.S., Choi Y., Allen, M.G., Prausnitz, M.R., 2005. In: Microneedle Electrode Array for Electroporation of Skin for Gene Therapy. 32nd Annual Meeting and Exposition. Controlled Release Society.

Cormier, M., Daddona, P.E., 2003. Macroflux Technology for Transdermal Delivery of Therapeutic Proteins and Vaccines. Marcel Dekker, New York, pp. 589–598.

Cormier, M., Johnson, B., Ameri, M., Nyam, K., Libiran, L., Zhang, D.D., Daddona, P., 2004. Transdermal delivery of desmopressin using a coated microneedle array patch system. J. Control. Release 97 (3), 503–511.

Davis, S.P., Martanto, W., Allen, M.G., Prausnitz, M.R., 2005. Hollow metal microneedles for insulin delivery to diabetic rats. IEEE Trans. Biomed. Eng. 52 (5), 909–915.

Douglas, J., Hugo, R. Jr, 2003. Micro infusion drug delivery device. US patent 6659982.

Gerstel, M., Place, V.A., 1976. US patent no. 3964482.

Gill, H.S., Prausnitz, M.R., 2007a. Coated microneedles for transdermal delivery. J. Control. Release 117 (2), 227–237.

Gill, H.S., Prausnitz, M.R., 2007b. Coating formulations for microneedles. Pharm. Res. 24 (7), 1369–1380.

Hashmi, S., Ling, P., Hashmi, G., Reed, M., Gaugler, R., Trimmer, W., 1995. Genetic transformation of nematodes using arrays of micromechanical piercing structures. Biotechniques 19 (5), 766–770.

Henry, S., McAllister, D.V., Allen, M.G., Prausnitz, M.R., 1998. Microfabricated microneedles: a novel approach to transdermal drug delivery. J. Pharm. Sci. 87 (8), 922–925.

Ito, Y., Yoshimitsu, J., Shiroyama, K., Sugioka, N., Takada, K., 2006a. Self-dissolving microneedles for the percutaneous absorption of EPO in mice. J. Drug Target. 14 (5), 255–261.

Ito, Y., Hagiwara, E., Saeki, A., Sugioka, N., Takada, K., 2006b. Feasibility of microneedles for percutaneous absorption of insulin. Eur. J. Pharm. Sci. 29 (1), 82–88.

Ji, J., Tay, F.E.H., Miao, J., Iliescu, C., 2006. Microfabricated silicon microneedle array for transdermal drug delivery. J. Phys. Conf. Ser. 34, 1127–1131.

Jiang, J., Gill, H.S., Ghate, D., McCarey, B.E., Patel, S.R., Edelhauser, H.F., Prausnitz, M.R., 2007. Coated microneedles for drug delivery to the eye. Invest. Ophthalmol. Vis. Sci. 48 (9), 4038–4043.

Kaushik, S., Hord, A.H., Denson, D.D., McAllister, D.V., Smitra, S., Allen, M.G., Prausnitz, M.R., 2001. Lack of pain associated with microfabricated microneedles. Anesth. Analg. 92 (2), 502–504.

Khumpuang, S., Horade, M., Fujioka, K., Sugiyama, S., 2006. Geometrical strengthening and tip-sharpening of a microneedle array fabricated by X-ray lithography. Microsyst. Technol. 13 (3), 209–214.

Kolli, C.S., Banga, A.K., 2008. Characterization of solid maltose microneedles and their use for transdermal delivery. Pharm. Res. 25, 104–113.

Lin, W., Cormier, M., Samiee, A., Griffin, A., Johnson, B., Teng, C.L., Hardee, G.E., Daddona, P.E., 2001. Transdermal delivery of antisense oligonucleotides with microprojection patch (Macroflux) technology. Pharm. Res. 18 (12), 1789–1793.

Martanto, W., Davis, S.P., Holiday, N.R., Wang, J., Gill, H.S., Prausnitz, M.R., 2004. Transdermal delivery of insulin using microneedles in vivo. Pharm. Res. 21 (6), 947–952.

Martanto, W., Moore, J.S., Couse, T., Prausnitz, M.R., 2006a. Mechanism of fluid infusion during microneedle insertion and retraction. J. Control. Release 112 (3), 357–361.

Martanto, W., Moore, J.S., Kashlan, O., Kamath, R., Wang, P.M., O'Neal, J.M., Prausnitz, M.R., 2006b. Microinfusion using hollow microneedles. Pharm. Res. 23 (1), 104–113.

Matriano, J.A., Cormier, M., Johnson, J., Young, W.A., Buttery, M., Nyam, K., Daddona, P.E., 2002. Macroflux microprojection array patch technology: a new and efficient approach for intracutaneous immunization. Pharm. Res. 19 (1), 63–70.

McAllister, D.V., Allen, M.G., Prausnitz, M.R., 2000. Microfabricated microneedles for gene and drug delivery. Annu. Rev. Biomed. Eng. 2, 289–313.

McAllister, D.V., Wang, P.M., Davis, S.P., Park, J.H., Canatella, P.J., Allen, M.G., Prausnitz, M.R., 2003. Microfabricated needles for transdermal delivery of macromolecules and nanoparticles: fabrication methods and transport studies. Proc. Natl. Acad. Sci. USA 100 (24), 13755–13760.

Mikszta, J.A., Dekker, J.P., 3rd, Harvey, N.G., Dean, C.H., Brittingham, J.M., Huang, J., Sullivan, V.J., Dyas, B., Roy, C.J., Ulrich, R.G., 2006. Microneedle-based intradermal delivery of the anthrax recombinant protective antigen vaccine. Infect. Immun. 74 (12), 6806–6810.

Moon, S.J., Lee, S.S., 2003. In: Fabrication of Microneedle Array Using Inclined LIGA Process, Transducers: 12th International Conference on Solid State Sensors, Actuators and Microsystems, Boston, p. 1546.

Moon, S.J., Lee, S.S., Lee, H.S., Kwon, T.H., 2005. Fabrication of microneedle array using LIGA and hot embossing process. Microsyst. Technol. 11 (4–5), 311–318.

Nordquist, L., Roxhed, N., Griss, P., Stemme, G., 2007. Novel microneedle patches for active insulin delivery are efficient in maintaining glycaemic control: an initial comparison with subcutaneous administration. Pharm. Res. 24 (7), 1381–1388.

Park, J.H., Allen, M.G., Prausnitz, M.R., 2006. Polymer microneedles for controlled-release drug delivery. Pharm. Res. 23 (5), 1008–1019.

Prausnitz, M.R., 2004. Microneedles for transdermal drug delivery. Adv. Drug Deliv. Rev. 56 (5), 581–587.

Raju, R., 2006. Method of fabricating microneedles. US patent 7097776.

Razzacki, S.Z., Thwar, P.K., Yang, M., Ugaz, V.M., Burns, M.A., 2004. Integrated microsystems for controlled drug delivery. Adv. Drug Deliv. Rev. 56 (2), 185–198.

Reed, M., Lye, W.K., 2004. Microsystems for drug and gene delivery. Proc. IEEE 92 (1), 56–75.

Reed, M.L., Wu, C., Kneller, J., Watkins, S., Vorp, D.A., Nadeem, A., Weiss, L.E., Rebello, K., Mescher, M., Smith, A.J., Rosenblum, W., Feldman, M.D., 1998. Micromechanical devices for intravascular drug delivery. J. Pharm. Sci. 87 (11), 1387–1394.

Sinko, P., 2005. Martin's Physical Pharmacy and Pharmaceutical Sciences, 5th ed. Lippincott Williams & Wilkins, p. 795.

Smart, W., Subramanian, K., Orloff, E., 2007. Silicon microprobe with integrated biosensor. US patent 7310543.

Tao, S.L., Desai, T.A., 2003. Microfabricated drug delivery systems: from particles to pores. Adv. Drug Deliv. Rev. 55 (3), 315–328.

Teo, M.A., Shearwood, C., Ng, K.C., Lu, J., Moochhala, S., 2005. In vitro and in vivo characterization of MEMS microneedles. Biomed. Microdevices 7 (1), 47–52.

Thakur, R., Michniak, B.B., 2007. In: Design of Microneedles – Polymer "Crown" System for Transdermal Delivery of Actives. AAPS Annual Meeting and Exposition. San Diego, CA.

Trimmer, W., Ling, P., Chin, C.K., Orton, P., Gaugler, R., 1995. Injection of DNA into plant and animal tissues with micromechanical piercing structures. Proc. 8th IEEE Microelectromech. Syst. Workshop. IEEE, Amsterdam, Piscataway, NJ, pp. 111–115.

Verbaan, F.J., Bal, S.M., Van den Berg, D.J., Groenink, W.H., Verpoorten, H., Luttge, R., Bouwstra, J.A., 2007. Assembled microneedle arrays enhance the transport of compounds varying over a large range of molecular weight across human dermatomed skin. J. Control. Release 117 (2), 238–245.

Wilke, N., Mulcahy, A., Ye, S.-R., Morrissey, A., 2005. Process optimization and characterization of silicon microneedles fabricated by wet etch technology. Microelectr. J. 36 (7), 650–656.

Williams, A.C., Barry, B.W., 2004. Penetration enhancers. Adv. Drug Deliv. Rev. 56 (5), 603–618.

Yeshurun, Y., Hefetz, M., de Boer, M., Berenschot, H.W., Gardeniers, J.G.E., 2003. Microneedle structure and production method therefor. US patent 6533949.

CHAPTER 7

Recent Advances in Ophthalmic Drug Delivery

James N. Chang

Chief Executive Officer, TaiMed Biologics, 5251 California Ave, Suite 230, Irvine, CA 92617, USA email: jchang@taimedbiologics.com

Contents

7.1 Introduction 165
7.2 The anatomy of the eye 167
7.3 Delivery systems for the anterior segment 169
 7.3.1 Viscous solutions 170
 7.3.2 Gels 171
 7.3.3 Bioadhesive polymers 173
 7.3.4 Nanosuspensions 175
 7.3.5 Inserts 176
 7.3.6 Liposomes 177
 7.3.7 Others 178
7.4 Retinal drug delivery systems 179
 7.4.1 Non-erodible implants 180
 7.4.2 Erodible implants 181
 7.4.3 Bolus injections 182
 7.4.4 Periocular drug delivery 182
 7.4.5 Others 186
7.5 Conclusions 186
References 186

7.1 INTRODUCTION

The eye presents unique challenges for drug delivery due to its anatomy and natural protective mechanisms. The constant tear turnover, drainage, and sensitivity to foreign bodies, as well as the low permeability of the cornea result in low bioavailability of ophthalmic medications administered topically using conventional dosage forms such as eye drops. Ophthalmic drugs delivered via systemic routes have proven even less successful due to the blood–aqueous and blood–retinal barriers, and significant side-effects associated with systemic exposure to large doses of drugs. Research in both industry and academia has been active over the last 30 years looking for ways

Handbook of Non-Invasive Drug Delivery Systems ISBN 9780815520252

to better deliver drugs to both the anterior and posterior segments of the eye, to achieve adequate bioavailability for better management of ocular diseases.

Most efforts in conventional formulations are directed toward lengthening the pre-cornea retention time of the medication, or increasing the cornea permeability, or both. Popular approaches such as the use of viscous solutions and gels to slow down nasolacrimal drainage have been widely used in formulating ophthalmic products with measured success. Other research efforts include the use of bioadhesive polymers to increase cornea contact time and lipids/liposome formulations to enhance cornea permeability.

In recent years, research activities have increasingly focused on the periocular route (Acharya and Young, 2004; Raghava et al., 2004; Ghate et al., 2007; Kansara et al., 2007; Amrite et al., 2008). Indeed, given the corneal barrier to drug penetration, it seems logical to consider the conjunctival–scleral route as an alternative for delivering drugs to both the anterior and posterior segments of the eye, since the drug permeability in the sclera has been reported to be approximately 20 times greater compared to the cornea. Numerous publications have reported on the mechanism of scleral drug delivery and pharmacokinetic modeling, and the relationship between the physicochemical properties of a drug and its permeability in the sclera (Ranta and Urtti, 2006). Passive diffusion through an interfibrillar aqueous pathway is likely to be the primary mechanism of drug permeation across the sclera due to its high water content. Moreover, it has been reported that the trans-scleral route is important for some drugs, even following topical dosing (Ahmed and Patton, 1985). For delivering drugs to the posterior segment of the eye, in particular, the trans-scleral route appears to be a promising alternative to the intravitreal route in light of the large scleral surface area, relatively few protein-binding sites, and its less invasive nature.

Nevertheless, trans-scleral drug delivery presents several challenges. To reach the neural retina by direct penetration, the drug molecules placed periocularly overlying the sclera must withstand the clearance by blood and lymphatic circulations in the episclera (the superficial layer of the sclera) and conjunctiva. Then, the drugs need to penetrate the sclera stroma, cross the choroidal vascular system, where they can be further removed or diluted, and then diffuse through the retinal pigment epithelium (RPE). The RPE is the outer part of the blood–retinal barrier, and it is known to express many transporters including efflux proteins that help restrict drug diffusion (Mannermaa et al., 2006). The physicochemical properties of the drug, its size and shape, as well as the circulation and elimination pathways all play an important role in the bioavailability of the drug delivered via this route (Pitkanen et al., 2005).

In this chapter, the many formulations, dosage forms, and delivery devices studied by the pharmaceutical industry and research institutions, and the various approaches for designing tomorrow's ophthalmic products, as well as recent advances in understanding the different ocular drug delivery routes are reviewed and, where appropriate, comments are made from an industrial perspective.

7.2 THE ANATOMY OF THE EYE

The eye (Figure 7.1) is a unique structure that is relatively isolated from the rest of the body due to the blood–ocular barriers, and yet well nourished and protected by the vascular and lacrimal circulations. A few tissues of the eye that play a more significant role in ocular drug delivery are briefly described below. Interested readers are encouraged to read more details in many excellent publications (Kaufman and Alm, 2003; Riordan-Eva and Whitcher, 2007; Saude, 1993).

The cornea is an optically transparent tissue that is a protective barrier to the interior of the eye. The shape of the cornea provides strong refractive power that, along with the lens, allows focusing of the light and image on to the retina. The cornea has a small surface area (~1 cm^2) and is composed of several layers. On the outside is the stratified, multicellular epithelium. These epithelial tight junctions present a natural barrier to drug penetration.

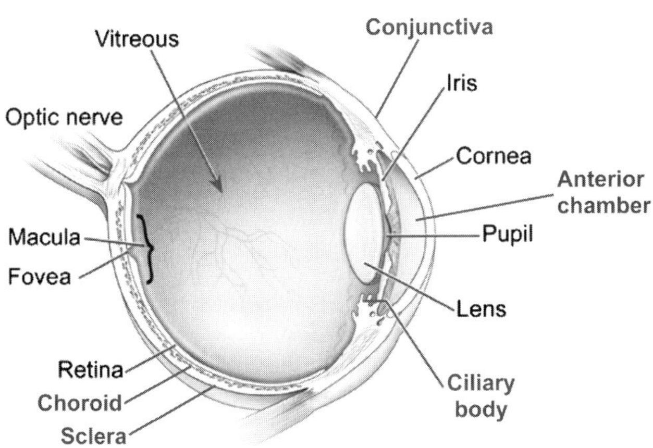

Figure 7.1 The eye in transverse section. *modified from a drawing from National Eye Institute (http://www.nei.nih.gov/diabetes/content/english/faq.asp)

Inside the cornea and in front of the iris and the lens is the anterior chamber filled with the aqueous humor. The aqueous humor is a transparent, watery substance secreted by the ciliary body. It maintains the intraocular pressure, for structural integrity, and provides diopteric power to the cornea as well as transports nutrients to, and metabolites from, the avascular tissues of the anterior segment.

The highly vascularized conjunctival membrane that lines the white part of the eye covers the rest of the anterior surface of the eye. The conjunctiva helps lubricate the eye by producing mucus from its goblet cells.

The sclera is a sponge-like, fibroblastic connective tissue that provides the structural integrity of the eye. The human sclera contains 70% water and has an average surface area of 17 cm^2 with an uneven thickness. The sclera is approximately 0.5 mm in thickness at the corneoscleral limbus, decreasing to 0.4 mm at the vertical equator of the eye where it is thinnest, and then gets substantially thicker to 1 mm near the optic nerve at the back.

The retina lines the back of eyeball, and is made up of the RPE and the neural retina, which consists of rod- and cone-shaped photoreceptor cells, second-order cells, and ganglion cells. The RPE is the outermost layer of the retina and, along with endothelial cells of retinal vessels, forms the blood–retinal barrier (BRB) that restricts transport of drugs, especially hydrophilic molecules, in both directions (Cunha-Vaz, 2004). The internal limiting membrane, the innermost layer of the retina that borders the vitreous cavity, prevents the transport of substances with size greater than approximately 100 kDa.

Positioned inside the sclera and outside the retina is the choroid, which is the vascular layer of the eye. The choroid provides oxygen and nourishment to the outer layers of the retina. Bruch's membrane is the inner layer of choroids and it thickens with age, slowing the transport of nutrients and metabolites. While the choroidal circulation nourishes the outer portion of the retina, the retinal vascular circulation nourishes the inner retina.

The central cavity of the eye globe is filled with the vitreous body, which is an aqueous gel of collagen and hyaluronic acid. The vitreous body has many important functions, including supporting the retina and lens that border it. In addition, it is a reservoir for nutrients, and serves as a barrier between the anterior and posterior segments of the eye. Drugs such as ranibizumab, triamcinolone acetonide and ganciclovir are frequently deposited in the vitreous body via injection or implantation for the treatment of retinal diseases. The physiology and functions of the vitreous body are important to the design of an intravitreal drug delivery system, its optimum placement in the vitreous body, and the movement of the drug.

7.3 DELIVERY SYSTEMS FOR THE ANTERIOR SEGMENT

The cornea is the primary route for anterior drug absorption, which is important for the treatment of diseases including infections, dry eye, glaucoma, and inflammations. Medications delivered in conventional eye drops are quickly removed from the eye by tear turnover and reflex blinking. The volume of a drop dispensed from a typical eye drop bottle varies significantly based on the tip design, formulation surface tension and viscosity, and the angle of instillation, but is on average approximately 40 μL. This volume is well in excess of the 20 μL holding capacity of the pre-corneal space, which is regularly filled with ~7 μL tear film in an average person. Accordingly, while some of the instilled 40 μL drop is absorbed systemically through the conjunctival stroma, much of the drug spills over onto the cheeks or drains into the nasolacrimal duct, where it flows down the throat and gets swallowed, or enters the systemic circulation through the nasopharyngeal mucosa. As a result, most topical eye drops have low bioavailability in the aqueous humor in the range of 1–5%. It has been suggested, based on pharmacokinetic considerations, that reducing the drop volume can increase bioavailability (Keister et al., 1991).

Despite decades of research efforts in novel ophthalmic drug delivery formulations and alternative dosage forms, the vast majority of the ophthalmic products currently available on the market, and those being developed by pharmaceutical companies worldwide, are simple aqueous solutions that do not address the issue of low bioavailability. There are several important reasons for this seemingly slow progress. First, the convenience and ease of use of an eye drop are unmatched by any other dosage form. Second, the manufacturability of pharmaceutical products is a critical factor in designing formulations, as all ophthalmic products need to be sterile. Most aqueous solutions can be readily sterilized by passing through 0.2-μm filters and therefore avoiding using terminal sterilization by autoclave, ethylene oxide gas, e-beam, or gamma irradiation. These terminal sterilization methods may cause drug decomposition, leave unwanted residues in the finished product, or in some cases require the use of new polymeric materials for the container to better withstand terminal sterilization. The sterile-filtered aqueous solution is aseptically filled into an eye drop bottle. While this manufacturing procedure is not as desirable as one that employs terminal sterilization in terms of sterility assurance, regulatory agencies around the world recognize the challenges faced by the pharmaceutical manufacturers and generally accept the sterile filtration and aseptic filling manufacturing method as long as bioburden is properly controlled. Third, a novel delivery system needs to demonstrate sufficient shelf stability, preferably at least 18 months, to be commercially viable. Finally, as the human eye is a delicate organ, novel formulations or delivery devices aimed at improving bioavailability or patient compliance must not cause excessive irritation such as stinging, foreign body sensation, or vision blurring. Given these constraints, even though the science for drug delivery and ocular pharmacokinetics has advanced a great deal to allow innovative design of new systems, improvement over the years has been incremental without major breakthroughs. A few interesting approaches have been undertaken by the pharmaceutical industry and developed into marketed products, while others are being actively pursued by both academia and industry. Further discussion of these approaches follows.

7.3.1 Viscous solutions

The addition of a biocompatible viscosity agent in the eye drop to slow down drainage into the nasolacrimal duct was one of the oldest methods, and was widely used. Robinson et al. (Chrai and Robinson, 1974; Patton and Robinson, 1975; Li and Robinson, 1989) demonstrated in the rabbit that drug bioavailability based on aqueous humor concentration increased as the viscosity of the solution became higher, presumably by increasing the drug–cornea contact time. The improvement in bioavailabity was only moderate, however, even as the viscosity was increased to a few thousand centipoises.

Many polymers have been studied over the last 30 years and used in ophthalmic products due to their proven drug compatibility and ocular safety profile. Hydroxypropylmethylcellulose (HPMC), sodium carboxymethylcellulose (CMC), carbomers, hyaluronic acid, and polyvinyl alcohol are among the most widely used in eye drop formulations. High-viscosity formulations are particularly popular for over-the-counter tear products. These polymers, to a varying degree, exhibit shear-thinning behavior that mimics the tear film. With shear thinning, the eye drop can more easily spread over the corneal surface by blinking. The viscosity of a polymeric solution depends not only on the amount of shear exerted, but also on the type, concentration, and molecular weight of the polymer. The viscosity profiles of several tear products are shown in Figure 7.2.

For products of moderate to high viscosities, it is generally not feasible for them to be manufactured using sterile filtration, as it would require too

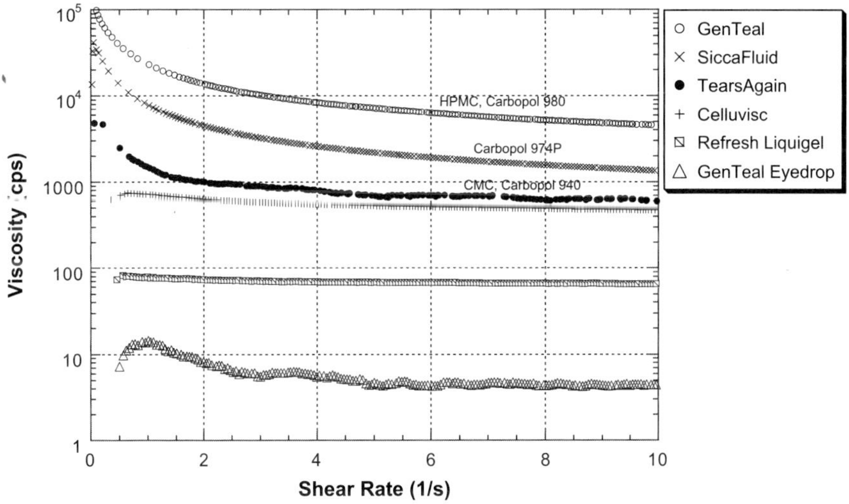

Figure 7.2 Viscosity measured by a cone-and-plate viscometer.

much pressure to pass through 0.2-μm filter cartridges, and they generally have to be manufactured aseptically by moving the sterile filtration step upstream to before the polymer(s) are added. The polymer is usually separately sterilized by dry heat or autoclave, or the finished product would have to be autoclaved prior to filling.

7.3.2 Gels

Ophthalmic gels have gained popularity as a replacement for ointment formulations as more synthetic, biocompatible polymers become available. Depending on how the gels are packaged and administered, their viscosities can vary between 1000 and 100,000 centipoises. There are a handful of ophthalmic gels on the market. Merck's Timoptic XE® gel-forming drops of timolol maleate contain Gelrite® gellan gum, a linear polysaccharide. This a low-viscosity aqueous solution that gels on contact with the cations in the tear film following instillation. Insite Vision recently received Food and Drug Administation (FDA) approval for AzaSite®, a polycarbophil gel-forming system for azithromycin for the treatment of bacterial conjunctivitis. Alcon's Pilopine HS® (pilocarpine hydrochloride) contains carbomer 940, and Laboratoires Thea's Virgan® (ganciclovir) contains carbomer 974. The carbomers are polyacrylic acid derivatives that impart high viscosity to their aqueous solutions at neutral pH (above their pK_a values) due to ionization and hydration of the carboxyl groups. Moreover, Akorn's Akten is an HPMC-based gel of lidocaine hydrochloride for ocular surface anesthesia.

Valley Forge and Novartis briefly collaborated on the development of a pirenzepine gel for the treatment of myopia in 2003. Such gels achieved prolonged ocular retention like ophthalmic ointments, but caused less vision blurring because they were optically clear and more easily diluted by the tear film.

Droppable gels such as Timoptic XE are easier for self-administration, which presents a significant marketing advantage, and have received much attention in recent years (Yoo et al., 2005). The *in situ* forming gels are typically aqueous solutions with a medium viscosity in the range of several hundred centipoises, but they undergo a phase transition to a gel that is far more viscous after being instilled in the eye. The phase transition can be triggered by several different mechanisms, including a sudden change in pH, temperature, ionic strength, or combination thereof when the eye drop encounters the physiological conditions in the eye (Miyazaki et al., 2001; Nanjawade et al., 2007). The polymers frequently used in pH-transition gels are carbomers and cellulose acetate phthalate (Gurny et al., 1985). These polymers typically have a pK_a around pH 4, and will hydrate and swell when they are exposed to an environmental pH that is above their pK_a values, causing the carboxyl groups on the polymers to become ionized. Lin and Sung (2000) reported that a mixture of carbopol and pluronic polymers works synergistically and achieves a better gel strength than each individually.

One major consideration in the design of pH gels is the need for them to be formulated at a substantially acidic pH that is below the pK_a of the polymer. The eye drop following instillation undergoes gel transition in the eye. The low formulation pH, however, often renders the drug product irritating. This is a particular problem when one considers that the carboxyl groups on the polymer present an intrinsic buffer capacity that resists a swift recovery of the pH back to the normal physiological condition following instillation, and thus further compounds the stinging sensation.

Another widely studied sol-to-gel system undergoes gelation when the temperature increases. For practical application, such thermal gels need to have a transition temperature somewhere between 32 and 35°C to ensure proper performance, as well as product quality during storage. They also need to be kept in the refrigerator in regions where the ambient temperature can reach 30°C or higher, which limits their marketability. Poloxamer block copolymers of polyethylene and polypropylene (Pluronic® and Tetronic® from BASF) are most studied for making thermal gels (gel Edsman et al., 1998). Vadnere et al. (1984), and separately Wei et al. (2002)

reported successful tuning of the transition temperature by mixing different poloxamers at a specific ratio. Using gamma scintigraphy, Wei et al. (2002) showed a more-than-threefold increase in ocular retention of their thermal gel system compared to a phosphate buffer solution.

Several studies have been reported where combination technologies were employed (Qi et al., 2007). Doijad et al. (2006) recently reported using sodium alginate as an ion-activated gelling agent in combination with HPMC to prepare a gatifloxacin eye drop, and achieved substantially prolonged, slow drug release compared to the marketed gatifloxacin formulation. In their *in vitro* study using a dialysis membrane, the cumulative drug release after 2 hours was 45% for the droppable gel system, compared to 92% for the commercial product. At the end of 8 hours, the cumulative release of gatifloxacin from the droppable gel is approximately 80%. While it can be argued that the diffusion of gatifloxacin through the dialysis membrane may be responsible for some of the slow release, the dramatic differences observed indicate the effect of the gel system. Ding (1998) discovered that a synergistic effect was achieved when a thermal gel (HPMC) and a pH gel (carbomer) were mixed together. Lin and Sung (2000) also reported use of a combination of carbopol and pluronic polymers. Gulsen and Chauhan (2004) showed that drug release from dimyristoyl phosphatidylcholine liposomes incorporated in a polyhydroxyethylmethacrylate hydrogel lasted for 8 days. Other gelling systems including the use of hydrogels (Park et al., 1993; Coviello et al., 2006) or xyloglucan polysaccharides derived from tamarind seed have also been proposed.

The concept of aqueous gel formulations, especially droppable gels, is appealing, as it reduces the vision-blurring problem associated with ointments. However, the drug release characteristics of the gel formulations generally show only marginal improvement when compared with solutions because the dissolved drug molecules can freely diffuse out of the gel, as in the solution, with little retardation from the polymer gel network. The ocular retention of the drug is considerably shorter than that of the polymers that are used to trap the drug in place. The use of a crosslinked gel network such as a hydrogel, which will be discussed further in section 7.3.5 later in the chapter, may lend more control of the drug diffusion and release rates.

7.3.3 Bioadhesive polymers

Bioadhesive formulations for ophthalmic drug delivery can be prepared as viscous solutions, gels, microspheres, and inserts (Saettone et al., 1994; Vasir et al., 2003; Weyenberg et al., 2003; Bernkop-Schnurch and Greimel,

2005). They are widely studied for enhancing ocular retention time, and have been the subject of many well-written reviews. Ionic, cellulose-based polymers such as sodium carboxymethylcellulose, carbomers, and chitosan are used frequently to increase the viscosity of aqueous ophthalmic solutions or to form gels. These polymers also exhibit strong adhesion to epithelial tissues. In the eye, mucus is secreted by conjunctival goblet cells. Mucus glycoproteins overlap and interpenetrate each other to form an extended gel network that is believed to be crosslinked by disulfide bonds. The bioadhesion of the polymers comes primarily from hydrogen bonding with the glycoprotein in the mucus and ionic interaction with the negatively charged sialic acids in the muco-glycoprotein. Physical entanglement of these macromolecules in the mucus, their hydrophobic interactions, and even van der Waals forces may also play an important role (Saettone, 2002).

Several methods have been developed for quantitative determination of bioadhesion (Pecosky and Robinson, 1990). Hui and Robinson (1985) reported that polycarbophil, a crosslinked polyacrylic acid, has good bioadhesion to conjunctival mucus. When compared with an aqueous suspension formulation without a bioadhesive polymer, their polycarbophil solution showed a nearly four-fold increase in bioavailability as measured by the area under the curve. Lehr et al. (1994) further demonstrated improved corneal penetration of gentamicin by the presence of polycarbophil. The use of collagen and fibrin as bioadhesive carriers for ophthalmic drug delivery has also been reported. Bonferoni et al. (2004) investigated the effect on timolol release rate of microspheres made with a carrageenan–gelatin ion pair that showed bioadhesive properties. A substantial increase in the aqueous humor concentration in rabbits was found compared to the gel-forming Timoptic XE®.

Chitosan has received broad attention as a promising bioadhesive excipient for ophthalmic application (Felt et al., 1999; Fini and Orienti, 2003). It is derived from chitin by deacetylation to form a cationic polysaccharide copolymer. Chitosan or chitosan-coated nanoparticles have also been investigated in recent years as researchers began to appreciate their potential for ophthalmic application given the tenability of the widely available polymeric carriers and the proven safety, compatibility, and bioadhesion of chitosan (deCampos et al., 2004; Barbu et al., 2006; de Salamanca et al., 2006; Kao et al., 2006). Chitosan, in addition to its bioadhesion, is also known as a penetration enhancer. Chondroitin sulfate is another modified polysaccharide that has been proposed as a potentially useful bioadhesive polymer. Kao et al. (2006) studied the release characteristics of a pilocarpine-loaded chitosan nanoparticle suspension containing carbomers, and showed more prolonged release when compared with gel and liposome formulations.

7.3.4 Nanosuspensions

The use of nanoparticles in ophthalmic formulations has received more attention with the recent advances in manufacturing technology, and appears to hold promise as a better dosage form than the conventional solutions and suspensions (Zimmer and Kreuter, 1995; Vandervoort and Ludwig, 2007). As more and more of the new chemical entities introduced into pharmaceutical development pipelines in recent years are less water soluble, formulating these drugs as a nanosuspension will likely gain popularity. Aqueous nanoparticle suspensions of hydrophobic drugs or a polymeric system embedded with a drug typically have a particle size distribution in the 100–400 nm range. Unlike microparticle suspensions, which tend to settle during storage and require shaking before use, these nanoparticle colloidal systems can stay suspended throughout their entire shelf-life without settling, when formulated properly. After being instilled in the eye, the fine particles tend to adhere to the eye tissues to form a depot and release drug over a desired period of time. In addition, due to the large surface area of the nanoparticles, the drug release rate may be sufficient to maintain an effective drug concentration in the tear film to achieve meaningful bioavailability. It is of particular interest if the nanoparticles are also mucoadhesive. De et al. (2004) reported that brimonidine-containing nanoparticles made with polycarboxylic acid show a sustained release profile through human cornea *ex vivo* when compared with a solution formulation. Pignatello et al. (2002) reported results that show improvement over a solution dosage form of ibuprofen loaded in Eudragit polymer nanoparticles with a positive surface charge (40–60 mV in z-potential) and approximately 100 nm in size.

As discussed in the previous section, chitosan has shown excellent bioadhesive properties that can help retain the nanoparticles in the eye, due at least in part to its cationic charge that can interact with cell membranes. Cui et al. (2006) have recently reported a detailed preparation procedure for nanoparticles coated with mucoadhesive polymers, including positively charged chitosan and negatively charged polyacrylic acid and carbopols. Motwani et al. (2007) further reported a poly(lactic-co-glycolic acid) (PLGA) nanoparticle delivery system for gatifloxacin that is coated with chitosan and polyvinyl alcohol to impart cationic charge for mucoadhesion. Lin et al. (2007) also investigated a nanosuspension of pilocarpine-loaded chitosan/acrylic acid copolymer, and found prolonged release both *in vitro* and *in vivo* compared to commercial eye drops.

In addition to chitosan, dextran, heparin, and hyaluronic acid have also been used for coating nanoparticles (Lemarchand et al., 2004).

7.3.5 Inserts

Of all the ophthalmic drug delivery systems that have been investigated, ocular inserts remain the most precise in terms of controlling the drug release profile through the use of permeable membranes or polymer matrices. Alza's Ocusert® is a non-erodible ethylene vinyl acetate membrane delivery system for weekly dosing of pilocarpine (Ali and Lehmussaari, 2006). Merck's Lacrisert®, on the other hand, is an erodible rod made with hydroxypropylcellulose for the treatment of dry eye. Both Ocusert and Lacrisert, developed some 30 years ago, have been poorly received in the marketplace however, due to the difficulty of applying, premature expulsion, and eye irritation.

Collagen shields and contact lenses are also used for delivering medications (Hwang et al., 1989; Friedberg et al., 1991; Friess, 1998). Several commercial collagen shields were developed in the 1980s as corneal bandages following eye surgery (Poland and Kaufman, 1988; Weissman and Lee, 1988). Bausch & Lomb's Bio-Cor® is made from porcine scleral tissue, and Alcon's Proshield® and Chiron's Medilens® are made from bovine corium tissue. These shields need to be hydrated before use. The shields swell by several hundred percent, typically, in aqueous antibiotic solution. When applied on the cornea, they keep the cornea moist and lubricated, promote wound healing, and also prevent possible infection. Kleinmann et al. (2007) reported that collagen shields can be an effective tool for delivering gatifloxacin and moxifloxacin to the anterior chamber, and help prevent bacterial postoperative endophthalmitis. Hwang et al. (1989) showed in rabbit eyes that a presoaked shield yielded equivalent or better drug delivery of dexamethasone compared with a regimen of hourly dosing by eye drops in a 24-hour period. Reidy et al. (1990) also reported a 10-fold increase in cyclosporine concentrations in both the cornea and aqueous humor when the drug was delivered from a collagen shield compared to an corn oil eye drop. These studies further confirm the utility of collagen shields for delivering the drugs previously reported.

Chow and Yang (2004) at Singapore's Institute of Bioengineering and Nanotechnology recently reported that long-term controlled release of drugs can be achieved by directly incorporating the drug into a nanostructured polymer matrix during microemulsion polymerization, and then molding the resulting polymer–drug mixture into a contact lens.

The drug-containing device made this way is different from a physical process of soaking up the drug with a collagen shield in that the drug rests in the microchannels of the polymer to allow for better control of drug release.

7.3.6 Liposomes

Liposomes have been investigated for several decades for use in ophthalmic drug delivery (Schaeffer and Krohn, 1982; Shek and Barber, 1987; Yanagawa et al., 1987; Velpandian et al., 1999). Most liposome formulations are prepared using phosphatidylcholine, cholesterol, and stearylamine. In theory, the drug may gain better penetration into and across the cornea when incorporated into a liposome as its lipid bilayer structure blends readily with the tear film and may also have higher affinity with the corneal epithelial cell membrane. There are barriers to overcome, however. Liposome formulations tend to clear from the precornea following instillation as fast as simple solutions, and therefore there is little time for the entrapped drug to be absorbed into the cornea. The surfactants used in liposome preparations may also cause eye irritation. Because the cornea and conjunctival epithelium cells and the mucus are negatively charged due to the presence of sialic acid-rich glycoproteins, some have suggested using cationic surfactants to impart a positive charge to enhance tissue uptake of the liposome and have achieved limited success. Other potential benefits of using liposomes include solubilizing lipophilic drugs and protecting the drug from premature enzyme degradation. Hathout et al. (2007) recently reported using both reverse-phase evaporation and multilamellar liposomes of acetazolamide, and achieved *in vitro* drug release rates over several hours. Budai et al. (2007) used phosphatidylcholine liposome formulations to prolong the release of ciprofloxacin.

While liposomes have been frequently referred to as being nontoxic (Kaur et al., 2004), it needs to be noted that overcoming local ocular irritation is a major consideration in bringing an ophthalmic product to commercialization. The surfactant(s) used in liposome formulations can be a major source for eye irritation; therefore, it is important to screen for cytotoxicity. It is believed that non-ionic surfactants are less irritating than cationic surfactants, which in turn are less irritating than anionic surfactants. This is why polysorbate 80 and poloxamers are most often used in ophthalmic formulations. Accordingly, there are many factors to consider in preparing liposome formulations, including the more complex manufacturing process, limited product stability, and the amount of surfactant to be used. It is a challenge to formulators to enhance drug bioavailability without inducing ocular irritation.

With the advent of gene therapy, DNA and RNA delivery has received much attention. These macromolecules, however, are unstable in blood and extracellular fluid as they are subject to enzymatic attack. In addition, their size and chemical structure pose further obstacles for intracellular uptake, therefore making it difficult for these therapeutic nucleotides to reach their target receptors. Liposomes have been studied for delivering plasmid DNAs, antisense oligonucleotides, and siRNAs to target cells given their unique lipid bilayer structure and the ability to design and modify the surface charge to meet the specific requirements for different cell types. Delivering these macromolecules in a liposome formulation for the treatment of retinal diseases has been reported (Fattal and Bochot, 2006).

7.3.7 Others

In addition to the more popular delivery systems discussed above, several other approaches have also been investigated such as microspheres, iontophoresis, and punctal plugs, to name a few. While nanosuspensions were the focus of discussion in section 7.3.4, microspheres continue to receive wide popularity, particularly for drug delivery to the posterior segment. This will be further discussed later in section 7.4. Another worthwhile approach is the use of penetration enhancers. It is logical to investigate their utility, given the low bioavailability delivered from eye drops and the progress made in transdermal drug delivery. However, the eye is much more sensitive than the skin and has proven to be far less tolerant to stinging and irritation, which are frequently associated with using penetration enhancers. It has been reported that surfactants and chelating agents such as polyoxyethylene stearyl ether, commonly used preservative benzalkonium chloride, editate disodium, and even chitosan enhance corneal penetration. While each of these penetration enhancers may have a different mechanism in promoting drug permeation across the cornea, all of them cause tissue damage when used at moderate concentrations. Burgalassi et al. (2001) studied some of these penetration enhancers using cultures of rabbit and human corneal epithelial cells and reported different levels of cytotoxicity from these molecules. They further investigated the ocular toxicity of these penetration enhancers using electrophysiology measurements on isolated rabbit corneas and showed correlations between ocular toxicity and transcorneal permeation enhancement (Chetoni et al., 2003).

Iontophoresis is another idea that has been investigated and has met with only limited success (Hill et al., 1978; Choi and Lee, 1988; Grossman and Lee, 1989; Eljarrat-Binstock and Domb, 2006). The process involves using low electric current density (~5 mA/cm^2) to drive ionic molecules across tissues to the target receptor site. Iomed and Eyegate are investigating retinal drug delivery with their respective iontophoretic systems. The iontophoretic procedure needs to be performed in a doctor's office by a trained technician. Iomed's Ocuphor® system requires placing a hydrogel applicator in the lower cul-de-sac and in direct contact with the conjunctiva to deliver drug through the sclera to the back of the eye. Both systems will require topical anesthesia. Much work remains to demonstrate clinically that iontophoresis is able to deliver substantial amounts of drug to the retina without damaging the eye tissues following repeated applications before it is given major consideration for commercial application.

Recently, the use of dendrimers as drug delivery vehicles has been investigated (Cunliffe et al., 2005; Patri et al., 2005; Svenson and Tomalia, 2005; Vandamme and Brobeck, 2005). Dendrimers are macromolecules with repeated branching characterized by their orderly and symmetrical structure. Their physical properties tend to be dominated by the functional groups on the molecular surface, which can be synthesized with anionic carboxyl, cationic amine, or non-ionic hydroxyl groups (Loutsch et al., 2003; Gillies and Frechet, 2005). The hydrophobicity and bioadhesive properties of a dendrimer can be tailor-made to suit specific drug delivery needs. While this appears interesting and warrants further research, many issues remain. Specifically, in addition to understanding their ocular retention and drug release characteristics, the ocular toxicity of dendrimers needs to be thoroughly investigated, as well as their long-term stability and compatibility with drugs.

Punctal plugs are used primarily for the treatment of dry eye conditions that do not respond to artificial tears (Balaram et al., 2001; Tai, 2002). For dry eye treatment, the punctal plugs block or reduce the drainage down the nasolacrimal ducts and keep the tears in the eye longer. Punctal plugs can be made with silicones, biodegradable polyesters and collagen, and they should be sterile. There are many commercial suppliers of punctal plugs. The idea may be useful for delivering potent drugs as the drug-loading capacities of these plugs are very limited given their small sizes.

7.4 RETINAL DRUG DELIVERY SYSTEMS

As the aging population continues to increase, treatment of retinal diseases has become a major issue and there is clearly an unmet medical need (Velez and Whitcup, 1999). Among the many debilitating retinal diseases that greatly impact on quality of life, age-related macular degeneration is the leading cause of loss of vision in the elderly, with an incidence of 200,000+

cases per year in the USA. Treatment options are often limited as therapeutic drug levels at the retina are difficult to attain by eye drops. It has proven to be equally challenging by the systemic route due to the blood–retinal barrier at the RPE. A large systemic drug dose is usually required in order to reach therapeutically effective drug levels at the retina, which invariably are accompanied with unwanted side-effects. Currently approved drugs on the market, including Lucentis® and Macugen®, require frequent bolus injections on a monthly basis into the vitreous body to manage chronic retinal diseases. These repeated injections can have severe side-effects, including retinal detachment and endophthalmitis. Accordingly, there is a need to develop a localized and controlled drug delivery system to efficiently deliver the many therapeutic agents, including steroids, and anti-angiogenic and antiviral compounds.

The eye anatomy, despite its many limitations, offers unique opportunities for designing posterior drug delivery systems as the eye is a relatively isolated tissue compartment. The use of mini-implants that can be inserted as an office procedure, or preferably injected, to deliver drugs to the posterior segment over an extended period of time is a practical approach, and has been the subject of intense research and development efforts (Cheng et al., 1995; Pearson et al., 1996; Alarcon and Martinez, 2006). The handful of marketed products for the treatment of retinal diseases and projects under development by the pharmaceutical industry are primarily these minimally invasive implants and injectables.

7.4.1 Non-erodible implants

The first approved product for slow drug release to the back of the eye was the Vitrasert® implant developed in the early 1990s by pSivida (formerly Control Delivery Systems) and Chiron Vision. Vitrasert, currently marketed by Bausch & Lomb through acquisition, contains the antiviral agent ganciclovir for the treatment of cytomegalovirus (CMV) retinitis in people with AIDS as a first-line therapy (Sanborn et al., 1992). The non-erodible implant is a polymer-based system made with an ethylene vinyl acetate and polyvinyl alcohol membrane. It needs to be surgically placed in the vitreous cavity and sutured on the sclera. The implant releases the embedded drug at a rate of approximately 2 μg/hr over several months by diffusion and can be removed at the time of the next insertion. A few years later Bausch & Lomb introduced Retisert® (also licensed from pSivida), a similar intravitreal implant containing fluocinolone acetonide for the treatment of uveitis. Retisert, smaller in size than Vitrasert, contains a drug core situated in a semipermeable membrane of polyvinyl alcohol, encased in a silicone elastomer cup. The drug is released from a small orifice after being hydrated. Like Vitrasert, Retisert is surgically placed in the vitreous cavity and sutured on the eyewall, where it releases fluocinolone acetonide over a period of 30 months at a desired daily dose. The delivery system allows accurate control of the drug release profile, which is important for the delivery of potent steroids such as fluocinolone acetonide, for a long period of time in order to minimize side-effects.

Both Vitrasert and Retisert require surgical implantation with a sizeable incision of a few millimeters. A less invasive implant that can be placed in the eye in a simpler procedure is desirable. In addition, it should be noted that use of these novel drug delivery systems may have unintended consequences. Long-term complications including retinal detachment have been reported following intraocular implantation (Guembel et al., 1999). Some patients receiving Retisert may develop elevated intraocular pressures that might not be immediately detected since the implant can effectively control uveitis inflammation over a long period of time, so these patients do not make frequent office visits. Alimera Sciences and pSivida are co-developing a similar delivery system called Medidur®, currently in late phase clinical trials. It is a non-erodible implant designed to deliver a very low dose of fluocinolone acetonide for up to 3 years for the treatment of diabetic macular edema. Unlike Retisert, Medidur is small in size and can be injected into the vitreous body during an office visit.

Neurotech is currently in clinical trials with an implant for delivering ciliary neurotrophic factor (CNTF) based on their encapsulated cell technology (ECT), for the treatment of retinitis pigmentosa. The ECT implant, a semipermeable hollow fiber about 6 mm in length, is inserted through the pars plana into the vitreous cavity and sutured on to the sclera. The semipermeable membrane design allows both the diffusion out of CNTF and diffusion in of oxygen and other nutrients to support cell viability while preventing any interaction between the host immune system with the encapsulated cells. This innovative drug delivery platform can also be applied to treatment of angiogenic and inflammatory diseases.

7.4.2 Erodible implants

Allergan's Ozurdex® is a polylactide-co-glycolide implant that delivers dexamethasone for the treatment of macular edema associated with retinal vein occlusion. Ozurdex was recently approved by the FDA. The rod-shaped erodible implant is hot-melt extruded. It is only about 1 mg in weight and can be injected into the vitreous cavity through a 22-gauge needle with a proprietary injector. The implant releases dexamethosone over approximately 4 weeks as the polymer matrix slowly hydrolyzes to lactic acid and glycolic acid, which are then cleared metabolically, obviating the need for implant removal.

Choonara et al. (2007) reported a donut-shaped minitablet for intravitreal delivery of ganciclovir and forscarnet using biodegradable PLGA polymers. The advantages of their system are the ease of manufacture of the minitablet using a standard tablet press, and the use of biodegradable polymers. Similar to many erodible systems, however, their system gave a nonlinear drug release profile with a substantial burst.

Biodegradable implants have the obvious advantage of leaving no residue in the eye. It is usually more challenging to control the drug release profile with an erodible system however, because as the implant erodes, the microenvironment surrounding the drug molecules also changes significantly with time.

7.4.3 Bolus injections

Recently introduced products, and those currently under development, have seen substantial improvement. Macugen® (pegaptanib sodium injection) is an antivascular endothelial growth factor inhibitor from Eyetech (acquired by OSI Pharmaceuticals) and Pfizer. It is injected intravitreally using a 27-gauge needle. The dosing frequency is less than ideal however, as it requires repeated injection every 6 weeks to manage the wet age-related macular degeneration (AMD) and slow vision loss. Genentech's Lucentis® (ranibizumab injection) is made from a human antibody fragment also for the treatment of wet AMD. It can be injected with a smaller 30-gauge needle and, while it has enjoyed much greater success than Macugen, it also requires frequent injections.

Both Macugen and Lucentis are bolus injections that do not employ a drug delivery system to achieve a more sustained release profile, leaving room for further improvement. In the foreseeable future, using anti-vascular endothelial growth factor molecules to treat retinal vascular diseases with a delivery system to achieve a long-term maintenance dose holds great promise.

7.4.4 Periocular drug delivery

Given the many risks associated with intravitreal injections and implants, placing the drug in the periocular space for the drug to penetrate across the sclera to reach the choroid and retina appears to be safer and less invasive

(Geroski and Edelhauser, 2001; Ogura, 2001). In addition, linear molecules larger than 40 kDa and globular molecules larger than 70 kDa do not readily cross the internal limiting membrane of the retina, precluding intravitreous delivery for many of the anti-angiogenic compounds. There has been considerable interest in recent years surrounding retina–choroid drug delivery via the scleral route (Cruysberg et al., 2005). This approach usually involves subconjunctival, sub-Tenon's or retrobulbar injections, or implantation. The injectable formulations range from solutions and microspheres to both erodible and non-erodible micro-implants. The objective is to deliver a significant and consistent amount of drug to the posterior segment over a prolonged period of time for the treatment of chronic diseases with better patient tolerance and less risk than drug delivery via the intravitreous route. It has been shown that the sclera is much more permeable than the cornea. While the permeability of the sclera depends on many properties such as the drug's lipophilicity, surface charge and shape, as well as tissue hydration (which is affected by pH and ionic strength), it is generally believed that the size of the drug molecule is the dominant factor (Lee et al., 1999). Prausnitz and Noonan (1998) compiled a list of the trans-scleral permeabilities of various compounds (Table 7.1). It has also been shown that the molecular radius of the drug, rather than its molecular weight, is a better predictor of scleral permeability (Ambati et al., 2000a).

Trans-scleral diffusion is generally not an issue with low-molecular-weight compounds, up to several thousand daltons, with the permeability roughly inversely proportional to the molecular weight. For macromolecules, however, meaningful permeability across the sclera is not guaranteed, and this is an important consideration for delivering certain anti-angiogenic compounds. Ambati et al. (2000a) have reported that the sclera is permeable to large molecules such as immunoglobulin G, whose molecular weight is approximately 150 kDa, possibly due to specific binding. The hydration level of the scleral tissue and the lipophilicity of the macromolecules are critical for their permeability. Moreover, it has been reported that the RPE is a major barrier, and may be the rate-limiting factor for the transport of hydrophilic drugs and macromolecules due to its tight cell junctions. On the other hand, the sclera and RPE–choroid are equal barriers for the penetration of lipophilic drugs.

While the sclera itself may not be a major barrier to the penetration of many drugs, the episclera and conjunctival lymphatics and blood vessels, as well as the blood flow of the choroids, can clear the drug molecules before they reach the retina, and therefore may present some challenges. Robinson

Table 7.1 Relative trans-scleral permeabilities of various compounds*

Molecule name	Molecular weight	K_{perm} (cm/s)	Reference
Polymyxin B–BODIPY	1800	$3.90 \pm 0.59 \times 10^{-7}$	Kao et al. (2005)
Vancomycin–BODIPY	1723	$6.66 \pm 1.46 \times 10^{-7}$	
Dexamethasone–fluorescein	841	$1.64 \pm 0.17 \times 10^{-6}$	Klein et al. (1996)
Rhodamine	479	$1.86 \pm 0.39 \times 10^{-6}$	
Penicillin–BODIPY	661	$1.89 \pm 0.21 \times 10^{-6}$	
Methotrexate–fluorescein	979	$3.36 \pm 0.62 \times 10^{-6}$	
Carboxyamidotriazole	423	$2.8–5.5 \pm 1.1 \times 10^{-6}$	
Fluorescein	332	$5.21 \pm 0.71 \times 10^{-6}$	
Dexamethasone	392	$8.94 \pm 1.5 \times 10^{-6}$	Bauer et al. (2000)
Carboxyfluorescein	317	$9.93 \pm 3.5 \times 10^{-6}$	
2-Methoxyestradiol	302	$9.96 \pm 1.1 \times 10^{-6}$	
Carboplatin	371	$27.0 \pm 1.7 \times 10^{-6}$	Fotsis et al. (1994)
Water	18	$51.8 \pm 18 \times 10^{-6}$	

* Human scleral permeability.

and coworkers investigated the effects of these lymph and blood flows by measuring the triamcinolone acetonide concentration in the rabbit vitreous following sub-Tenon's injection (Robinson et al., 2006; Csaky, 2007). In their study, the conjunctival circulations of one group of rabbits were removed by cuts made in the conjunctiva to bare the sclera. In a second group, the choroidal circulation was stopped by forming a scar on the choroidal–retinal tissue using cryosurgery. In a third group, both conjunctival and choroidal circulations were halted by euthanasia. They reported that the drug concentrations in the vitreous were substantially higher when cuts were made or when the rabbit was euthanized, but not when cryosurgery was performed, suggesting that the conjunctival lymphatic and vascular circulations are a more significant barrier than the choroid in trans-scleral drug delivery. Another study employing a dynamic three-dimensional magnetic resonance imaging technique by the same group further confirmed that episcleral circulation presents a major hurdle for trans-scleral drug delivery (Kim et al., 2007).

Subjconjuctival injections have been shown to be an efficient and less invasive way of achieving drug delivery to the posterior segment of the eye. In one study reported by Kompella et al. (2003), it was shown that a subconjunctival injection of 50 μg budesonide solution in rats resulted in measurable drug levels in the retina and vitreous for up to 3 days, indicating a slow elimination rate of the drug from the subconjunctival space, possibly due to specific tissue binding of the lipophilic drug. No budesonide drug level was detectable in the contralateral eye, suggesting direct penetration of the drug across the sclera to reach the intraocular tissues rather than through systemic circulation. Lee and Robinson (2001) drew a similar conclusion about direct penetration being the dominant pathway following subconjunctival injection. In Kompella et al.'s study, it was also reported that a PLGA microsphere preparation of budesonide delivered a sustained level of the drug to the back of the eye for 14 days following subconjunctival injection. In a separate study reported by Cheruvu et al. (2003), subconjunctival injections of cyclodextrin aqueous solutions of sodium fluorescein, budesonide, and celecoxib in rats were performed, and it was found that all three drugs readily reached the retina with no apparent relationship to the drug lipophilicity. The use of a collagen matrix gel for prolonged delivery of cisplatin for up to 2 weeks by subconjunctival injection has also been reported (Gilbert et al., 2003).

In another study, Kato et al. (2004) constructed an implant for betamethasone coated with a non-erodible ethylene vinyl acetate membrane. The episcleral implant released betamethasone in a remarkably linear fashion both *in vitro* and *in vivo*. In a separate study, Sakurai et al. (2003) demonstrated that experimental uveitis in a rabbit model can be successfully suppressed by tacrolimus released from a biodegradable scleral plug made with PLGA.

It has also been reported that thermally reversible gels, discussed earlier in this chapter, can be used successfully for periocular drug delivery to ensure contact between the drug and the sclera. Lee et al. (1999) prepared a thermal gel using Pluronic® F-127 and their *in vitro* study demonstrated slow release of dexamethasone across the human sclera.

7.4.5 Others

In addition to the aforementioned approaches for delivering drugs to the back of the eye, new techniques continue to be explored. Jiang et al. (2007) recently reported coating pilocarpine and sodium fluorescein onto stainless steel microneedles, which are inserted halfway into the upper region of the rabbit cornea for 20 seconds under anesthesia. Tissue damage is minimal. More studies will be needed to assess the long-term effect of the microneedle concept for transcorneal or trans-scleral delivery.

Ambati et al. (2000b) reported the use of an osmotic pump to deliver IgG and anti-intercellular adhesion molecule-1 monoclonal antibody transsclerally to the choroid and retina in pigmented rabbits. Their results indicated that antibodies can be delivered across the sclera into the choroid and retina without significant proteolytic degradation, loss of bioactivity, or systemic absorption. They also suggested that the gradual delivery achieved by their osmotic pump system resulted in more complete scleral absorption because of the long drug–sclera contact time.

7.5 CONCLUSIONS

The advancement in formulation technology for conventional eye drop dosage forms has slowed in recent years. While there is always a possibility for breakthroughs, it is unlikely to see dramatic improvement in the foreseeable future. With pharmaceutical development scientists turning their focus more toward injectable and implant systems for intravitreal and periocular delivery, however, there could be a paradigm shift in how anterior segment chronic diseases such as glaucoma are treated. For some patients who may find the daily use of chronic medications challenging, there could be an alternative in receiving quarterly injections instead of putting eye drops in their eyes each day.

As the world population ages and more people face debilitating visual diseases and their quality of life is severely threatened, the development of a drug delivery system that is effective, convenient, and of low risk will surely benefit many. The ever-advancing polymer delivery systems, especially those in periocular drug delivery, for the treatment of age-related macular degeneration, retinitis pigmentosa, and the many other retinal diseases, hold great promise as new therapeutic agents are discovered.

REFERENCES

Acharya, N., Young, L., 2004. Sustained-release drug implants for the treatment of intraocular disease. Int. Ophthalmol. Clinics 44 (3), 33–39.

Ahmed, I., Patton, T.F., 1985. Importance of the noncorneal absorption route in topical ophthalmic drug delivery. Invest. Ophthalmol. Vis. Sci. 26 (4), 584–587.
Alarcon, R., Martinez, M., 2006. Intraocular drug delivery systems (Sistemas de cesión controlada por vía intraocular). Arch. Soc. Esp. Oftalmol. 81 (2), 57–59.
Ali, Y., Lehmussaari, K., 2006. Industrial perspective in ocular drug delivery. Adv. Drug Deliv. Rev. 58 (11), 1258–1268.
Ambati, J., Canakis, C.S., Miller, J.W., Gragoudas, E.S., Edwards, A., Weissgold, D.J., Kim, I., Delori, F.C., Adamis, A.P., 2000a. Diffusion of high molecular weight compounds through sclera. Invest. Ophthalmol. Vis. Sci. 41, 1181–1185.
Ambati, J., Canakis, C.S., Gragoudas, E.S., Miller, J.W., You, T.T., Miyamoto, K., Delori, F. C., Adamis, A.P., 2000b. Transscleral delivery of bioactive protein to the choroid and retina. Invest. Ophthalmol. Vis. Sci. 41, 1186–1191.
Amrite, A., Edelhauser, H., Kompella, U., 2008. Modeling of corneal and retinal pharmacokinetics after periocular drug administration. Invest. Ophthalmol. Vis. Sci. 49 (1), 320–332.
Balaram, M., Schaumberg, D.A., Dana, M.R., 2001. Efficacy and tolerability outcomes after punctal occlusion with silicone plugs in dry eye syndrome. Am. J. Ophthalmol. 131 (1), 30–36.
Barbu, E., Verestiuc, L., Nevell, T.G., Tsibouklis, J., 2006. Polymeric materials for ophthalmic drug delivery: trends and perspectives. J. Mater. Chem. 16, 3439–3443.
Bauer, K.S., Cude, K.J., Dixon, S.C., Kruger, E.A., Figg, W.D., 2000. Carboxyamidotriazole inhibits angiogenesis by blocking the calcium-mediated nitric-oxide synthase–vascular endothelial growth factor pathway. J. Pharmacol. Exp. Ther. 292, 31–37.
Bernkop-Schnurch, A., Greimel, A., 2005. Thiomers: the next generation of mucoadhesive polymers. Am. J. Drug Deliv. 3 (3), 141–154.
Bonferoni, M.C., Chetoni, P., Giunchedi, P., Rossi, S., Ferrari, F., Burgallassi, S., Caramella, C., 2004. Carrageenan–gelatin mucoadhesive systems for ion-exchange based ophthalmic delivery: in vitro and preliminary in vivo studies. Eur. J. Pharm. Biopharm. 57 (3), 465–472.
Budai, L., Hajdu, M., Budai, M., Grof, P., Beni, S., Noszal, B., Klebovich, I., Antal, I., 2007. Gels and liposomes in optimized ocular drug delivery: studies on ciprofloxacin formulations. Int. J. Pharm. 343 (1–2), 34–40.
Burgalassi, S., Chetoni, P., Monti, D., Saettone, M., 2001. Cytotoxicity of potential ocular permeation enhancers evaluated on rabbit and human corneal epithelial cell lines. Toxicol. Lett. 122 (1), 1–8.
Cheng, C., Berger, A., Pearson, P., Ashton, P., Jaffe, G., 1995. Intravitreal sustained-release dexamethasone device in the treatment of experimental uveitis. Invest. Ophthalmol. Vis. Sci. 36, 442–453.
Cheruvu, N., Ayalasomayajula, S., Kompella, U., 2003. Retinal delivery of sodium fluorescein, budesonide and celecoxib following subconjunctival injection. Drug Deliv. Technol. 3 (6), 52–58.
Chetoni, P., Burgalassi, S., Monti, D., Saettone, M.F., 2003. Ocular toxicity of some corneal penetration enhancers evaluated by electrophysiology measurements on isolated rabbit corneas. Toxicol. In Vitro 17 (4), 497–504.
Choi, T., Lee, D., 1988. Transscleral and transcorneal iontophoresis of vancomycin in rabbit eyes. J. Ocul. Pharmacol. 4, 153–164.
Choonara, Y., Pillay, V., Carmichael, T., Danckwerts, M., 2007. Studies on a novel doughnut-shaped minitablet for intraocular drug delivery. AAPS Pharm. Sci. Tech. 8 (4), article 118.
Chow, E., Yang, Y., 2004. Available from: <www.news-medical.net>, accessed 29 October.
Chrai, S., Robinson, J., 1974. Ocular evaluation of methylcellulose vehicle in albino rabbits. J. Pharm. Sci. 63, 1218–1223.

Coviello, T., Matricardi, P., Alhaique, F., 2006. Drug delivery strategies using polysaccharidic gels. Expert Opin. Drug Del. 3 (3), 395–404.

Cruysberg, L.P., Franklin, A.J., Sanders, J., Self, C., Yuan, P., Csaky, K.G., Robinson, M.R., Kohn, E.C., Edelhauser, H.F., 2005. Effective transscleral delivery of two retinal anti-angiogenic molecules: carboxyamido-triazole (CAI) and 2-methoxyestradiol (2ME2). Retina 25 (8), 1022–1031.

Csaky, K.C., 2007. New developments in the transscleral delivery of ophthalmic agents. Retina Today, March/April, 32.

Cui, F., Qian, F., Yin, C., 2006. Preparation and characterization of mucoadhesive polymer-coated nanoparticles. Int. J. Pharm 316 (1–2), 154–161.

Cunha-Vaz, J., 2004, The blood–retinal barriers system. Basic concepts and clinical evaluation. Exp. Eye Res. 78(3), 715–721.

Cunliffe, D., Kirby, A., Alexander, C., 2005. Molecularly imprinted drug delivery systems. Adv. Drug Deliv. Rev. 57, 1836–1853.

De, T.K., Bergey, E.J., Chung, S.J., Rodman, D.J., Bharali, D.J., Prasad, P.N., 2004. Polycarboxylic acid nanoparticles for ophthalmic drug delivery: an ex vivo evaluation with human cornea. J. Microencaps 21 (8), 841–855.

deCampos, A.M., Diebold, Y., Carvalho, E.L.S., Sanchez, A., Alonso, M.J., 2004. Chitosan nanoparticles as new ocular drug delivery systems: in vitro stability, in vivo fate, and cellular toxicity. Pharm. Res. 21 (5), 803–810.

de Salamanca, A.E., Diebold, Y., Calonge, M., Garcia-Vazquez, C., Callejo, S., Vila, A., Alonso, M.J., 2006. Chitosan nanoparticles as a potential drug delivery system for the ocular surface: toxicity, uptake mechanism and in vivo tolerance. Invest. Ophthalmol. Vis. Sci. 47, 1416–1425.

Ding, S. 1998. Recent developments in ophthalmic drug delivery. Pharm. Sci. Tech. Today. 1 (8), 328–335.

Doijad, R.C., Manvi, F.V., Malleswara Rao, V.S.N., Alase, P., 2006. Sustained ophthalmic delivery of gatifloxacin from in situ gelling system. Indian J. Pharm. Sci. 68 (3), 814–818.

Edsman, K., Carlfors, J., Petersson, R., 1998. Rheological evaluation of poloxamer as an in situ gel for ophthalmic use. Eur. J. Pharm. Sci. 6, 105–119.

Eljarrat-Binstock, E.E., Domb, A.J., 2006. Iontophoresis: a non-invasive ocular drug delivery. J. Control. Release 110 (3), 479–489.

Fattal, E., Bochot, A., 2006. Ocular delivery of nucleic acids: antisense oligonucleotides, aptamers and siRNA. Adv. Drug Deliv. Rev. 58, 1203–1223.

Felt, O., Furrer, P., Mayer, J.M., Plazonnet, B., Buri, P., Gurny, R., 1999. Topical use of chitosan in ophthalmology: tolerance assessment and evaluation of precorneal retention. Int. J. Pharm. 180 (2), 185–193.

Fini, A., Orienti, I., 2003. The role of chitosan in drug delivery: current and potential applications. Am. J. Drug Deliv. 1 (1), 43–59.

Fotsis, T., Zhang, Y., Pepper, M.S., et al., 1994. The endogenous oestrogen metabolite 2-methoxyoestradiol inhibits angiogenesis and suppresses tumour growth. Nature 368, 237–239.

Friedberg, M.L., Pleyer, U., Mondino, B.J., 1991. Device drug delivery to the eye: collagen shields, iontophoresis, and pumps. Ophthalmology 98, 725–732.

Friess, W., 1998. Collagen – biomaterial for drug delivery. Eur. J. Pharm. Biopharm. 45 (2), 113–136.

Geroski, D.H., Edelhauser, H.F., 2001. Transscleral drug delivery for posterior segment disease. Adv. Drug Deliv. Rev. 52, 37–48.

Ghate, D., Brooks, W., McCarey, B., Edelhauser, H., 2007. Pharmacokinetics of intraocular drug delivery by periocular injections using ocular fluorophotometry. Invest. Ophthalmol. Vis. Sci. 48, 2230–2237.

Gilbert, J.A., Simpson, A.F., Rudnick, D.E., Geroski, D.H., Aaberg Jr., T.M., Edelhauser, H.F., 2003. Transscleral permeability and intraocular concentrations of cisplatin from a collagen matrix. J. Control. Release 89 (3), 409–417.

Gillies, E.R., Frechet, J.M.J., 2005. Dendrimers and dendritic polymers in drug delivery. Drug Discovery Today 10 (1), 35–43.

Grossman, R., Lee, D., 1989. Transscleral and transcorneal iontophoresis of ketoconazole in the rabbit eye. Ophthalmology 96, 724–729.

Guembel, H., Krieglsteiner, S., Rosenkranz, C., Hattenbach, L., Koch, F., Ohrloff, C., 1999. Complications after implantation of intraocular devices in patients with cytomegalovirus retinitis. Graefe's Arch. Clin. Exp. Ophthalmol. 237 (10), 824–829.

Gulsen, D., Chauhan, A., 2004. Ophthalmic drug delivery through contact lenses. Invest. Ophthalmol. Vis. Sci. 45, 2342–2347.

Gurny, R., Boye, T., Ibrahim, H.J., 1985. Ocular therapy with nanoparticular systems for controlled drug delivery. J. Control. Release 2, 353–361.

Hathout, R., Mansour, S., Mortada, N., Guinedi, A., 2007. Liposomes as an ocular delivery system for acetazolamide: in vitro and in vivo studies. AAPS Pharm. Sci. Tech. 891, article 1.

Hill, J., Park, N.-H., Gangarosa, L., et al., 1978. Iontophoresis of vidarabine monophosphate into rabbit eyes. Invest. Ophthalmol. Vis. Sci. 17, 473–476.

Hui, H.W., Robinson, J.R., 1985. Ocular drug delivery of progesterone using a bioadhesive polymer. Int. J. Pharm. 26, 203–213.

Hwang, D., Stern, W., Hwang, P., MacGowan-Smith, L., 1989. Collagen shield enhancement of topical dexamethasone penetration. Arch. Ophthalmol. 107, 1375–1380.

Jiang, J., Gill, H., Ghate, D., McCarey, B., Patel, S., Edelhauser, H., Prausnitz, M., 2007. Coated microneedles for drug delivery to the eye. Invest. Ophthalmol. Vis. Sci. 48 (9), 4038.

Kansara, V., Hao, Y., Mitra, A., 2007. Dipeptide monoester ganciclovir prodrugs for transscleral drug delivery: targeting the oligopeptide transporter on rabbit retina. J. Ocul. Pharmacol. Ther. 23 (4), 321–334.

Kao, H.J., Lin, H.R., Lo, Y.L., Yu, S.P., 2006. Characterization of pilocarpine-loaded chitosan/carbopol nanoparticles. J. Pharm. Pharmacol. 58 (2), 179–186.

Kao, J.C., Geroski, D.H., Edelhauser, H.F., 2005. Transscleral permeability of fluorescent-labeled antibiotics. J. Ocul. Pharmacol. Ther. 21, 1–10.

Kato, A., Kimura, H., Okabe, K., Okabe, J., Kunou, N., Ogura, Y., 2004. Feasibility of drug delivery to the posterior pole of the rabbit eye with an episcleral implant. Invest. Ophthalmol. Vis. Sci. 45, 238.

Kaufman, P., Alm, A., 2003. Adler's Physiology of the Eye, 10th ed. Mosby, St. Louis, MO.

Kaur, I., Garg, A., Singla, A., Aggarwal, D., 2004. Vesicular systems in ocular drug delivery: an overview. Int. J. Pharm 269 (1), 1–14.

Keister, J., Cooper, E., Missel, P., Lang, J., Hager, D., 1991. Limits on optimizing ocular drug delivery. J. Pharm. Sci. 80 (1), 50–53.

Kim, S., Lutz, R., Wang, N., Robinson, M., 2007. Transport barriers in transscleral drug delivery for retinal diseases. Ophthal. Res. 39, 244–254.

Klein, R., Klein, B.E., Lee, K.E., 1996. Changes in visual acuity in a population. The Beaver Dam Eye Study. Ophthalmology 103, 1169–1178.

Kleinmann, G., Larson, S., Huter, B., Stevens, S., Mamalis, N., Olson, R., 2007. Collagen shields as a drug delivery system for the fourth-generation fluoroquinolones. Ophthalmologica 221 (1), 51–56.

Kompella, U., Bandi, N., Ayalasomayajula, S., 2003. Subconjunctival nano- and microparticles sustain retinal delivery of budesonide, a corticosteroid capable of inhibiting VEGF expression. Invest. Ophthalmol. Vis. Sci. 44, 1192–1201.

Lee, S., Noonan, J., Geroski, D., Prausnitz, M., Edelhauser, H., 1999. Drug delivery through the sclera: the influence of thickness, hydration, and sustained release systems. Invest. Ophthalmol. Vis. Sci. 40, S83.

Lee, T., Robinson, J., 2001. Drug delivery to the posterior segment of the eye: some insights on the penetration pathways after subconjunctival injection. J. Ocul. Pharmacol. Ther. 17 (6), 565–572.

Lehr, C., Lee, Y., Lee, V.H.L., 1994. Improved ocular penetration of gentamicin by mucoadhesive polymer polycarbophil in the pigmented rabbit. Invest. Ophthalmol. Vis. Sci. 35 (6), 2809.

Lemarchand, C., Gref, R., Couvreur, P., 2004. Polysaccharide-decorated nanoparticles. Eur. J. Pharm. Biopharm. 58, 327–341.

Li, V., Robinson, J., 1989. Solution viscosity effects on the ocular disposition of cromolyn sodium in the albino rabbit. Int. J. Pharm. 53, 219–225.

Lin, H.R., Sung, K.C., 2000. Carbopol/pluronic phase change solutions for ophthalmic drug delivery. J. Control. Release 69 (3), 379–388.

Lin, H., Yu, S., Kuo, C., Kao, H., Lo, Y., Lin, Y., 2007. Pilocarpine-loaded chitosan–PAA nanosuspension for ophthalmic delivery. J. Biomater. Sci. Polym. Ed. 18 (2), 205–221.

Loutsch, J., Ong, D., Hill, J., 2003. Dendrimers: an innovative and enhanced ocular drug delivery system. In: Mitra, A. (Ed.), Ophthalmic Drug Delivery Systems, 2nd ed. Marcel Dekker, New York, Chapter 15, p. 467.

Mannermaa, E., Vellonen, K., Urtti, A., 2006. Drug transport in corneal epithelium and blood–retina barrier: emerging role of transporters in ocular pharmacokinetics. Adv. Drug Deliv. Rev. 58 (11), 1136–1163.

Miyazaki, S., Suzuki, S., Kawasaki, N., Endo, K., Takahashi, A., Attwood, D., 2001. In situ gelling xyloglucan formulations for sustained release ocular delivery of pilocarpine hydrochloride. Int. J. Pharm. 229, 29–36.

Motwani, S.K., Ahmad, F.J., Iqbal, Z., Talegaonkar, S., Khar, R.K., 2007. Gatifloxacin nanoparticles for ophthalmic delivery. Nanotech 2007, vol. 2, Technical Proceedings of the 2007 NSTI Nanotechnology Conference and Trade Show, Chapter 4, pp. 310–312.

Nanjawade, B., Manvi, F., Manjappa, A., 2007. In situ-forming hydrogels for sustained ophthalmic drug delivery. J. Control. Release 122 (2), 119–134.

Ogura, Y., 2001. Drug delivery to the posterior segments of the eye. Adv. Drug Deliv. Rev. 52, 1–3.

Park, K., Shalaby, W., Park, H., 1993. Biodegradable Hydogels for Drug Delivery. Technomic, Lancaster, PA.

Patri, A., Kukowska-Latallo, J., Baker, J., 2005. Targeted drug delivery with dendrimers: comparison of the release kinetics of covalently conjugated drug and non-covalent drug inclusion complex. Adv. Drug Deliv. Rev. 57, 2203–2214.

Patton, T., Robinson, J., 1975. Ocular evaluation of polyvinyl alcohol vehicle in rabbits. J. Pharm. Sci. 64, 1312–1316.

Pearson, P., Jaffe, G., Martin, D., Cordahi, G., Grossniklaus, H., Schmeisser, E., Ashton, P., 1996. Evaluation of a delivery system providing long-term release of cyclosporine. Arch. Ophthalmol. 114, 311–317.

Pecosky, D., Robinson, J., 1990. Bioadhesive Polymers and Drug Delivery in Polymers for Controlled Drug Delivery. CRC, Boca Raton, FL.

Pignatello, R., Bucolo, C., Ferrara, P., Maltese, A., Puleo, A., Puglisi, G., 2002. Eudragit RS100(r) nanosuspensions for the ophthalmic controlled delivery of ibuprofen. Eur. J. Pharm. Sci. 16 (1–2), 53–61.

Pitkanen, L., Ranta, V., Moilanen, H., Urtti, A., 2005. Permeability of retinal pigment epithelium: effects of permeant molecular weight and lipophilicity. Invest. Ophthalmol. Vis. Sci. 46, 641–646.

Poland, D., Kaufman, H., 1988. Clinical uses of collagen shields. J. Cataract Refract. Surg. 14, 492–495.

Prausnitz, M., Noonan, J., 1998. Permeability of cornea, sclera, and conjunctiva: a literature analysis for drug delivery to the eye. J. Pharm. Sci. 87 (12), 1479–1488.

Qi, H., Chen, W., Huang, C., Li, L., Chen, C., Li, W., Wu, C., 2007. Development of a poloxamer analogs/carbopol-based in situ gelling and mucoadhesive ophthalmic delivery system for puerarin. Int. J. Pharm. 337 (1–2), 178–187.

Raghava, S., Hammond, M., Kompella, U., 2004. Periocular routes for retinal drug delivery. Expert Opin. Drug Deliv. 1 (1), 99–114.

Ranta, V., Urtti, A., 2006. Transscleral drug delivery to the posterior eye: prospects of pharmacokinetic modeling. Adv. Drug Deliv. Rev. 58 (11), 1164–1181.

Reidy, J., Gebhardt, B., Kaufman, H., 1990. The collagen shield. A new vehicle for delivery of cyclosporin A to the eye. Cornea 9, 196–199.

Riordan-Eva, P., Whitcher, J., 2007. Vaughan & Asbury's General Ophthalmology, 17th ed. McGraw-Hill, New York.

Robinson, M., Lee, S., Kim, H., Kim, S., Lutz, R., Galban, C., Bungay, P., Yuan, P., Wang, N., Kim, J., Csaky, G., 2006. A rabbit model for assessing the ocular barriers to the transscleral delivery of triamcinolone acetonide. Exp. Eye Res. 82, 479–487.

Saettone, M., 2002. Progress and problems in ophthalmic drug delivery. Bus. Brief. Pharmatech. 161–171.

Saettone, M.F., Monti, D., Torracca, M.T., Chetoni, P., 1994. Mucoadhesive ophthalmic vehicles: evaluation of polymeric low-viscosity formulations. J. Ocul. Pharmacol. Ther. 10 (1), 83–92.

Sakurai, E., Nozaki, M., Okabe, K., Kunou, N., Kimura, H., Ogura, Y., 2003. Scleral plug of biodegradable polymers containing tacrolimus (FK506) for experimental uveitis. Invest. Ophthalmol. Vis. Sci. 44, 4845–4852.

Sanborn, G., Anand, R., Torti, R., Nightingale, S., Cal, S., Yates, B., Ashton, P., Smith, T., 1992. Sustained-release ganciclovir therapy for treatment of cytomegalovirus retinitis. Arch. Ophthalmol. 110, 188.

Saude, T., 1993. Ocular Anatomy and Physiology. Blackwell Science.

Schaeffer, H., Krohn, D., 1982. Liposomes in topical drug delivery. Invest. Ophthalmol. Vis. Sci. 22, 220–227.

Shek, P., Barber, R., 1987. Liposomes are effective carriers for the ocular delivery of prophylactics. Biochim. Biophys. Acta 902, 229–236.

Svenson, S., Tomalia, D., 2005. Dendrimers in biomedical applications – reflections on the field. Adv. Drug Deliv. Rev. 57, 2106–2129.

Tai, M., Cosar, C., Cohen, E., Rapuano, C., Laibson, P., 2002. The clinical efficacy of silicone punctal plug therapy. Cornea 21 (2), 135–139.

Vadnere, M., Amidon, G., Lindenbaum, S., Haslam, J.L., 1984. Thermodynamic studies on the gel–sol transition of some pluronic polyols. Int. J. Pharm. 22, 207–218.

Vandamme, Th., Brobeck, L., 2005. Poly(amidoamine) dendrimers as ophthalmic vehicles for ocular delivery of pilocarpine nitrate and tropicamide. J. Control. Release 102 (1), 23–38.

Vandervoort, J., Ludwig, A., 2007. Ocular drug delivery: nanomedicine applications. Nanomedicine 2 (1), 11–21.

Vasir, J.K., Tambwekar, K., Garg, S., 2003. Bioadhesive microspheres as a controlled drug delivery system. Int. J. Pharm. 255 (1–2), 13–32.

Velez, G., Whitcup, S., 1999. New developments in sustained release drug delivery for the treatment of intraocular disease. Br. J. Ophthalmol. 83, 1225–1229.

Velpandian, T., Gupta, S., Gupta, Y., et al., 1999. Ocular drug targeting by liposomes and their corneal interactions. J. Microencaps. 16, 243–250.

Wei, G., Xu, H., Ding, P.T., Li, S.M., Zheng, J.M., 2002. Thermosetting gels with modulated gelation temperature for ophthalmic use: the rheological and gamma scintigraphic studies. J. Control. Release 83 (1), 65–74.

Weissman, B., Lee, D., 1988. Oxygen transmissibility, thickness, and water content of three types of collagen shields. Arch. Ophthalmol. 106, 1706–1708.

Weyenberg, W., Vermeire, A., Remon, J., Ludwig, A., 2003. Characterization and in vivo evaluation of ocular bioadhesive minitablets compressed at different forces. J. Control. Release 89 (2), 329–340.

Yanagawa, A., Mizushima, Y., Komatsu, A., Horiuchi, M., Shirasawa, E., Igarashi, R., 1987. Application of a drug delivery system to a steroidal ophthalmic preparation with lipid microspheres. J. Microencaps. 4, 329–331.
Yoo, M.K., Cho, K.Y., Song, H.H., Choi, Y.J., 2005. Release of ciprofloxacin from chondroitin 6-sulfate-graft-Poloxamer hydrogel in vitro for ophthalmic drug delivery. Drug Dev. Ind. Pharm. 31 (4–5), 455–463.
Zimmer, A., Kreuter, J., 1995. Microspheres and nanoparticles used in ocular delivery systems. Adv. Drug Deliv. Rev. 16 (1), 61–73.

CHAPTER 8

Nasal Delivery of Micro- and Nano-encapsulated Drugs

Joe McDonough, Hong Dixon* *and* **Michael Ladika**
Southwest Research Institute, 6220 Culebra Rd., San Antonio, TX 78238, USA

Contents

8.1 Introduction	193
8.2 Physiology of the nose	198
8.3 Drug delivery formulation	199
8.4 Micro- and nano-encapsulation	201
8.4.1 PLGA	201
8.4.2 Chitosan	203
8.4.3 Other polymers	205
8.5 Summary	206
References	206

8.1 INTRODUCTION

The benefits of the nasal route of drug delivery are being increasingly recognized and reaped. Current share of the nasally delivered drug products is almost $2 billion of the domestic pharmaceutical market, up from $1 billion in 1997 (Langer, 1998; Law, 2000), suggesting a new trend in drug delivery route that has obvious benefits. Many pharmaceutical products have been developed for the nasal delivery route that now take advantages of the benefits, including convenience and ease of use, rapid onset of action, avoidance of immediate hepatic metabolism, possibilities to minimize systemic exposure, higher bioavailability, and rapid brain access through the nerves that perforate the cribriform plate into the olfactory bulb. Some of the products that have successfully leveraged the benefits of the nasal route are listed in Table 8.1. These products are usually delivered as a nasal spray, though some offer the options of direct liquid or powder presentation. Recently, the issues associated with this route of administration were thoroughly reviewed (Illum, 2003).

* Author to whom correspondence should be addressed.

Handbook of Non-Invasive Drug Delivery Systems
ISBN 9780815520252

Table 8.1 Commercially available nasal formulations*

Trade Name	Active	Delivery Mode	Marketing Company	Indication	Device Type/ Description
Miacalcin® NS	Salmon Calcitronin	Nasal Spray	Novartis	Postmenopausal Osteoporosis	3.7 mL fill clear glass bottle fitted with a screw-on metered nasal spray pump
Astelin®	Azelastine	Nasal Spray	Medpointe Pharmaceuticals	Rhinorrhea	HDPE bottle fitted with a metered-dose spray pump unit with blue saftey cap and blue plastic dust cover
DDAVP®	Desmopressin Acetate	Rhinal Tube or Nasal Spray	Sanofi-Aventis (Rhone-Poulenc Rorer)	Central Cranial Diabetus Insidpidus	5 mL bottle with Compression Spray Pump or Rhinal Tube
Suprefact®	Buserelin Acetate	Nasal Spray	Aventis	Prostate Cancer	10 mL amber glass bottle with metered dose pump (nebulizer) – mechanical action/ no propellant

Imitrex®	Sumatriptan	Nasal Spray	GlaxoSmithKline	Migraine	Single use mechanical plunger atomizer
Synarel®	Nafarelin	Nasal Spray	Searle	Endometriosis	Metered spray pump
Migranal®	Dihyrdro-ergotamine	Nasal Spray	Novartis Pharmaceuticals Inc.	Migraine	3.5 mL amber glass bottle with metered spray pump
Stadol NS®	Butorphanol Tartrate	Nasal Spray	Mylan	Pain Management	2.5 mL amber glass bottle with metered spray pump (mechanical)
Flumist™	Attenuated Flu Virus	Nasal Spray	MedImmune Inc. and Wyeth	Flu	Single use sprayer – plunger type
Nasonex®	Mometasone Furoate	Nasal Spray	Schering Corp	Allergic Rhinitis	White/HDPE bottle with metered-dose manual spray pump
Flonase®	Fluticasone Propionate	Nasal Spray	GlaxoSmithKline	Allergic Rhinitis	Amber glass bottle fitted with a white metering atomizing pump w/nasal adapter

(Continued)

Table 8.1 Commercially available nasal formulations*—Cont'd

Trade Name	Active	Delivery Mode	Marketing Company	Indication	Device Type/ Description
Beconase AQ®	Beclomethasone	Nasal Spray	GlaxoSmithKline	Allergic and Non-allergic Rhinitis	Amber gloss bottle fitted with a metering atomizing pump w/nasal adapter
Nasarel®	Flunisolide	Nasal Spray	IVAX Laboratories	Seasonal Rhinitis	White, HDPE spray bottle fitted with a meter pump, nasal adapter and a protective cap
Rhinocort AQUA®	Budesonide	Nasal Spray	Astrazeneca LP	Allergic Rhinitis	Amber glass bottle with a metered-dose pump spray and a green protection cap
Bactroban Nasal®	Mupirocin	Nasal Ointment	GlaxoSmithKline	*Staph. Aureus* Nasal Colonization	22 gram tube

Atrovent® Nasal	Ipratropium Bromide	Nasal Spray	Boehringer Ingelheim	Rhinorrhea	White HDPE bottle fitted with a white and clear metered nasal spray pump, a green safety clip to prevent accidental discharge of the spray, and a clear plastic dust cap
Vicks® Sinex	Oxymetazoline HCl	Nasal Spray	Procter and Gamble	Sinusitis	Squeeze bottle nasal spray dispenser
Nasacort AQ®	Tramcinolone Acetonide	Nasal Spray	Sanofi Aventis	Allergic Rhinitis	White HDPE container with a metered-dose pump unit, white nasal adapter
Nasalcrom®	Cromolyn Sodium	Nasal Spray	Pharmacia Upjohn	Allergic Rhinitis	Metered-dose nebulizer

* Abbreviation: HDPE, high-density polyethylene.

8.2 PHYSIOLOGY OF THE NOSE

There are many challenges to successful nasal delivery that arise due to the complexity of the nose's physiology and function (Chien et al., 1989). Physiology of the nose is shown in Figure 8.1. The nose provides a means of warming, humidifying, and filtering the inhaled air before it reaches the lungs. It serves these functions by passing inhaled and expired air over the large surface area of the nasal cavity (approximately 180 cm^2).

More eloquently, the main function of the nose is to be a first line of defense, because the mucus-coated membranes lining the nasal cavity contain immunoglobulin A, proteases, esterases, and many other enzymes that help break down contaminants before they can be inspired and enter the body. Unfortunately, these defensive systems also serve to prevent the delivery of drugs (Hochestrasser, 1983; Scott and McKenna, 1985; Lee, 1988; Reed, 1993). Airflow into the nasal cavity comes in contact with the olfactory nerves that penetrate the cribriform plate above the concha. The olfactory neural bundle provides the sense of smell; consequently, it is also a potential route for delivering therapeutic agents directly to the brain via intra- or extraneuronal pathways (Thorne and Frey, 2001).

The nasal cavity is segmented by the superior, middle, and inferior concha (turbinates), which have ciliated epithelial linings. These cilia, 4–6 μm in length, beat with a frequency of the order of 1000 strokes/min, propelling

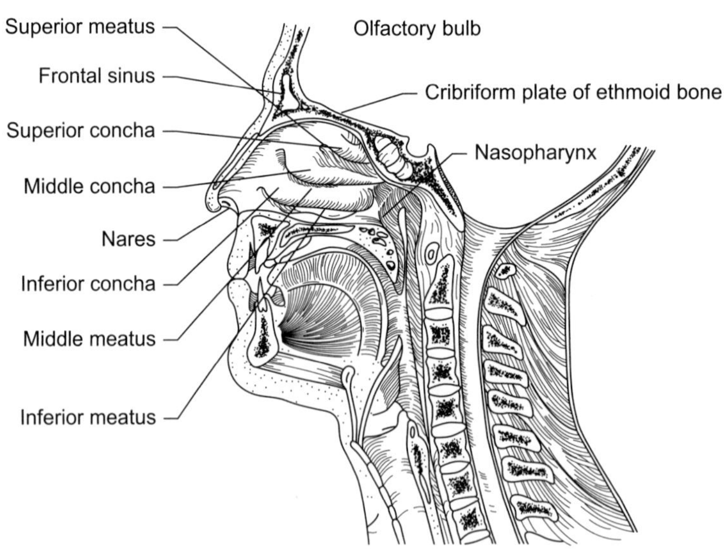

Figure 8.1 Physiology of the nasal cavity.

mucus secreted from the secretory glands and foreign contaminants to the nasopharynx, where it drains further down the esophagus and the gastrointestinal system. These cilia propel the mucus through the nose at a rate of 5–6 mm/min (Mygind and Anggard, 1983). The secretion volumes vary from person to person, but it is generally accepted that 1 liter/day is secreted into the nose then swept posterior. This secretion rate can more than double when the nose is inflamed. The volume of the nasal cavity is roughly 16–19 cm^3. Therefore, a mucosal turnover of more than 2.5 times/hour can be assumed with mucus layer renewal occurring every 15 min (Chien and Chang, 1987). Thus, most of a drug dosed in the nasal cavity generally will be carried completely away in 10–80 min (Washington and McGlashan, 2000). Complicating matters further, the patency of each nasal passageway changes with time, creating another challenge to delivering drugs. This cycle can last from 2 to 7 hours and determines the airflow into each side of the nose during inspiration and exhalation (Lund, 1996; Washington and McGlashan, 2000). The temperature in the nasal cavity is generally 36°C, and the pH generally varies between 5.5 and 6.5 in adults and 5.0 and 6.7 in infants and young children, and tends to become more alkaline at lower temperatures (Hehar et al., 1999). At lower temperatures, where alkaline pHs can occur, the lysozyme activity decreases, which can increase survival of bacteria in the mucus (Jackson and Turner, 1966).

8.3 DRUG DELIVERY FORMULATION

Regardless of its physiological challenges, the nasal route offers advantageous administration compared to subcutaneous routes. Drug administration by the nasal route has been examined by many investigators because this route allows for high absorption of low-molecular-weight, hydrophobic drugs, avoidance of first-pass effects, and ease in administration by patients. Mucosal delivery is also a well-documented and highly effective route for the stimulation of local and systemic immunity.

Because of the complex physiological function of the nose, which is balanced to protect against infection while allowing for respiration and sensing, drug delivery formulations must be appropriately formulated so as not to upset this delicate stasis. One measure of a drug delivery technology's compatibility with nose and nasal cavity is the response of the cilia after dosing. If the cilial beat remains constant or if the cilia are not exfoliated by the dosing activity, the dose is compatible. Upsetting the cilia by exfoliation or perturbing the beat frequency with an incompatible formulation partially compromises this defense system and creates mucosal irritation. The cilial beat frequency or condition can be measured by scanning electron microscopy or by optical microscopy *in vitro* (Dalhanm and Rylander, 1962; Jian et al., 1995; Martin et al., 1999). Additionally, the presentation of mucosal irritation is also an indication of bioincompatibility. Mucosal irritation can be observed in larger test species, such as the beagle, through visual observation (Ramanthan et al., 1998). In smaller species, the visualization is more difficult, so the observation can be made through histopathological means (Young, 1981). This method was carried out successfully in our laboratories for studies using ethopropazine and promethazine (McDonough et al., 2004a, b). With this technique, nasal irritation was recognizable by the absence of nasal epithelium or leukocyte infiltration.

Delivery of nonbiocompatible, unprotected drugs, and certain excipients directly to the delicate nasal epithelium is likely to upset delicate nasal stasis, resulting in nasal irritation. Therefore, drug formulations must be compatible with nasal physiology if the nasal route of administration is to be used successfully. Prudent use of absorption enhancers as part of the nasal formulation decreases the risk of upsetting the stasis of the nasal physiology (Martin et al., 1998). By facilitating transport through the nasal epithelium, the effective drug concentration on the cilia can remain low enough such that no damage occurs. Furthermore, with rapid absorption, the drug is less likely to be swept back to the nasal pharynx and the esophagus. Absorption enhancement, however, increases the absorption of the intended as well as the unintended components and must be used judiciously.

Vaccination with antigen alone at mucosal sites generally induces weak mucosal and systemic immune responses and protection, thereby requiring the need for a mucosal adjuvant. Nasal vaccines also have shown promise to elicit a more protective immune response than intramuscular injection of vaccines using equivalent doses. Most side-effects observed with mucosally delivered experimental vaccines seem to be associated with the chosen adjuvant rather than with the vaccine itself.

There is a possibility for nasally administered therapeutic agents to enter into the brain via olfactory nerves and provide extra benefits for this administration route over other delivery routes in treating central nervous system disorders. The mechanism by which large molecule plasmid DNA is substantially absorbed in the brain is not clear; perhaps it is through the interaction of plasmid DNA and the nasal epithelial membranes, therefore transiently affecting the tight junctions. The localization of plasmid to the brain appears to be higher after intranasal administration as compared to intravenous administration. The intranasal route providing the substantial absorption of DNA might have potential as a promising non-invasive delivery route of DNA vaccines (Oh et al., 2001).

A recent review evaluated various mouse models for the study of mucosal vaccination against otitis media, a common infectious disease in humans (Sabirov and Metzger, 2008). Murine antibody responses are similar to those in humans and, therefore, the mouse model is useful for testing hypotheses regarding immune responses. The authors discussed the mechanisms for mucosal immunization at one site leading to the secretion of specific IgA antibodies at other mucosal tissues and the preferential distribution of responding cells to sites that are anatomically related to the inductive sites that received the original antigenic stimulation. Although various rodents have been used in the induction of otitis media, the mouse is a more advantageous study animal because mice are inexpensive and there is an extensive genetic toolkit available for manipulating the mouse genome. The mouse model has been used to assess the role of T- and B-cell-dependent mechanisms in protection against respiratory pathogens following intranasal exposure to either live or killed whole-cell vaccine. Several mucosal adjuvants and mucosal delivery systems were briefly described (Sabirov and Metzger, 2008).

8.4 MICRO- AND NANO-ENCAPSULATION

One method to prepare safe nasal formulations is to use micro- or nano-encapsulation. The size of the particles is a critical factor that affects their interaction and transport across mucosal surfaces. In general, particles smaller than 1 μm are more efficient in penetrating through mucosal barriers. In addition, the surface compositions of the particles (their surface charge and hydrophobicity or hydrophilicity) are known to influence the interaction of the nanoparticles with the biological environment.

With encapsulation and controlled release, low effective concentrations of active agents can be achieved in nasal epithelium and increase the likelihood of the formulation being biocompatible. The slow release can result in reduced exposure of the mucosal epithelial cells, especially if absorption enhancers are dosed with the microcapsule so that the active drug can be absorbed as quickly as it is released. Two biocompatible and biodegradable polymers, namely poly(D,L-lactide-co-glycolide) (PLGA) and chitosan, have been extensively used in encapsulation of active substances for nasal drug delivery.

8.4.1 PLGA

PLGA polymers have been widely used in the encapsulation of peptides and enhanced immune responses have been shown in animal studies with these microparticles (Carcaboso et al., 2004; Stanley et al., 2004). PLGA particles appear to be efficient immunological adjuvants and the nasal route is a safe mode of vaccination, suitable for application to a variety of vaccines. However, micro- or nanoparticles made of PLGA often need to be optimized to preserve the biological activity of the encapsulated plasmid DNA and improve the interaction with the biological environment. Two major concerns with PLGA polymers are preventing protein deactivation and aggregation during the harsh encapsulation process and suppressing protein degradation during release caused by the increased acidity due to degrading PLGA polymer. There are several approaches to mitigate the deactivation problems due to inherent hydrophobicity and free carboxylic groups of PLGA. One is the use of poly(ethylene glycol) (PEG)-coated PLGA particles (Vila et al., 2004). The other is to use composites of PLGA with either polyoxyethylene derivatives or polyanhydrides (Li et al., 2005; Csaba et al., 2006). Yet another is to modify PLGA by grafting it to cyclodextrins (H. Gao et al., 2006).

Intranasal administration of PLGA microparticles (about 1 μm size) encapsulated with synthetic malaria vaccine SPf66 in Balb/c mice elicited higher antibody levels compared to the conventional alum adjuvant and to the administration of the particles by either the subcutaneous or the oral route. Systemic immune responses were demonstrated without requiring boosting doses and the responses have been associated with protective immunity in malaria (Carcaboso et al., 2004). Also, sheep immunized with PLGA micro- and nanoparticles containing toxoplasma antigen were shown to produce enhanced levels of both local and systemic antibody and showed increased cellular immune responses after intranasal immunization (Stanley et al., 2004).

PEG coating helps prevent the aggregation of the nanoparticles upon contact with the nasal mucosa and facilitates the transport across the nasal epithelia. In addition, the transport of nanoparticles made by double emulsion was more pronounced compared to that of the nanoparticles prepared by the nanoprecipitation method, possibly due to the larger particle size and lower PEG coating of the latter sample. It was found that the PLA–PEG particles can cross the rat mucosa and that this transport is greatly affected not only by the size of the particles, but also by PEG coating density (Vila et al., 2004).

Nanoparticles in the 100–200 nm size range consisting of PLGA and polyoxyethylene blends exhibiting sustained release of the encapsulated plasmid were able to enter the nasal epithelium by a transcellular pathway throughout the depth of the tissue (about 25 μm). These nanoparticles elicited higher systemic immune responses than those corresponding to the naked plasmid and the maximum response was observed at 4 weeks post-administration in mice (Csaba et al., 2006). Bioadhesive properties of PLGA microparticles can be enhanced by blending PLGA with polyanhydrides. PLGA and poly[1,3 bis(*p*-carboxy-phenoxy)propane-co-*p*-(carboxyethyl-formamido)benzoic anhydride] (P(CPP:CEFB)) composite microspheres were used to microencapsulate α-cobratoxin by an emulsion solvent evaporation method. The microparticles (about 25 μm size) were intranasally administered to rats; the presence of P(CPP:CEFB) in the microspheres increased their residence time at the surface of the rat nasal mucosa. Compared with PLGA microspheres and free α-cobratoxin, PLGA/P(CPP:CEFB) microspheres showed significant increase in the intensity and duration of analgesic effect after intranasal administration and exhibited only mild toxicity to rat nasal mucosa (Li et al., 2005).

Cyclodextrins can serve as absorption enhancers through mucous membrane in nasal administration. Biodegradable amphiphilic copolymers have been synthesized by conjugating poly(D,L-lactic acid) (PLA) on to ethylenediamino- or diethylenetriamino-bridged bis-β-cyclodextrins. Bovine serum albumin (BSA) encapsulated inside these copolymers showed controlled release without significantly aggregating or degrading during *in vitro* release tests (H. Gao et al., 2006).

An advantage of nasal delivery is that it provides increased potential for access to the central nervous system (CNS). To improve the absorption of nanoparticles in the CNS following nasal administration, wheatgerm agglutinin (WGA) was conjugated to the surface of PEG–PLA nanoparticles. WGA specifically binds to *N*-acetyl-D-glucosamine and sialic acid, which are abundant in the nasal cavity. The surfaces of the nanoparticles of PEG–PLA containing vasoactive intestinal peptide (VIP), a 28-amino-acid neuropeptide of the glucagon–secretin family, were also modified with WGA and the nanoparticles were evaluated for their ability to mediate drug delivery to the brain following intranasal administration. After intranasal administration, it was shown that WGA modification enhanced the brain uptake of the drug entrapped in the nanoparticles in rats (X. Gao et al., 2006, 2007).

8.4.2 Chitosan

Among bioadhesive materials proposed for nasal delivery, chitosan has received particular interest because of its favorable biological properties, low toxicity, and biodegradability. Chitosan is known to be bioadhesive and has been used as an absorption enhancer as well as an adjuvant to enhance both systemic and local immune responses against a variety of antigens after nasal administration. It significantly increases the residence time of the formulations in the nasal cavity. After nasal administration, drugs can be transported directly from the nasal cavity to the CNS via the olfactory epithelium and/or the trigeminal nerve system. An *in vitro* cell permeability study indicated that the chitosan has a larger effect on opening the tight junctions between the cells than the pectins for hydrophilic drugs (Charlton et al., 2007).

Chitosan microparticles loaded with *Bordetella bronchiseptica* multiple antigens have immune-stimulating activities both *in vitro* and *in vivo*. But the microparticles showed limited storage stability. Pluronic® F127 was added to the formulation for its protein-stabilizing and high solubilizing properties to overcome the aggregation of the chitosan microparticles and to enhance the intranasal vaccine delivery. The antibody responses elicited by nasal immunization with chitosan/F127 microparticles were higher than those achieved by chitosan microparticles in mice (Kang et al., 2007). Targeted delivery based on mannosylated chitosan microspheres (MCMs) to macrophages with the hope of improvement of receptor-mediated uptake upon binding macrophage mannose receptors (MMRs) did not significantly increase protective immunity in mice (Jiang et al., 2008).

Chitosan can function as an adjuvant itself, as well as an enhancer to secondary adjuvants that are effective mucosally with *Helicobacter pylori* urease. The combination of chitosan and muramyl dipeptide (MDP) following intranasal administration resulted in elevated serum IgG responses and strongly proliferating memory splenocytes releasing elevated quantities of Th1 responses in mice (Moschos et al., 2005). It was also found that the immunogenicity of the vaccine was enhanced by formulation with chitosan when cross-reacting material $(CRM)_{197}$, a nontoxic mutant of diphtheria toxin, was delivered nasally in adult volunteers. The responses were enhanced following the second intranasal immunization (McNeela et al., 2004). In a clinical trial of inactivated, trivalent influenza virus vaccine administered with chitosan glutamate as a nasal spray, the serum HI antibody responses obtained were similar following administration of the non-adjuvanted, commercial influenza vaccine by the intramuscular route. Contrary to earlier literature results (Read et al., 2005), the second nasal dose did not enhance further serum HI antibody responses.

A modified chitosan, *N*-trimethylchitosan (TMC), that is soluble at physiological pH, was developed for its mucoadhesive properties and absorption-enhancing effects at neutral pH. A monovalent influenza subunit vaccine was loaded in TMC nanoparticles and administered intranasally in mice; the TMC nanoparticles were able to cross the nasal mucosal layer and elicited strong systemic as well as local antibody responses. Intranasally administered antigen-loaded TMC nanoparticles can induce an S-IgA response at the site of influenza virus entry (Amidi et al., 2006, 2007).

To prolong the contact time of the drug with the nasal mucosa and enhance its permeation into the nasal mucosa, a nasal gel can be used. Like encapsulation, nasal gels also release drug slowly and continuously, but their high viscosity makes it hard to apply accurate dosage and distribute it uniformly in the nasal cavity. A new thermosensitive hydrogel was developed from *N*-[(2-hydroxy-3-trimethylammonium)propyl]chitosan chloride (HTCC) and PEG with a small amount of α,β-glycerophosphate (α,β-GP). The formulation has a thermal transition temperature around 37°C, changing from a liquid at ambient temperature to a viscous gel at body temperature. The enhancement of absorption of fluorescein isothiocyanate (FITC)-labeled insulin in rat nasal cavity by this formulation has been demonstrated. The hydrogel formulation decreased the blood glucose concentration to about 50% of initial blood glucose concentration for 4–5 h after administration (Wu et al., 2007).

8.4.3 Other polymers

Different strategies have been explored to increase the bioavailability following nasal administration for hydrophilic and high-molecular-weight drugs. Besides PLGA and chitosan, other polymer systems have been developed for nasal drug delivery. For example, aminated gelatin, a positively charged gelatin derivative, was shown to significantly increase the nasal absorption of dextran and insulin with slight perturbation to rat nasal mucosa. Dry powders of crosslinked aminated gelatin microparticles with insulin administered nasally in rats showed enhanced insulin absorption effect; the enhanced absorption was attributed to the temporary dehydration of the epithelium membrane and opening of the tight junctions, as well as the positive charge of the microparticles and its mucoadhesive properties (Wang et al., 2006). Ethylcellulose was also used as an inert excipient with a mucolytic agent, *N*-acetyl-L-cysteine (NAC), to study the nasal absorption of a powder form of salmon calcitonin (SCT) in rats and dogs. The powder formulation with NAC gave significant nasal absorption of SCT in both animal models and gave a quicker absorption rate than subcutaneous administration of SCT. The powder SCT/NAC/ethylcellulose formulation did not induce irritation or histological damage to the nasal membrane in rabbits (Matsuyama et al., 2006). A nasal microparticulate delivery system based on the thiomer polycarbophil-cysteine (PCP-Cys) in combination with the permeation mediator glutathione (GSH) was developed for delivery of human growth hormone (hGH). The intranasal administration of the microparticulate formulation resulted in threefold increased bioavailability compared to that of the formulation without GSH (Leitner et al., 2004).

8.5 SUMMARY

Even though the challenges to develop a successful nasal formulation can be onerous, appropriate use of absorption enhancement, adjuvants, mucosal adhesives, and encapsulation can help ameliorate the problems. Using the controlled-release aspects of the encapsulated drug ensures low concentrations of the drug at the nasal epithelium and is a consistent approach to develop biocompatible and effective nasal formulations. The nasal route has been demonstrated as a safe and promising new needle-free mode of delivering a variety of small-molecule drugs as well as for vaccination against respiratory diseases with DNA-based vaccines. This route of administration also has potential for delivering therapeutics to the brain; therefore, when delivering drugs via the nasal route, only nontoxic mucosal adjuvants are to be used to avoid neurotoxic potential.

REFERENCES

Amidi, M., Romeijn, S.G., Borchard, G., Junginger, H.E., Hennink, W.E., Jiskoot, W., 2006. Preparation and characterization of protein-loaded *N*-trimethyl chitosan nanoparticles as nasal delivery system. J. Control. Release 111, 107–116.

Amidi, M., Romeijn, S.G., Verhoef, J.C., Junginger, H.E., Bungener, L., Huckriede, A., Crommelin, D.J.A., Jiskoot, W., 2007. *N*-Trimethyl chitosan (TMC) nanoparticles loaded with influenza subunit antigen for intranasal vaccination: biological properties and immunogenicity in a mouse model. Vaccine 25, 144–153.

Carcaboso, A.M., Hernández, R.M., Igartua, M., Rosas, J.E., Patarroyo, M.E., Pedraz Potent, J.L., 2004. Long lasting systemic antibody levels and mixed Th1/Th2 immune response after nasal immunization with malaria antigen loaded PLGA microparticles. Vaccine 22, 1423–1432.

Charlton, S.T., Davis, S.S., Illum, L., 2007. Evaluation of bioadhesive polymers as delivery systems for nose to brain delivery: in vitro characterisation studies. J. Control. Release 118, 225–234.

Chien, Y.W., Chang, S., 1987. Intranasal drug delivery for systemic medications. Crit. Rev. Ther. Drug Carrier Syst. 4, 67–194.

Chien, Y.W., Chang, S., Su, K., 1989. Nasal Systemic Drug Delivery, 39th ed. Marcel Dekker, New York.

Csaba, N., Sánchez, A., Alonso, M.J., 2006. PLGA:poloxamer and PLGA:poloxamine blend nanostructures as carriers for nasal gene delivery. J. Control. Release 113, 164–172.

Dalhanm, T., Rylander, R., 1962. Frequency of ciliary beat measured with a photo-sensitive cell. Nature 196, 592–593.
Gao, H., Yang, Y.-w., Fan, Y.-g., Ma, J.-b., 2006. Conjugates of poly(DL-lactic acid) with ethylenediamino or diethylenetriamino bridged bis(β-cyclodextrin)s and their nanoparticles as protein delivery systems. J. Control. Release 112, 301–311.
Gao, X., Tao, W., Lu, W., Zhang, Q., Zhang, Y., Jiang, X., Fu, S., 2006. Lectin-conjugated PEG–PLA nanoparticles: preparation and brain delivery after intranasal administration. Biomaterials 27, 3482–3490.
Gao, X., Wu, B., Zhang, Q., Chen, J., Zhu, J., Zhang, W., Rong, Z., Chen, H., Jiang, X., 2007. Brain delivery of vasoactive intestinal peptide enhanced with the nanoparticles conjugated with wheat germ agglutinin following intranasal administration. J. Control. Release 121, 156–167.
Hehar, S.S., Masan, J.D.T., Stephen, A.B., 1999. Twenty-four hour ambulatory nasal pH monitoring. Clin. Otolaryngol. 24, 24–25.
Hochestrasser, K., 1983. Proteinases and their inhibitors in human nasal mucus. Rhinology 21, 217.
Illum, L., 2003. Nasal drug delivery: possibilities, problems and solutions. J. Control. Release 87, 187–198.
Jackson, R.T., Turner, J.S., 1966. Some observations on nasal pH. Arch. Otolaryngol. 84, 446.
Jian, X.G., Cui, J.B., Fang, X.L., 1995. Toxicity of drugs on nasal mucocilia and the method of its evaluation. Acta Pharm. Sinica 30, 848–853.
Jiang, H.-L., Kang, M.L., Quan, J.-S., Kang, S.G., Akaike, T., Yoo, H.S., Cho, C.-S., 2008. The potential of mannosylated chitosan microspheres to target macrophage mannose receptors in an adjuvant-delivery system for intranasal immunization. Biomaterials 29, 1931–1939.
Kang, M.L., Jiang, H.-L., Kang, S.G., Guo, D.D., Lee, D.Y., Cho, C.-S., Yoo, H.S., 2007. Pluronic® F127 enhances the effect as an adjuvant of chitosan microspheres in the intranasal delivery of *Bordetella bronchiseptica* antigens containing dermonecrotoxin. Vaccine 25, 4602–4610.
Langer, R., 1998. Drug delivery and targeting. Nature 392, 5–10.
Law, J., 2000. Not to be sniffed at. Scrip 5, 35–38.
Lee, V.H.L., 1988. Enzymatic barriers to peptide and protein absorption. CRC Crit. Rev. Ther. Drug Carrier Syst. 5, 69–97.
Leitner, V.M., Guggi, D., Krauland, A.H., Bernkop-Schnqrch, A., 2004. Nasal delivery of human growth hormone: in vitro and in vivo evaluation of a thiomer/glutathione microparticulate delivery system. J. Control. Release 100, 87–95.
Li, Y., Jiang, H.L., Zhu, K.J., Liu, J.H., Hao, Y.L., 2005. Preparation, characterization and nasal delivery of α-cobrotoxin-loaded poly(lactide-co-glycolide)/polyanhydride microspheres. J. Control. Release 108, 10–20.
Lund, V.J., 1996. Nasal physiology: neurochemical receptors, nasal cycle and ciliary action. Allergy Asthma Proc. 17, 179–184.
Martin, E., Schipper, N.G.M., Verhoef, J.C., Merkus, F.W.H.M., 1998. Nasal mucociliary clearance as a factor in nasal drug delivery. Adv. Drug Deliv. Rev. 29, 13–38.
Martin, E., Verhoef, J.C., Spies, F., 1999. The effect of methylated beta-cyclodextrin on the tight junctions of the rat nasal respiratory epithelium: electron microscope and confocal laser scanning microscope visualization studies. J. Control. Release 57, 205–213.
Matsuyama, T., Morita, T., Horikiri, Y., Yamahara, H., Yoshino, H., 2006. Improved nasal absorption of salmon calcitonin by powdery formulation with *N*-acetyl-L-cysteine as a mucolytic agent. J. Control. Release 115, 183–188.
McDonough, J.A., Persyn, J., Nino, J., 2004a. Mucosal delivery of cytotoxic agents. Proc. Control. Release Society, Abstract 318, June.
McDonough, J.A., Putcha, L., Persyn, J., Nino, J., 2004b. Controlled nasal delivery of promethazine hydrochloride, a motion sickness medication. Proc. Control. Release Society, Abstract 295, June.

McNeela, E.A., Jabbal-Gill, I., Illum, L., Pizza, M., Rappuoli, R., Podda, A., Lewis, D.J.M., Mills, K.H.G., 2004. Intranasal immunization with genetically detoxified diphtheria toxin induces T cell responses in humans: enhancement of Th2 responses and toxin-neutralizing antibodies by formulation with chitosan. Vaccine 22, 909–914.

Moschos, S.A., Bramwell, V.W., Somavarapu, S., Alpar, H.O., 2005. Comparative immunomodulatory properties of a chitosan–MDP adjuvant combination following intranasal or intramuscular immunization. Vaccine 23, 1923–1930.

Mygind, N., Anggard, A., 1983. Anatomy and physiology of the nose: pathophysiologic alterations in allergic rhinitis. Clin. Rev. Allergy 2, 173–188.

Oh, Y.-K., Kim, J.-P., Hwang, T.S., Ko, J.J., Kim, J.M., Yang, J.-S., Kim, C.-K., 2001. Nasal absorption and biodistribution of plasmid DNA: an alternative route of DNA vaccine delivery. Vaccine 19, 4519–4525.

Ramanthan, R., Geary, R., Bourne, D.W., Putcha, L., 1998. Bioavailability of intranasal promethazine dosage forms in dogs. Pharm. Res. 38 (1), 35–39.

Read, R.C., Naylor, S.C., Potter, C.W., Bond, J., Jabbal-Gill, I., Fisher, A., Illum, L., Jennings, R., 2005. Effective nasal influenza vaccine delivery using chitosan. Vaccine 23, 4367–4374.

Reed, C.J., 1993. Drug metabolisim in the nasal cavity: relevance to toxicology. Drug Metab. Rev. 25, 173–205.

Sabirov, A., Metzger, D.W., 2008. Mouse models for the study of mucosal vaccination against otitis media. Vaccine 26, 1501–1524.

Scott, W.T., McKenna, M.J., 1985. Hydrolysis of several glycol ether acetates and acrylate esters by nasal mucosal carboxylesterase in vitro. Fundam. Appl. Toxicol. 5, 399.

Stanley, A.C., Buxton, D., Innes, E.A., Huntley, J.F., 2004. Intranasal immunisation with Toxoplasma gondii tachyzoite antigen encapsulated into PLG microspheres induces humoral and cell-mediated immunity in sheep. Vaccine 22, 3929–3941.

Thorne, R.G., Frey, W.H., 2001. Delivery of neurotrophic factors to the central nervous system. Clin. Pharmacokinet. 40, 907–946.

Vila, A., Gill, H., McCallion, O., Alonso, M.J., 2004. Transport of PLA–PEG particles across the nasal mucosa: effect of particle size and PEG coating density. J. Control. Release 98, 231–244.

Wang, J., Tabata, Y., Morimoto, K., 2006. Aminated gelatin microspheres as a nasal delivery system for peptide drugs: evaluation of in vitro release and in vivo insulin absorption in rats. J. Control. Release 113, 31–37.

Washington, N., McGlashan, J.A., 2000. The effect of nasal patency on the clearance of radiolabeled saline in healthy volunteers. Pharm. Res. 17 (6), 733–736.

Wu, J., Wei, W., Wang, L.-Y., Su, Z.-G., Ma, G.-H., 2007. A thermosensitive hydrogel based on quaternized chitosan and poly(ethylene glycol) for nasal drug delivery system. Biomaterials 28, 2220–2232.

Young, J.T., 1981. Histopathologic examination of rat nasal cavity. Fundam. Appl. Toxicol. 1, 309–312.

CHAPTER 9

Pulmonary Drug Delivery

Mohammed Ali *

Assistant Professor, Department of Technology, Jackson State University, 1400 John R. Lynch Street, P.O. Box 18480, Jackson, MS 39217, USA; email: mohammed.ali@jsums.edu

Contents

9.1 General introduction 209
9.1.1 Non-invasive drug delivery 209
9.1.2 The human respiratory system 211
9.1.3 Physical properties of drug aerosol particles 212
9.2 Pulmonary drug deposition mechanisms 214
9.3 Pulmonary drug delivery methods 218
9.3.1 Metered-dose inhalers 219
9.3.2 Dry powder inhalers 225
9.3.3 Nebulizers 227
9.3.4 Comparable and contrasting features of MDIs, DPIs, and nebulizers 229
9.4 Analytical methods 231
9.4.1 Non-invasive scintigraphic techniques 232
9.4.2 Inertial impaction 233
9.4.3 Optical microscopy 237
9.4.4 Laser-induced tools 237
9.5 Challenges and advantages 239
9.6 Conclusions 241
References 242

9.1 GENERAL INTRODUCTION

9.1.1 Non-invasive drug delivery

Non-invasive drug delivery through the pulmonary airway tree has been a long-standing pharmaceutical objective. The history of inhalation therapy can be traced back to when the leaves of the *Atropa belladonna* plant were smoked in the Indian subcontinent as a cough suppressant 4000 years ago (Grossman, 1994). This method of treatment of various respiratory infections and tuberculosis was discovered in Europe (UK) in the seventeeth century (O'Callaghan et al., 2002). An aromatic plant such as datura (*Stramonium*) tobacco mixture in Potter's asthma cigarette is one of the best-

* Address for correspondence: 106 Somerset Cove, Madison, MS 39110, USA.

Handbook of Non-Invasive Drug Delivery Systems ISBN 9780815520252

known examples in this regard. In the early nineteenth century, inhaled vapors were replaced by wet aerosols, and the first atomizer, a device for the pulverization of liquids, was constructed in 1829, followed by a portable device in 1860. A century later in 1938, the handheld bulb nebulizer became popular (Grossman, 1994). During the mid-twentieth century inhaler technology rapidly improved and diverged. The introduction of electric pumps enabled continuous airflow through jet nebulizers, which combined with baffles and reduced the width of the size distribution of the aerosols. The second breakthrough in inhalation therapy took place with the introduction of the first pressurized metered-dose inhaler (MDI; Medihaler[†] by Riker Pharma) utilizing chlorofluorocarbon propellants in 1956. Nearly 15 years later the first dry powder inhaler (DPI; Spinhaler by Fisons) reached the market (Ashurst et al., 2000). All three drug delivery devices (DDDs) provide convenient ways of administering pulmonary medication, and will be discussed in more detail in section 9.3.

The therapeutic advantages of selective targeting of drugs to the airways and lungs for the treatment of respiratory diseases are enormous, as are those of the emerging field of systemic delivery via the lung. There are seven non-invasive drug delivery routes: (1) pulmonary; (2) oral; (3) buccal (the cheek-facing side of a postcanine tooth); (4) nasal; (5) vaginal/uterine; (6) ocular (eye); (7) transdermal (skin). Among these, the pulmonary route is one of the preferred drug delivery routes since it provides (i) a large surface area (about 100 m^2) for rapid absorption provided by terminal bronchioles and alveoli, (ii) a thin (0.1–0.2 μm of alveolar epithelium) physical barrier for absorption, promoting rapid uptake into the bloodstream, (iii) the absence of extreme pH and metabolism, (iv) no first-pass liver with minimum reduction of bioavailability, (v) rich blood supply, (vi) rapid systemic delivery from alveolar region to lung, and (vii) minimal extracellular enzyme levels for metabolic breakdown compared with the gastrointestinal (GI) tract. Because of the large surface area and rich blood supply, the pulmonary airway enables rapid onset of therapeutic action even with reduced doses compared to the oral (in case of poorly bioavailable oral drugs) and other systemic routes; therefore, systemic side-effects are minimal. For example, sodium cromoglycate is well absorbed from the lung whereas less than 5% is absorbed from the GI tract (Kellaway, 1988). Many examples are also summarized by de Boer et al. (2001). Nevertheless, it is well accepted that for any drug delivery route target selectivity via differential sensitivity is virtually impossible to achieve.

[†] For the convenience of the reader, the registered trademark® has not been used in this chapter.

As an invasive drug delivery method, injections are usually more expensive than tablets and more cumbersome to use, especially when considering children, the elderly, and the infirm. If blockage occurs in the needle, more than one-third of a microparticulate formulation can remain in the syringe, and in some cases blockages are not infrequent (Minter, 2003). There is growing interest in using the surface of the lung as an absorptive membrane to allow uptake of a medication directly into the bloodstream. While most systemic drugs are yet to be approved for pulmonary drug delivery, DPI technology for delivering drugs like morphine for pain and insulin for diabetes is at the forefront of development or already marketed.

In this chapter, a brief outline of the human respiratory system is provided, and mechanisms of drug particle deposition in human lungs, various pulmonary drug delivery devices, and analytical methods associated with characterization of aerosols are reviewed. An attempt has also been made to assess the challenges and opportunities of pulmonary drug delivery.

9.1.2 The human respiratory system

The human respiratory tract can be divided into 24 generations (Weibel, 1963). Regarding physiological functions, the contiguous airway from the trachea to the terminal bronchioles (generation index 0–16) is called the conducting zone, and the area from the respiratory bronchioles to the alveolar sacs (generation index 17–23) is called the transitional and respiratory zone.

Figure 9.1 is a schematic representation of the respiratory tract. The human respiratory system could also be said to include the nasal cavity, associated sinuses, nasopharynx, oral cavity, oropharynx, and larynx. The respiratory system of a normal adult processes 10–25 m^3 air per day (140–290 mL/s) and the average breathing condition is defined by a tidal flow rate of 500 mL/s (Hinds, 1998). The inhalation flow rate for light to various levels of physical activities are: (i) sedentary, 233.3 mL/s; (ii) low, 333.1 mL/s; (iii) light, 667 mL/s; (iv) heavy, 2000 mL/s; and (v) maximal, 4066.7 mL/s (Hofmann et al., 2003). However, the inhalation rate significantly varies in the asthma and other chronic obstructive pulmonary disease (COPD)-hampered lung. A study of the obstructive spirometry patterns in the lungs of 6167 adult patients from northern India showed considerable variability between patients' forced expiratory volume in 1 second (FEV_1), expressed as a percentage of predicted, and peak expiratory flow (PEF), expressed as a percentage of predicted (Aggarwal et al., 2006). Analysis also showed that PEF expressed in this way overestimated the FEV_1 percentage in patients with less severe obstruction and underestimated the FEV_1 percentage in

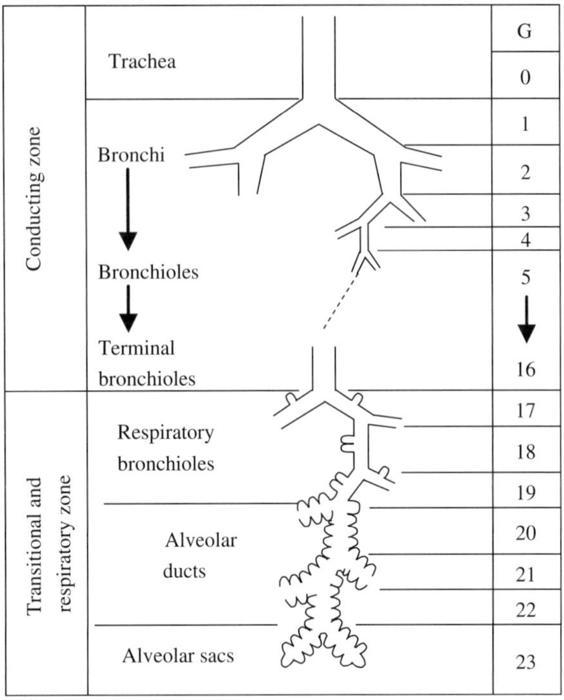

Figure 9.1 The human respiratory airway can be divided into 24 various lung generations. Generation index 0–16 functions as the conducting zone, and generation index 17–23 functions as the transitional and respiratory zone. From Weibel (1963).

those with more severe obstruction, and inhalation rate differed in many patients with airflow limitation. Therefore, it is unjustified to assume general parity between PEF and FEV_1 expressed as percentages and these terms should not be used interchangeably.

9.1.3 Physical properties of drug aerosol particles

The major physical properties of respiratory drug aerosol particles include particle size (diameter), shape, diffusivity, density, and electrostatic charge (Suarez and Hickey, 2000). In order to achieve non-invasive methods to treat asthma, cystic fibrosis and COPD, and for the efficacy of administering therapeutic aerosols, understanding and controlling the physical and chemical properties are topics of current interest (Peart et al., 2003). In fact, the aerosol particle size influences all of the other physical properties. The gravitational force, terminal settling velocity, inertia force, and electrostatic forces are approximately proportional to the square of the particle diameter (US EPA, 2006). Therefore, one of the most commonly used terminologies in aerosol medicine is the aerodynamic diameter (d_a), representing the size of a drug particle. This is defined, for a particular particle, as the diameter of the spherical particle with density of 1000 kg/m^3 (the density of a water droplet) that has the same settling velocity as the particle. Mathematically, it was derived from Stokes' law and is defined by:

$$d_a = (\rho_p/\rho_0)^{0.5} d_p$$

where d_p is the particle diameter, ρ_p is particle density, and ρ_0 is the standard particle density (1000 kg/m^3). The aerodynamic diameter can be perceived as the diameter of a water droplet having the same aerodynamic behavior as the particle (Hinds, 1998). If a particle has an aerodynamic diameter of 1 μm, it behaves in an aerodynamic sense like a 1-μm water droplet regardless of its shape, density, or physical size. The aerodynamic diameter is the key property of a particle that helps to determine its respiratory deposition, coulombic attraction/repulsion force, diffusivity, and velocity while airborne and suspended in the aerosol. This property can be engineered to deliver particles to different areas of the lung.

Therapeutic aerosols known at present or anticipated to be of primary practical importance for lung deposition have aerodynamic diameters in the range 0.5–10 μm (Swift, 1996). Medicinal aerosols containing particles per droplet of the drug should ideally be in the size range of 2–5 μm in diameter (Smith and Williams, 1998). To achieve clinical efficacy and efficient systemic delivery using inhalation therapy requires aerosol particles to be designed with an aerodynamic diameter between 1 and 5 μm in order to maximize deposition in the alveolar region (Lippmann et al., 1977; Rees et al., 1982; Newman et al., 1991; Hickey, 1992). In the case of systemic absorption of proteins and peptides, aggregation of particles themselves of respirable size and binding of drug molecules to an excipient lower the deposition efficiency and reproducibility of aerosol delivery to the deep lungs (Vanbever, 2003).

Particles acquire electrostatic charges during generation by the diffusion of ions in aerosol streams, which is caused by the collisions resulting from Brownian motion of the ions and particles (Hinds, 1998). Acquisition of charges by a particle as a function of time and the charge state of the particle can be calculated in simplified and dimensionless form using the following equation (Fjeld et al., 1990; Lawless, 1996):

$$q = \frac{ne^2}{2\pi\varepsilon d_p kT} \tag{9.1}$$

where q is the particle charge, n is the number of elementary charge units, e is the electronic charge (1.6×10^{-19} C), ε is the permittivity of air, d_p is the particle diameter, k is Boltzmann's constant, and T is the absolute temperature. If we know the density of the particle, ρ, then the charge on an individual particle can be described in terms of its average charge-to-mass ratio, q/m, which can be calculated by the following equation:

$$\frac{q}{m} = \frac{3\,ne^2}{\pi^2 \varepsilon \rho d_p^4 kT} \tag{9.2}$$

During the inhalation of an aerosol bolus, the complicated inlet velocity profile and flow conditions within the respiratory airway may affect the abilities of particles to gain electrostatic charges through tribo-electric charging processes (Ali et al., 2009). Tribo-electric charging is a process of generating static electricity that results from two materials coming into contact and then separating during device actuation (Staniforth, 1994). This phenomenon takes place when particles flow through a conduit. When this occurs, one of the materials typically gains a positive charge, while the other acquires a negative charge (Cross, 1987). The magnitude of the charge is affected by numerous factors, including drug propellant surfactants, metal surfaces of delivery devices, drug-carrier inhomogeneities, and excipient particle size distributions (Staniforth, 1994).

9.2 PULMONARY DRUG DEPOSITION MECHANISMS

The clinical efficacy of an inhaled drug is proportional to the amount of drug reaching the airways (Selroos et al., 1996; Pauwels et al., 1997). This statement clearly indicates that with a higher percentage lung deposition, the same clinical efficacy can be achieved using a smaller or nominal label of drug dose. Furthermore, if more drugs reach the lungs from a given dose, this means that there is less deposition in the oropharynx and thus fewer local or global side-effects. Human lungs have natural effective defense mechanisms against aerosol hazards. For example, macrophages in the lung engulf any foreign particle as soon as it arrives in the airway and carry it to the oropharynx for removal through coughing or swallowing through the GI tract. But the administration of pharmaceutical aerosol particles is accomplished by delivering them in a quantity that exceeds that clearing capacity. Overall, in the changing geometry of respiratory airways the primary mechanisms (basic theories) of deposition of inhaled respiratory drug aerosol particles consist of five electromechanical processes: impaction, diffusion, interception, gravitational settling, and electrostatic effects (Figure 9.2). Mathematical expressions that are widely accepted to calculate the lung deposition fraction of inhaled aerosols have been derived in the

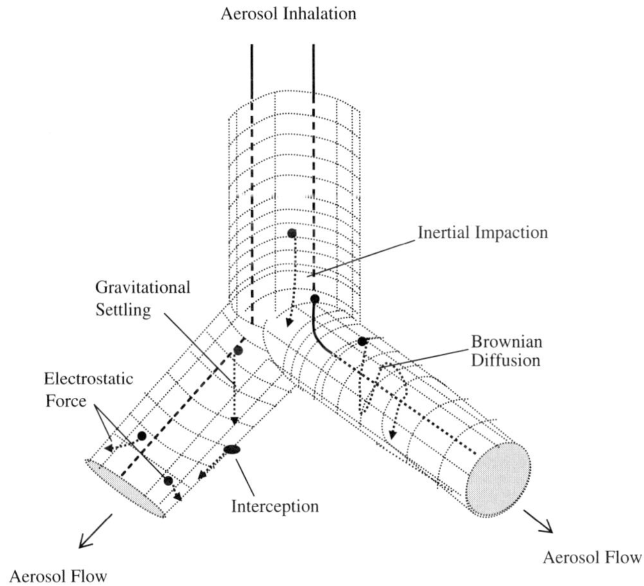

Figure 9.2 Respiratory drug particle deposition mechanisms in the human lung. Figure shows tracheobronchial region for an example.

following categories: (i) impaction (Yu and Diu, 1982); (ii) gravitational settling (Yu and Pich, 1972; Heyder, 1975); (iii) turbulent diffusion (Landahl, 1963); (iv) laminar diffusion (Ingham, 1975); and (v) space and image charge forces (Yu, 1977). A modeling study on spherical aerosol particles' deposition efficiency due to an interception mechanism did not show a significant impact on overall and regional deposition in the lung; Brownian diffusion and gravitational settling replace this effect (Liu and Wang, 1997). Another study showed that interception efficiency is prominent for fibrous particles (Su et al., 2008).

Predicting respiratory deposition realistically from the above-mentioned basic theories is much more difficult due to the hygroscopic nature of most therapeutic aerosols, which allows the particles to charge drastically during transport in the highly humid atmosphere of the respiratory tract. Other factors that directly influence aerosol deposition are (i) physical properties of the inhaled particles, such as size, shape, diffusivity, density, and electrical charge (Suarez and Hickey, 2000); (ii) structural properties of airways' mucus, temperature, and relative humidity (Oberdorster, 2001); (iii) the method of redistributing and removing particles (Kim, 2000); and (iv) the breathing pattern, such as tidal volume, mean flow rate, mean residence time, and frequency of inhalation and exhalation. Studying the effects of these properties on particle deposition in the various regions of the human lung is necessary (Orban and Peart, 2004). However, investigations on the relevance of electrostatic charge on respiratory drug delivery are quite rare and started in the early 1980s, though aerosol science is a decade-old discipline.

During inhalation, the aerosols undergo a series of direction changes while flowing from the mouth through the branching airways to the alveolar region. The probability of deposition due to impaction depends on the ratio of particle stopping distance to airway dimensions (Hinds, 1998). This ratio is higher for larger particles and in the bronchial region; therefore, the large particles tend to impact bronchial bifurcations (carina) due to inertial impaction. As a result, inhaled drug aerosols experience the highest inertial impaction deposition in the tracheobronchial bifurcation (Figure 9.3a).

Gravitational settling is a very important deposition mechanism in distal airways such as terminal and respiratory bronchioles, and favors greater deposition in the horizontal-oriented airways. The main reason is that this mechanism depends on the settling distance (terminal settling velocity × residence time) of a particle. Also, the gravitational settling is very effective when particles grow in size due to agglomeration by coulombic attraction among particles with opposite electrostatic charge. In addition, hygroscopic particles grow as they pass through the humid airways of the lung (Figure 9.3c). Mathematical models and experimental measurements demonstrated that large particles have high deposition efficiency in the alveolar region if they can reach there (Figure 9.3e) (Swift, 1996).

Brownian diffusion is very effective for submicrometer particles and in the terminal and respiratory bronchioles and alveolar region, where the particles travel short distances but experience a comparatively longer residence time. Particle deposition by diffusion is largely dependent upon the ratio of the root-mean-square displacement in the airways to the airway diameter (Hinds, 1998). As a result, this mechanism is the predominant means of deposition for particles smaller than 0.5 μm. Interception is a mechanism that occurs for particles that have not deviated from the aerosol streamline and come into contact with the airway surface because of their shape and size (for long fibers and large particles) (Hinds, 1998; Phalen et al., 2008). This mechanism is predominant for fibrous particles and those flowing through the tortuous path of the smaller airways. In the case of spherical particles, Brownian diffusion and settling methods cause the most deposition, and interception becomes negligible.

When the therapeutic aerosol particles move through the contiguous respiratory airway generations, and if they carry an electrical charge,

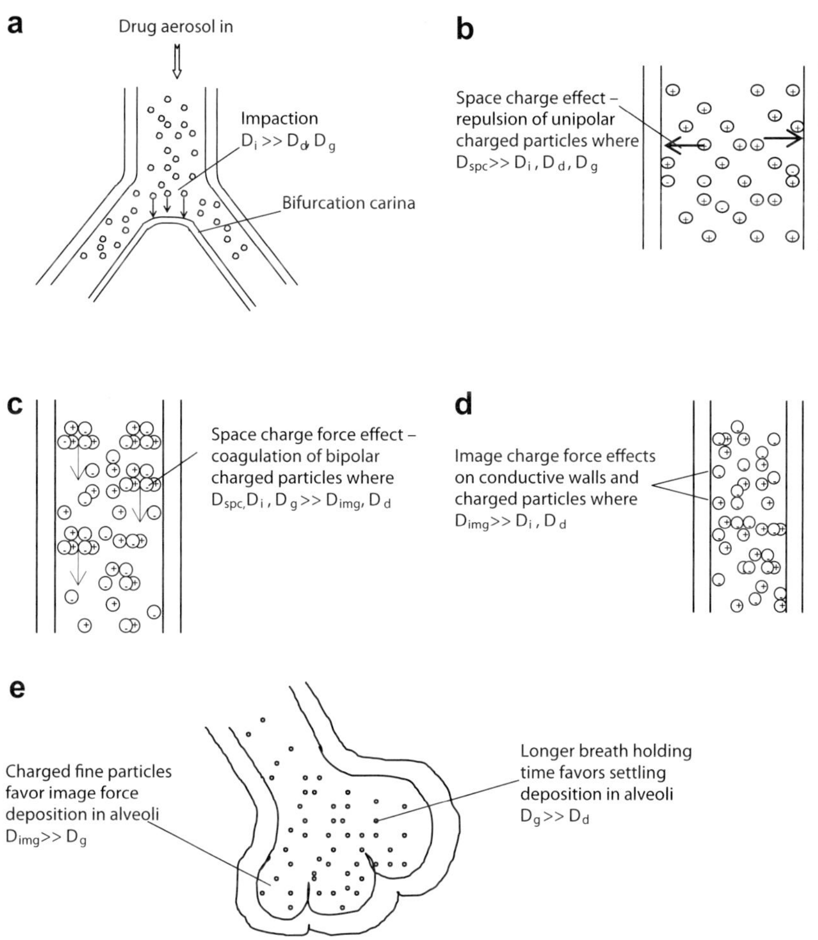

Figure 9.3 Relative deposition efficiencies among impaction, diffusion, gravitational settling, and electrostatic image and space charge forces with respect to various regions in the pulmonary airway. D_i = inertial impaction deposition efficiency; D_d = Brownian diffusion deposition efficiency; D_g = gravitational settling deposition efficiency; D_{img} = image charge force deposition efficiency; D_{spc} = space charge force deposition efficiency.

deposition due to electrostatic force plays an important role. There are several possible scenarios. Firstly, highly charged particles are attracted by the mirror image forces those particles to induce in the conductive lung airways. So the charged particles exhibit higher image force deposition compared to impaction and diffusion (Figure 9.3d). Secondly, a large number of unipolarly charged aerosol particles obtain space charge forces and mutual repulsions and, as a result, particles come closer to airway walls and are captured (Figure 9.3b). These particles therefore show greater deposition in the main, lobar, and segmental bronchi. Thirdly, bipolar charged particles agglomerate due to coulombic attraction forces, which allow smaller particles to grow larger and therefore deposit more efficiently due to gravitational settling and impaction (Figure 9.3c). This phenomenon usually takes place in the terminal bronchioles of the conducting and respiratory zone. When charged particles flow in the respiratory airways, axial particle flux occurs due to advection and electrostatic image charge forces, and diffusional flux takes place due to the particle density gradient (Barat et al., 2002). Therapeutic aerosols reside for a longer time in the alveolar sacs (compared to the traveling time in the tracheobronchial region), where gas exchange takes place; therefore, if patients can hold their breath for longer, deposition due to gravitational settling and diffusion will take place in the alveoli.

A systematic *in vitro* study was conducted to find out whether the electrostatic charge of the MDI and DPI aerosols influences loss in the oropharyngeal (OPL) region and to gain further insight into possible measures to mitigate this problem (Ali et al., 2009). Results showed that the electrostatic image force influenced deposition of particles on the OPL walls. They also revealed that if the particles are bipolar, OPL loss can be minimized due to the high space charge force and agglomeration. If the net charge of drug aerosols becomes negligible (positively charged particles are neutralized by negatively charged particles), the process helps particles to travel further deep to the lung and deposit by sedimentation and diffusion. The agglomeration due to electromagnetic forces is a physical phenomenon caused by attraction between oppositely charged drug particles (Finlay, 2001). Space charge effects favor coagulation of oppositely charged particles; then impaction and gravitational settling become the dominant methods of deposition.

9.3 PULMONARY DRUG DELIVERY METHODS

The pulmonary drug delivery (PDD) industry supplies approximately half a billion inhalers globally for the treatment of asthma (Dalby and Suman, 2003). The mainstay of treating asthma and COPD is inhalation therapy of medications to the site of the diseased region in the lung. This can be achieved by three major drug delivery methods. In method 1, the drug is dissolved or suspended in the propellant under pressure. When activated, a valve system releases a metered volume of drug and propellant. To administer the medication, the patient may either use direct inhalation or squeeze the canister of the inhaler. Devices that adopt this technique are called pressurized metered-dose inhalers (MDIs). In method 2, the drug is in dry powder form. The patient exhales a full breath, places the lips around the mouthpiece, and then quickly breathes in the powder. The powdered drug is dispersed into particles by inspiration instead of using chemical propellants. The strength of the inhalation required is stronger than that needed to release the medication in the MDI. Devices that adopt this technique are called dry powder inhalers (DPIs). In method 3, oxygen, compressed air, or ultrasonic power is used to break up solutions or suspensions of medication into droplets for inhalation, i.e. the device converts the medication into a fine mist. The aerosol is administered by mask or mouthpiece. These devices are called nebulizers. A number of different inhalation devices for each method have become commercially available; there are wide variations in cost to the health service.

Choice of device (nebulizer, MDI, or DPI) is based on ease of use for the patient and providing the required therapeutic dose with minimal adverse effects. There has been no previous systematic review of the evidence of the clinical effectiveness and cost-effectiveness of these different inhaler devices. However, to be accepted for clinical use, any respiratory drug delivery device must satisfy several criteria, including: (1) the ability to generate particles of ideal size (0.5–5 μm) because deposition of particles larger than 2–5 μm is very limited (≤12%) in the distal airways (generations 17–23) (Cheng, 2003); (2) very reliable dose reproducibility; (3) drug stability at room temperature; (4) ability to provide flexible dosing to patients; (5) portability and ease of use with minimum training; and (6) affordability (Dalby et al., 1996). The MDI, DPI, and nebulizer are commonly marketed as they satisfy these criteria, but they do not represent optimum drug delivery (Dalby et al., 1996). The ideal drug delivery system delivers drug at a rate dictated by the needs of the body over a period of treatment.

9.3.1 Metered-dose inhalers

The pressurized metered-dose inhaler (pMDI or MDI) is the most commonly used inhaler, and is an inexpensive, quick and convenient way to take asthma medicine, and the device may also be used in conjunction with a spacer. The MDI, often simply called an inhaler, is a small device incorporating a pressurized canister that contains aerosol medicine to be inhaled. Many asthma medications are taken with an inhaler. Every day more than 500 million patents in the world carry easy-to-use MDI devices to deliver medicines to their pulmonary airways (Colthorpe, 2003). MDIs are also used by people who suffer from other respiratory ailments such as emphysema, chronic lung disease, and bronchitis.

There are five parts to an MDI: the medication, the propellant, the canister, the metering valve, and the mouthpiece. In addition, surfactants are

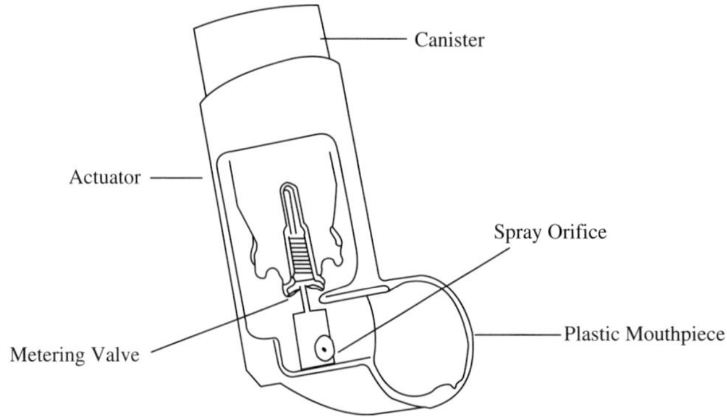

Figure 9.4 Design of primary functional parts of a classic metered-dose inhaler. Within the canister is the drug formulation, which is typically comprised of submicron-sized drug particles; the formulation is suspended in the propellant and stabilized by a surfactant.

used to aid dispersion or dissolution of partially soluble drug and to lubricate the metering valve mechanism. The propellant can be either a chlorofluorocarbon (CFC) or hydrofluoroalkane (HFA). Figure 9.4 shows the design of the primary functional parts of a classic MDI.

Each time a patient uses an MDI, a precisely measured, or "metered," amount of medicine is delivered. Inhalers can be used by all asthma patients aged 5 and older, according to the American Medical Association. Through inhalation, it takes only 5–15 minutes for short-acting bronchodilators (quick-relief medicine) to have an effect, compared to oral asthma medicines, which can take 1–3 hours. With an inhaler, there are also fewer side-effects because the medicine goes directly to the lungs and not to other parts of the body. Even though MDIs are the most widely prescribed product for the treatment of asthma, bronchitis, allergies, and COPD, it is estimated that up to 70% of patients have difficulty in coordinating their inspiratory flow rate with pressing of the canister (Larsen et al., 1994; Plaza et al., 1998). The inspiratory flow rate can affect the dose emitted from an inhaler, amount inhaled, OPL deposition, and regional lung deposition (Smyth, 2003).

9.3.1.1 Valve-holding chamber or spacer

It is important to administer an MDI in an appropriate way to get the necessary amount of medication into diseased lungs, such as proper co-ordination between the actuation of the device and the patient's inhalation. To ease this coordination, tubes are attached to the inhalers that act as a reservoir or valve-holding chamber (VHC). These tubes are also called spacers (Figure 9.5a). They serve to hold the medication that is sprayed by

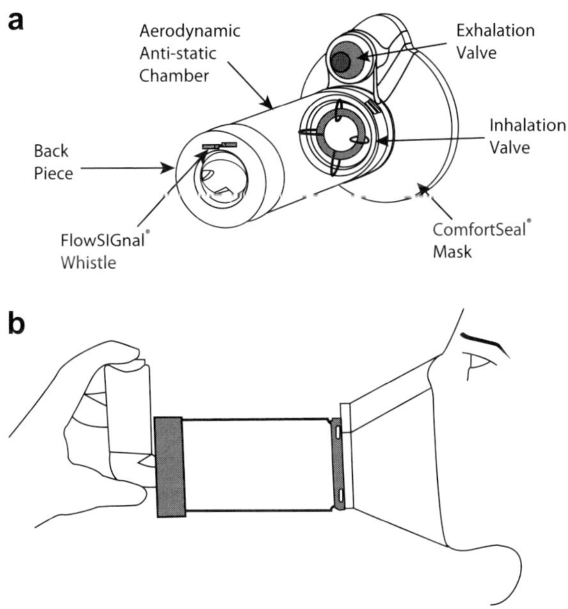

Figure 9.5 (a) Diagram of an AeroChamber Plus® Z STAT® anti-static valve-holding chamber (spacer). (b) Schematic of the proper administration of MDI with the spacer. The spacer has valves that prevent medication loss during exhalation by allowing air to flow through the chamber only during inspiration. Courtesy of Monaghan Medical Corporation, NY, USA; reproduced with permission.

the inhaler. This makes it easier to use the inhaler and helps patients to get more of the medication into the lungs instead of just into the mouth. With proper use, a spacer can make an inhaler 20% more effective in delivering medicine to the lungs. Furthermore, it decreases deposition of medication in the mouth and throat, and eliminates the need to coordinate activating the MDI with inhaling the medication. Spacers can be especially helpful to adults and children who find a regular inhaler hard to use. People who use corticosteroid inhalers should use a spacer to prevent getting the medicine in their mouth, which can cause an oral yeast infection. Therefore, physicians recommend the use of a spacer because it improves delivery of medication to the lungs and also reduces the risk of mouth infections (yeast).

A reservoir is sometimes used for actuation of the MDI prior to initiating an inhalation. Also, spacers are used to start inhaling prior to actuating the MDI. These auxiliary devices help patients with coordination problems. When the reservoir and spacer are smaller they facilitate larger drug particle deposition in the components due to inertial impaction and gravitational sedimentation. The delays between actuation and inhalation reduce drug delivery efficiency. MDIs may also generate some carcinogenic compounds in low concentrations extracted from the valve system. Clinical improvements in lung function for a given dose of inhaler are markedly increased when a spacer device is used correctly in conjunction with an MDI. Studies showed that the deposition values for terbutaline and budesonide ranged between 8.2% and 16.7% without a spacer (Pauwels et al., 1997). The deposition of terbutaline was determined by both gamma scintigraphy and the charcoal block pharmacokinetic method, and it was shown that when the Nebuhaler spacer was used (inspiratory flow rate 15 L/min), values were 31.6% and 33.8% respectively, approximately three times those obtained when a spacer was not used (Newman et al., 1995). Therefore, it is suggested that the correct use of a spacer with an MDI may result in substantially greater lung deposition.

The comparative advantages and disadvantages of various commonly marketed spacers have not been firmly established with substantial evidence; varying opinions have been expressed. *In vitro* aerosol deposition from a beclomethasone dipropionate MDI containing HFA propellant compared with that of the MDI in combination with two commonly marketed valve-holding chambers, namely OptiChamber VHC (Cardinal Health, Dublin, OH) and AeroChamber-Plus VHC (Invacare Corp., Elyria, OH), did not demonstrate equivalent performance (Asmus et al., 2003). In order to investigate the Proventil MDI (Schering-Plough Corp., Kenilworth, NJ) aerosol particle size distributions leaving the MediSpacer (Allegiance Healthcare Corp., McGaw Park, IL) and AeroChamber-Plus, Foss et al. (1997) found that both devices released particles of similar size distributions.

Young children in the age range of 5 months to 2 years form a special treatment group, as factors such as cooperation, acceptance, and the use of face masks may determine the success or failure of the inhalation therapy. In addition, when using metal or plastic spacers differences in results can be interpreted as being due to spacer design or to the presence or absence of electrostatic charge. Several studies have shown that plastic spacers can become electrostatically charged, which decreases drug delivery (Barry and O'Callaghan, 1995; Wildhaber et al., 1996). With metal spacers there is less possibility of holding static charges. Electrostatic charge can be minimized by coating the plastic spacer with a household detergent (Wildhaber et al., 1997). In an open crossover study of budesonide pMDI (Pulmicort, AstraZeneca, Lund, Sweden) drug delivery in 25 wheezy infants aged 5–26 months, a metal spacer (Nebuchamber, AstraZeneca, Lund, Sweden), a detergent-coated (DC) and a non-detergent-coated (non-DC) plastic spacer (Babyhaler of Glaxo Wellcome, London, UK) were tested at home for 7 days each (Janssens et al., 2000). The results showed that the DC spacer provided increased dose delivery compared to the metal and non-DC spacers as it reduced static charges in the aerosols. However, the electrostatic charge had no influence on the dose variability; rather, considerable within-subject dose variability was found for the metal spacer and it was speculated to be due to suboptimal fit of the face mask of the Nebuchamber to the infants. Another *in vitro* study of the electrostatic charge properties of aerosols leaving new and detergent-coated AeroChamber-Plus spacers showed that the former type releases much higher charge per mass (charge-to-mass ratio) drug particles than the latter (Kwok et al., 2006). However, in all cases it was clear that patient cooperation during administration of inhaled drugs leads to more effective medicinal aerosol treatment. Therefore, it is very helpful for clinicians to prescribe a spacer that will physically fit and be comfortable on the face of the patient.

9.3.1.2 Chlorofluorocarbon to hydrofluoroalkane propellants

A liquefied or compressed gas is used in the MDI as propellant. Most pulmonary drugs are of low propellant solubility and they are frequently formulated as micronized suspensions. Chlorofluorocarbons (CFCs) have been the propellant of choice because of their low pulmonary toxicity, high chemical stability, purity, nonflammability, and compatibility with other packaging materials. After the 1985 discovery of a hole in the ozone layer above Antarctica, governments around the world combined to address the growing problem of ozone depletion. The result was the 1987 Montreal Protocol on Substances that Deplete the Ozone Layer. Although the Montreal Protocol has been amended and updated over the years, the overall goal still remains the same: to protect the ozone layer by controlling the emission of harmful substances, such as CFCs (US FDA, 2006a). The ban on CFCs caused concern for many people with asthma and lung disease, since CFCs have long been an important component of MDIs. To comply with the Montreal Protocol, the US Food and Drug Administration (FDA) began a reformulation effort to find and approve non-CFC medical products. In March 2005, the FDA announced that it would ban the production and sale of all CFC albuterol MDIs by 31 December 2008 (US FDA, 2006b). Ironically, the most significant recent changes in MDI technology have not been the drug molecule characteristics, formulation chemistry or patient compliance factors, but the drive to change propellants triggered by environmental concerns.

The CFC propellants in MDIs are being replaced by hydrofluoroalkanes (HFAs). Two approved propellants of this group are HFA-134a and

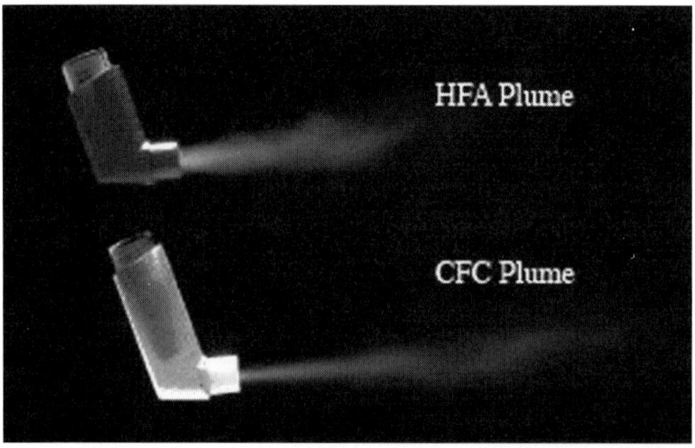

Figure 9.6 Comparison of the real-time aerodynamic pattern of the sprays generated from albuterol MDIs with CFC and HFA plumes. Reproduced with permission from ProAir HFA (2008).

HFA-227. Although the HFA products may taste or smell slightly different, the FDA has found them to be safe and effective alternatives to CFCs.

Figure 9.6 shows the real-time aerodynamic patterns of the sprays generated from albuterol MDIs with CFC and HFA plumes. The investigators estimated that HFA inhalers will cost three times as much as CFC inhalers, at least until generic versions become available. However, the HFA MDIs have many advantages. For example:

1. Albuterol MDIs with HFA propellant are softer, with mist-like spray, puffs are warmer, and they require lower spray force (Boccuzzi et al., 2000).
2. They are as safe and effective as CFC inhalers (US FDA, 2006b).
3. HFA MDIs are more ozone friendly than CFC inhalers (US FDA, 2006c).
4. ProAir HFA (albuterol sulfate) MDIs require no priming when used within 2 weeks of the last dose and can reliably deliver up to 200 actuations (ProAir HFA, 2008).

A clinical comparability study between CFC and HFA budesonide MDI (Pulmicort, AstraZeneca, Lund, Sweden) administration in random pediatric asthma patients (age 6–12 years) was conducted in five countries, namely France, Germany, Spain, Sweden, and Ukraine (Escribano et al., 2006). The results suggested that delivery of budesonide via the novel HFA MDI offered comparable efficacy and tolerability to the conventional budesonide CFC MDI.

9.3.2 Dry powder inhalers

A dry powder inhaler (DPI) is a breath-activated device that aerosolizes a set dose of micronized drug in the solid state on an airstream. The DPI provides certain advantages regarding drug stability, ease of handling, and range of possible dose weights to be administered. Since the drug is processed, weighed, and packed in powder form, decomposition, separation, and microbiological contamination hazards are minimal compared to wet formulations (Frijlink and de Boer, 2004). In most cases, the inspiratory air flow through the DPI delivers the energy for aerosol generation, meaning that this DDD can be small, portable, and relatively cheap. Every DPI powder consists of active pharmaceutical ingredient (API) or drug and the carrier(s) either mixed or co-precipitated together into dry powder form. The API is a disease-specific therapeutic agent, and the choice of carrier depends on the nature and safety of carrier, the type of drug (i.e. protein, peptides, hydrophilic, or hydrophobic), the device for delivery, the site of action, and the disease state. The dry powder particles should be near spherical in shape and monodispersed with an aerodynamic diameter range of 0.5–5 μm. Particles in the range 0.5–3 μm are appropriate for systemic absorption in the distal region including alveoli, and 3–5 μm for local action in the terminal bronchioles.

When designing a DPI, four basic features are extremely important: (i) dose metering; (ii) aerosolization; (iii) disaggregation; and (iv) adapter to direct aerosol. These features are achieved by incorporating certain primary and secondary functional parts. The primary parts are the same for all types of DPI currently on the market, and determine the performance of inhaler with respect to producing a fine particle fraction in aerodynamic size distribution and lung deposition, dosing mechanism, powder de-agglomeration principle, redispensing the powder into the inhaled air stream, and geometry of the mouthpiece (Frijlink and de Boer, 2004). All the primary parts are assembled in a suitable housing by mechanical means to access a premeasured single dose or multiple doses. The secondary functional parts are for enabling safety, ease of handling, signaling to patients, moisture protection of the drug formulation, etc. Figure 9.7 shows the design of primary functional parts of a classic DPI.

Development of the DPI involves powder recrystallization, formulation, dispersion, delivery, and deposition of medicinal agents in the respiratory bronchioles to the alveoli for treatment or prophylaxis of pulmonary and systemic disorders. Conventional production technology of DPI powders has many limitations, resulting in the development of alternative techniques to mitigate these problems. In the past decade remarkable performance of

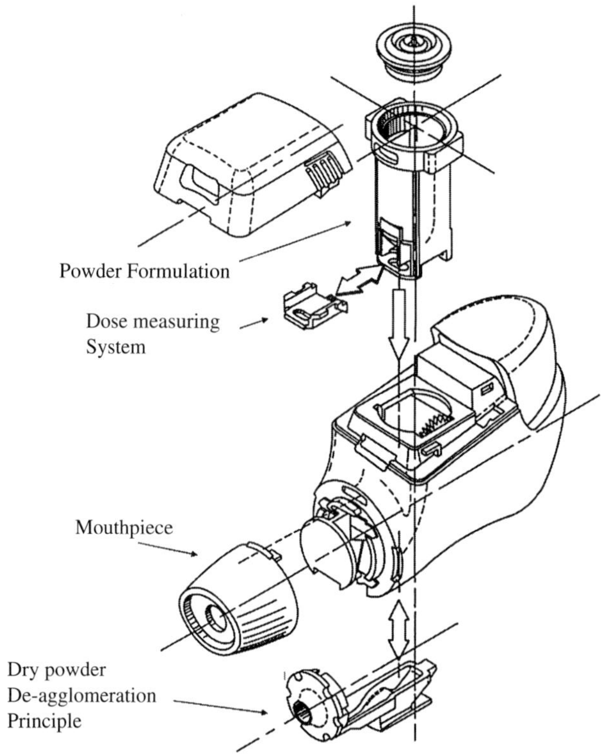

Figure 9.7 Design of primary functional parts of a classic dry powder inhaler. Reproduced with permission from de Boer et al. (2006).

DPIs has been achieved through the following: (i) incorporation of fine carrier powders to occupy active sites of the surface of the API, and hydrophobic carriers to facilitate reduced surface energy and particle interaction; (ii) reduction of particle aerodynamic size distributions through particle engineering and incorporating drug into porous or low density; (iii) creation of less cohesive and adhesive particles through corrugated surfaces, low bulk density, reduced surface energy, and particle interaction (Chougule et al., 2007). However, the breakthrough application of DPI technology is the inhalation of insulin (e.g. Exubera, AERx IDMS, Technosphere, etc.) to treat type I and type II diabetes.

In a DPI, the asthma medication comes in dry powder form – inside a small capsule, disk, or compartment inside the inhaler. The dry powder in some inhalers has no taste, while in others the medication is mixed with lactose (a simple sugar) to give it a sweet taste. Most adult patients find the DPI easy to use. DPIs are not designed for use with a spacer. There are many types of this kind of DDD commercially available, and each works differently.

Some have to be loaded each time they are used (e.g. Inhalator, Spinhaler, Rotahaler). Others have disks with a set number of doses (four or eight). Still others have as many as 200 doses stored in the device (Turbuhaler, Accuhaler).

It is understood that well-designed DPIs are highly efficient systems and can offer the greatest potential of any approach to PDD. Both quick-relief medicine (inhaled bronchodilators) and long-term control medicine (inhaled corticosteroids) can be delivered to the airways using a DPI. If the dry powders are engineered properly, they can meet the criteria for success while delivering a wide range of molecules from traditional asthma drugs and antibiotic molecules to proteins, peptides, and oligonucleotides. In particular, delivering biological therapeutics deeper into the lungs for systemic delivery provides the most promising inhalation therapy platform. However, the devices are also complicated systems, the performance of which relies on many aspects, including the design mechanisms for less resistance to airflow and de-agglomeration principles to generate the inhalation aerosol, the powder formulation, and the airflow generated by the patient (Frijlink and de Boer, 2004). The DPI approach attracted considerable investment in the 1990s, and still offers considerable potential for growth and value.

9.3.3 Nebulizers

Nebulizers have been available since the beginning of the twentieth century. Nebulization from a drug solution is a common method of medical aerosol generation. To deliver a drug by nebulization, the drug must first be dispersed in a liquid (usually aqueous) medium. After application of a dispersing force (either a jet of gas or ultrasonic waves), the drug particles are contained within the aerosol droplets, which are then inhaled. At present all commercially available nebulizers can be categorized into two types: (i) jet (or pneumatic) small-volume nebulizers and (ii) ultrasonic nebulizers. Jet nebulizers are based on the venturi principle, whereas ultrasonic nebulizers use the converse piezoelectric effect to convert alternating current to high-frequency acoustic energy (Rau, 2002). The major features of both types of nebulizer are duration of treatment at each time of use, particle size distributions produced, and aerosol drug output.

The formulation of drug solution is usually designed to optimize drug solubility and stability; small changes in formulation may also affect inhaled mass, particle size distribution, and treatment time. The differences between nebulizer brands probably has a greater impact than differences in formulation (O'Riordan, 2002). There are several advantages to jet nebulization, including the fact that effective use requires only simple, tidal breathing, and that dose modification and dose compounding are possible. Disadvantages include the duration of treatment time and equipment size. Design modifications to the constant-output nebulizer have resulted in breath-enhanced, open-vent nebulizers such as the Pari LC Plus and the dosimetric AeroEclipse. Figure 9.8 shows the design of the primary functional parts of a jet nebulizer.

The second type, ultrasonic nebulizers, usually generates a higher output rate than jet nebulizers; however, the average particle size is larger (Rau, 2002). Ultrasonic nebulizers can substantially increase reservoir solution temperature, which is the opposite of jet nebulizer cooling. Drug concentration in the reservoir does not increase with ultrasonic nebulization as it does with jet nebulization. Although ultrasonic nebulizers have the same advantages as jet nebulizers, the former is more expensive and fragile than the latter type; in addition, the ultrasonic nebulizer may cause drug degradation and may not nebulize suspensions well. It has been shown that neither type of nebulizer meets the criteria of an ideal inhaler, which would generate efficient and quick dose delivery with reproducibility, cost-effectiveness, and no ambient contamination by escaped aerosols during administration (Rau, 2002). However, the electrostatic charge per mass of the nebulized aerosols is negligible, which extensively reduces OPL deposition of drug particles and enhanced deposition into the deeper lung (Ali et al., 2008a). Another study showed that nebulized

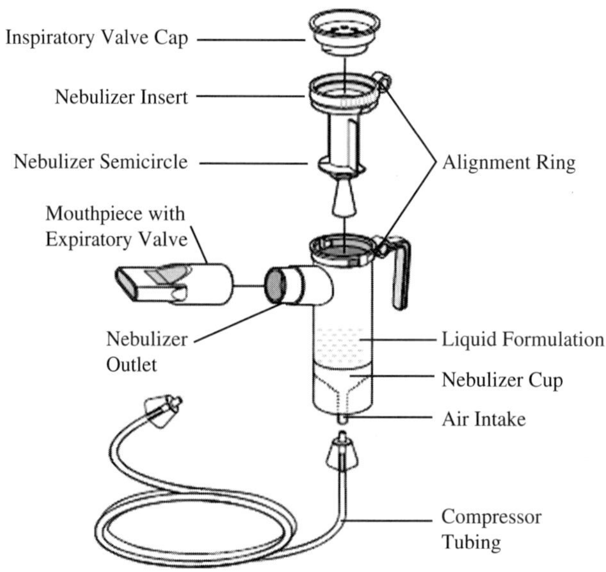

Figure 9.8 Design of primary functional parts of a classic jet nebulizer. Courtesy of PARI Respiratory Equipment Inc., Midlothian, VA, USA.

terbutaline aerosol, 1.8, 4.6, and 10.3 μm particles, deposited 80%, 60%, and 44% of drug respectively in the lung (Clay and Clarke, 1987).

9.3.4 Comparable and contrasting features of MDIs, DPIs, and nebulizers

The PDD industry has marketed a wide variety of inhaler systems, each of which has innate advantages and disadvantages. These have to be considered when making a choice of inhaler system for a specific clinical circumstance, as no currently available system is ideal in every clinical situation. The following are some comparable and contrasting features of the three PDD methods:

1. The MDI employs high-vapor-pressure propellants, the DPI uses a patient-independent energy source, and the nebulizer is electrically or pneumatically driven.
2. The typical method of PDD is via MDIs or DPIs. The former has been in use for around 50 years and is undergoing reformulation due to environmental concerns. DPIs are complicated devices requiring dose counters and breath-actuated or self-inhalation efforts in some cases. MDIs will be more convenient as they eliminate a patient's breath actuation or self-inhalation (especially for children and older patients) and, if they can exploit the physics of aerosol flow, their own source of energy and electrostatic forces.
3. The most popular advantages of MDIs are their portability, price, and ease of use for a properly instructed and adequately coordinated patient. A study showed that 30 of 36 patients exceeded the bioavailability of drug dose expectations when they received training and coordinated accordingly, thereby highlighting the benefit of using a training aid without any change to their prescription (Al-Showair et al., 2007).
4. Density of drug, propellant, and surfactant of the MDI dictates drug dosing. Since the density and solvency of drug and propellant have direct impact on vapor pressure, they can be manipulated to only a limited extent (Dalby and Suman, 2003). The novelty of various DPI devices is their mechanism for aerosolizing the powder rather than the packaged unit dose. Desiccant protects drug aggregation or chemical degradation in the reservoir.
5. DPIs do not have a standard protocol as in the case of MDIs due to the variety of operating mechanisms. They present the unique opportunity of dispersing dry particles as an aerosol in a patient's inspired airflow. The potential drawback of the DPI is that patients need to generate inspiratory flow rates of at least 30 L/min to obtain optimal drug

bioavailability. Therefore, the effectiveness of DPIs is more dependent on patients' age, gender, disease, and breathing cycle. Structured education of a patient's handling of the DPI does not provide an advantage over the usual education incidental to the prescribing or dispensing process; however, suboptimal breath holding is a common handling flaw, and does make a difference (Epstein et al., 2001).

6. Drugs that are susceptible to hydrolysis are good for MDIs and DPIs. Also, drugs that are less soluble and react with packaging material are good for DPIs. MDIs and DPIs are ideal for delivering drugs to the upper airways (e.g. asthmatics) where receptors are located. However, they are incapable of delivering a large percentage of drug to distal airways. As a result, nebulizers come into use in aerosol therapy.

7. MDIs and DPIs deliver concentrated drug in a bolus of single breath. During inhalation both drug delivery devices cause a significant deposition of drug particles in the oral–pharyngeal–laryngeal (OPL) region due to high inertial force (Kim et al., 1987; Fink, 2000). An *in vitro* study of an MDI (Atrovent, ipratropium bromide) and a DPI (Spiriva Handihaler, tiotropium bromide) where both devices were the trademarked brands of the same pharmaceutical company, and contained the same amounts of different drugs, was carried out. The deposition of aerosolized drug particles in the OPL region was higher for the DPI (21.5%) than for the MDI (3.0%) (Ali et al., 2009). Therefore, if the DPI employs higher doses, an equivalent therapeutic effect can be achieved.

8. Some DPIs (e.g. Turbohaler) have a series of tortuous channels. The dimensions of these channels dictate the pressure drops. Materials coming into contact with drug during the aerosolization process have a significant effect on its electrical properties (Staniforth, 1994). Pulmicort and Bricanyl Turbuhaler deliver approximately twice as much drug to the lungs as the corresponding CFC MDI, and Diskus/Accuhaler or Diskhaler (Thorsson et al., 1994, 2001; Tønnesen et al., 2004).

9. Unlike MDIs and DPIs, no established United States Pharmacopeia (USP) functionality standards are available for nebulizers. Nevertheless, biotechnologically developed drugs that stabilize or dry in organic solvents are safe to use in a nebulizer as a drug delivery device. Respiratory drug solution (e.g. budesonide in Europe) properties such as density, surface tension, viscosity, and nebulizer operating condition determine atomization.

10. Nebulizers require preparation of drug solutions and cleaning or sterilization after use. In contrast to nebulizers, high doses can be inhaled from a DPI with a single breath maneuver, and the cleaning process is simple, requiring only thorough wiping of the mouthpiece using a dry cloth.

11. The major advantages of nebulizers are: (i) generation of very small droplets capable of penetrating the deeper lung; (ii) capability of high dose delivery; (iii) propellant free; (iv) shorter treatment time and increased efficiency; (v) usefulness in treatment of children and elderly patients; (vi) easier to atomize aqueous solution than with MDIs and DPIs; and (vii) they may be used to deliver more complex dispersed drugs such as liposome.
12. The major disadvantages of nebulizers are: (i) large size, obtrusive dosing and cleaning requirements; (ii) a single-dose nature device – solution/tubing/mouthpiece; (iii) it is impossible to specify the actual dose of drug; (iv) inter-device variability; and (v) they require a power source at the pace of administering pulmonary drugs.
13. For nebulizers, drug aerosol particle loss in the head airways (OPL region) is less than with MDIs and DPIs, since nebulizers generate the smallest number of particles carrying electrostatic charge (Ali et al., 2008b, 2009). Nebulizers' small particles of antibiotics may reach the alveolar region for a systemic effect to cure viral infection. Conventional nebulizers tend to be large and require either compressed air or an electric power supply, but the new mesh technology is an improvement, delivering portable, silent, and battery-operated devices.

9.4 ANALYTICAL METHODS

For MDI and DPI drug products, the US Food and Drug Administration (FDA) recommends certain studies should be performed to determine appropriate stability test storage conditions and optimum performance properties (US FDA, 2006b). Devices may vary in both design and mode of operation, and these characteristics may be unique to a particular drug product. Drug product-specific information helps to define the appropriate storage conditions, facilitates correct use and maintenance of the inhaler, and contributes to patient compliance. Therefore, the sampling and analysis of an emitted dose of drug aerosol from the PDD devices depend upon the device used. Testing of medication delivered from the devices involves studying the interaction between metering valves, drug formulation, and actuators. In addition, there are several variables (e.g. inhalation flow rate, particle size, concentration) that contribute to the complexity of sampling; some of them are listed in Table 9.1. The sampling technique for each DDD must be adapted to account for all these variables (Lalor and Hickey, 1997).

As detailed in section 9.2, inhaled pharmaceutical aerosol deposition efficiency in the lung depends on the physical characteristics of the particles.

Table 9.1 Variables associated with metered-dose inhalers, dry powder inhalers, and nebulizers

Drug delivery device	Variables			
	Inhalation flow rate (L/min)	Particle size range (μm)	Particle concentration (number/cm^3)	Generation and sampling
MDI	28.3, 30, or 60	0.5–10	Variable	Active
DPI	Bolus, total 4 L	0.5–10	High	Passive
Nebulizer	Device output rate	0.5–5	Low	Passive

The particle or droplet size distribution in the spray discharged from MDIs, and the particle size distribution in the cloud discharged from DPIs, are important characteristics used in judging inhaler performance (USP 24 Aerosols, 2000). A typical respirable aerosol is produced in a fraction of a second with thousands of micronized particles of varying aerodynamic size and electrostatic charge distribution, mass concentration, and velocity. Knowledge of aerodynamic diameter of a particle allows determination of: (i) if and where the particle will be deposited in the human respiratory tract, and (ii) how long the particle will remain airborne in the emitted drug aerosol. There are various techniques and tools available to characterize and analyze these physical properties, as reported by others (Zhao, 2006; Ali et al., 2008b, 2009). In the next few subsections these techniques and tools will be discussed briefly with a few assumptions, such as: (i) particles are spherical with uniform composition; (ii) a uniform state of temperature and humidity within the instruments (tools) exists; and (iii) there is no growth or evaporation of particles in transit.

9.4.1 Non-invasive scintigraphic techniques

Scintigraphic techniques can be used for *in vivo* measurements of the dose delivered to animal and human lungs. These techniques include rectilinear scanners, two-dimensional planar gamma scintigraphy and three-dimensional imaging methods, single photon emission computed tomography, and positron emission tomography. By design, both planar and single photon emission computed tomography employ low-energy gamma emitters, usually 99m-technetium. On the other hand, positron emission tomography utilizes isotopes of basic biological molecules such as oxygen, carbon, nitrogen, fluorine, and neon, which permit functional imaging as well as deposition measurements. Adopting any of these methods requires labeling of the particles before actuation and inhalation.

The labeling quality must be confirmed before use, including consistency in the dose of radioactivity inhaled. This confirmation involves characterizing both radioactivity and the drug aerodynamic size distributions in order to make sure that the drug is not altered by the labeling procedure. Measurements of aerodynamic diameter or size distributions require *in vitro* study of the unlabeled and labeled drug products, and allow clinical response measurements obtained post-inhalation to be correlated with deposition.

High-speed direct laser imaging also precisely measures aerodynamic size distribution, shape, and velocity distribution of drug aerosol particles (Oxford Lasers, 2008). Figure 9.9 illustrates the imaging setup. A short flash of light from a diode laser illuminates a screen, which creates a bright background behind the subject aerosol plume. The short pulse from the laser freezes the motion, allowing blur-free visualization of particle size and shape. Images from the digital camera are transferred to a computer. The high-speed real-time particle sizing software analyzes the images obtained to generate the size distributions. This non-intrusive measurement method is very useful to help understand the nozzle effects, and formation of aerosols from MDIs and nasal sprays.

9.4.2 Inertial impaction

In a uniformly moving aerosol, when flowing air converges, changes direction, or passes an obstacle, curvilinear motion will result for particles responding to forces along two or more axes. Inertial impaction is a special case of curvilinear motion of suspended particles that is extensively applied for collection, measurement, and aerosol separation (Hinds, 1998).

9.4.2.1 Cascade impactor

Cascade impactors operate on the principle of curvilinear motion of particles in the aerosol stream. Figure 9.10a is a schematic representation of the principle of operation of an Andersen cascade impactor (ACI). A single jet per impactor stage is shown. Impactors with multiple jets in each stage function in the same manner.

The flat plate of the stage, called the impaction plate or collection plate, blocks the flow to form an abrupt bend in the streamlines. Particles whose inertia exceeds a certain value (cutoff size) will be unable to follow the streamlines and will impact on the collection plate and be deposited on it.

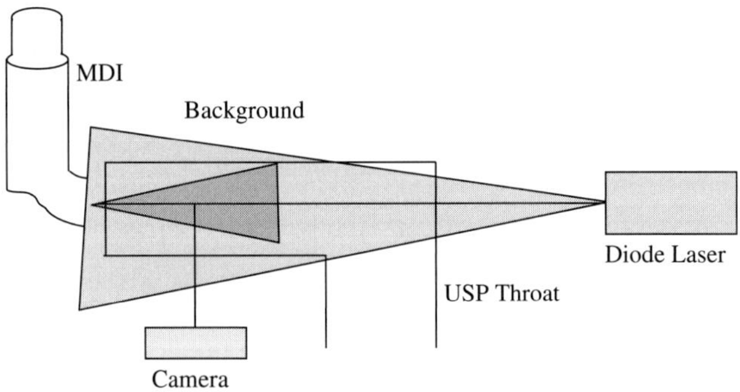

Figure 9.9 Illustration of high-speed direct laser imaging technique to measure the MDI actuated aerosol plume fired into the United States Pharmacopeia (USP) throat.

Smaller particles can follow the streamlines and avoid impacting the collection plate. They become airborne and will have to go through the same phenomenon while flowing through the next successive stages. In addition, particles will deposit on the bead surfaces due to electrostatic image force or space charge force. Since the residence time of the particle is very low, deposition due to diffusion, sedimentation, and interception will be negligible. Thus, each stage of the ACI will separate aerosol particles into two size ranges: particles larger than the cutoff size will be removed from the aerosol stream, and particles smaller than that size will remain airborne and pass through the next stages. Figure 9.10b illustrates the inertial impaction

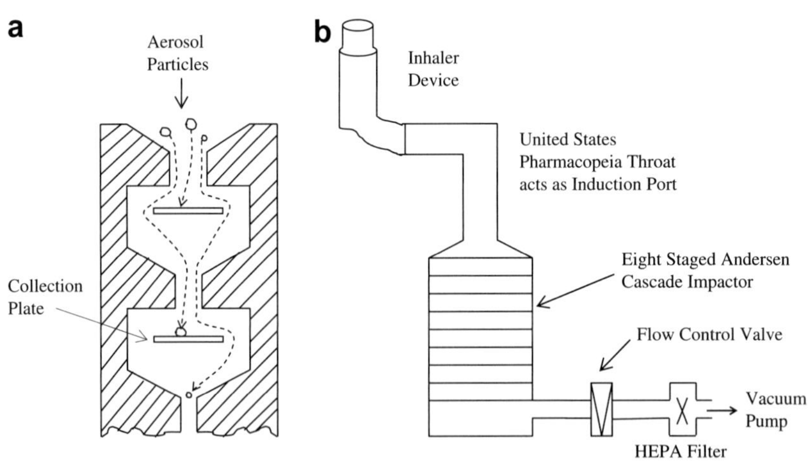

Figure 9.10 (a) Schematic representation of the principle of operation of an Andersen cascade impactor. (A single jet per impactor stage is shown. Impactors with multiple jets in each stage function in the same manner.) (b) Illustration of the inertial impaction setup for measurement of aerodynamic size distributions of respiratory drug aerosols in real time.

setup for measurement of aerodynamic size distributions of respiratory drug aerosols in real time.

The human respiratory tract is an aerodynamic classifying system for inhaled particles (US EPA, 2006). Therefore, the deposition of aerosol particles in each of the ACI stages is expected to show similar deposition patterns in various regions of the human lung. However, the ACI provides the aerodynamic size distribution, and apart from inertial impaction, it is unable to simulate other deposition mechanisms (Hinds, 1998). The correct operation of inertial impactors is crucial for accurate estimation of particle mass median aerodynamic diameter and geometric standard deviation. The aerosol inlet velocity of the impactor should be approximately equal to the velocity leaving the device (Hickey, 1990). Additionally, the inter-stage drug loss or wall losses should not exceed 5% of the inhaler's total delivered drug mass (USP 24 Aerosols, 2000). Particle re-entrainment takes place when particles bounce off the collector plate, which can be minimized by coating the plates with glycol, silicone fluid, or another particle adhesive substance (Byron et al., 1994).

9.4.2.2 Electrical low-pressure impactor

The electrical low-pressure impactor (ELPI) is an instrument to measure airborne submicrometer particle size distribution, concentration, and electrostatic charge of pharmaceutical aerosols in real time (Keskinen et al., 1992). It is developed based on the well-known impactor technology combined with particle charging and extra fittings to provide electrical characteristics such as particle charge distribution in real time, and to make gravimetric impactor measurements. It operates in the size range of 7 nm–10 μm, and can be applied to various different types of measurements where the requirements for the instrument include wide particle size range and fast response time.

Figure 9.11 illustrates the ELPI setup for measurement of both aerodynamic and electrical properties of the MDI emitted aerosols. Because of its rapid response, the ELPI is an effective measurement instrument for the analysis of unstable particulate concentrations and size distributions, or the evolution of size distributions. The operating principle of the ELPI can be divided into three major parts: particle charging in a unipolar corona charger, size classification in a cascade impactor, and electrical detection with sensitive electrometers. The particles are first charged to a known charge level in the charger. After charging, the particles enter a cascade low-pressure impactor with electrically insulated collection stages. The particles are collected in different impactor stages according to their aerodynamic

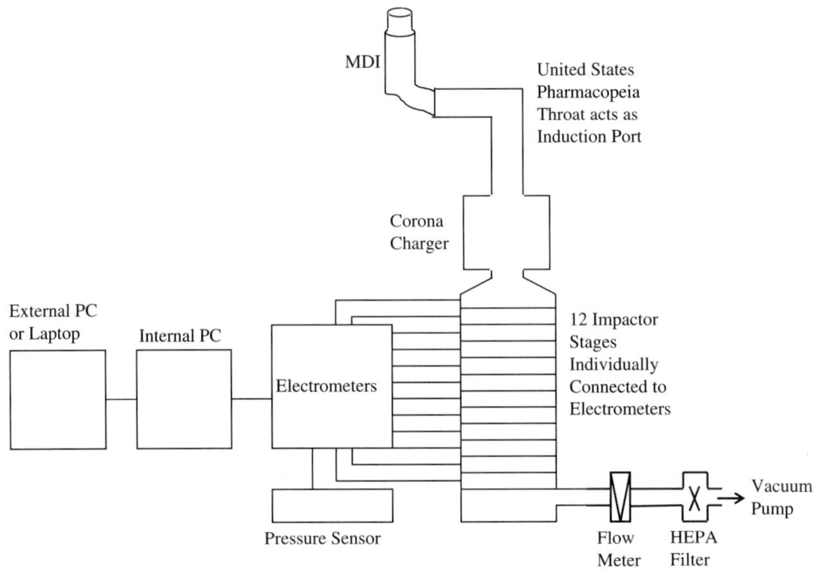

Figure 9.11 Illustration of the ELPI setup for measurement of aerodynamic size and electrostatic charge of respiratory drug aerosols in real time.

diameter, and the electric charge carried by particles into each impactor stage is measured in real time by sensitive multichannel electrometers. The measured current signal is directly proportional to particle number concentration and size. The particle collection into each impactor stage is dependent on the aerodynamic size of the particles. Measured current signals are converted to (aerodynamic) size distribution using particle size-dependent relations describing the properties of the charger and the impactor stages. The result is particle number concentration and size distribution. However, the ELPI has the limitation that it provides only the net charge of all particles deposited at each impactor plate, and does not provide data for each individual particle in real time.

Other unique tools adopting inertial impaction as an operating principle are also available for aerosol particle size fractionation, such as virtual impactors and the PC-2 Air Particle Analyzer (California Measurements, Inc., Sierra Madre, CA). The former does not have particle collector plates; rather the aerosol remains airborne after inertial separation has occurred. The latter uses real-time measurements by piezoelectric quartz crystal microbalance mass sensors to electronically weigh particles in each stage (Hickey, 1992). USP official monographs recommend a Marple Miller Impactor with USP induction port to characterize DPIs. The impactor has five stages and an after-filter. The inhalation flow rate should be 60 L/min and the collector plates should be coated to ensure that particles that impact on a given stage are not re-entrained in the flowing airstream.

9.4.3 Optical microscopy

USP official monographs recommended optical microscopy for suspension-based inhalation aerosols but not for solution-based MDIs or DPIs (USP 22, 1992). The MDI is actuated on to a clean microscope slide held 5 cm from the end of the actuator, perpendicular to the direction of the spray (Byron et al., 1994). The slide is examined under a microscope equipped with a calibrated ocular micrometer, using ×500 magnification. At least 25 fields of view should be examined to acquire accurate particle size information. The advisory panel dealing with the aerosols suggested that these tests can be replaced with a more acceptable method of determining particle aerodynamic diameter, such as use of a cascade impactor (Byron et al., 1994).

9.4.4 Laser-induced tools

There are various laser-induced tools and techniques that are used for performance characterization of therapeutic aerosol DDDs. These tools operate on the basis of interaction of particles with lights, preventing the necessity for tedious manual analysis. One drawback of these instruments is that the technique provides the aerodynamic size of each particle in real time but the particle may not contain an API. A twin-beam phase Doppler analyzer has been used to determine the size and velocity of particles in the cloud delivered from commercially available MDIs (Clifford et al., 1990). It works by laser Doppler anemometry. Laser diffraction techniques are also used to find aerosol particle size, velocity, and dynamics of atomization, in the size range of 0.6 nm–10 μm (Mastersizer, Zetasizer, Spraytec, Sysmex, Malvern Instruments, Southborough, MA). Two other widely known tools, the aerodynamic particle sizer spectrometer and electronic single-particle aerodynamic relaxation time analyzer, will be discussed in the following subsections.

9.4.4.1 Aerodynamic particle sizer

The aerodynamic particle sizer (APS) spectrometer sizes particles in the range from 0.5 to 20 μm using a sophisticated time-of-flight technique that measures aerodynamic diameter in real time. The advantage of time-of-flight aerodynamic sizing over normal light scattering is that it accounts for particle shape and is unaffected by index of refraction or Mie scattering. In addition, the monotonic response curve of the time-of-flight measurement ensures high-resolution sizing over the entire particle size range. The APS detects particles from size 0.37 to 20 μm using a light-scattering technique.

While light-scattering intensity is not always a reliable indicator of particle size, it remains a parameter of interest. The APS keeps this second measurement separate and distinct from aerodynamic size. The ability to provide two measurements of each particle using the same sensor allows us to gain exciting new insights into the makeup of an aerosol. The APS uses a double-crest optical system to detect the occurrence of particle coincidence (when more than one particle is in the detection area) and to identify poor signals near the instrument's lower detection threshold. This ensures robust and high-quality measurements. Other features, such as microprocessor-controlled volumetric flow control, barometric pressure correction, and separate pumps for sheath and total flows, enable the APS to operate under a wide range of conditions and remain in calibration.

9.4.4.2 Electronic single-particle aerodynamic relaxation time analyzer

The electronic single-particle aerodynamic relaxation time (ESPART) analyzer determines the aerosol particles' aerodynamic size and electrostatic charge in real time (Mazumder et al., 1989). Figure 9.12 depicts its working principle. The ESPART samples aerosol at a rate of 1 L/min, which flows downwards through the sensing volume of focused beams of laser radiation. As particles move downwards, they come across two converging laser beams. They also experience AC electric excitation, which makes them oscillate horizontally. A photomultiplier is used to measure the intensity of the scattered light generated by each particle as it passes through the sensing volume. An electronic signal processor and data processor analyze the phase lag of the motion of the particle with respect to the AC electric field driving the particles. The aerodynamic diameter is derived from the phase lag value. The direction and amplitude of the electrical migration velocity of the particles with respect to the electric field provide the polarity and magnitude of the electrostatic charges.

The ESPART analyzer operates in two modes. In mode 1, it measures the aerodynamic diameter of each particle whether the particle is charged or uncharged. Acquired aerosol data in this mode represent all charged and uncharged inhaled particles. In mode 2, it measures the aerodynamic diameter and the electrostatic charge of each charged particle only. In mode 2, it ignores the uncharged particles by design. The ESPART has certain limitations too. It is unable to detect and characterize aerosol particles smaller than 0.5 μm. Its system parameters were set for a particle size range from 0.5 to 20 μm. Particles larger or smaller than this operating range are ignored. The maximum counting limit of this instrument is 200 particles per second.

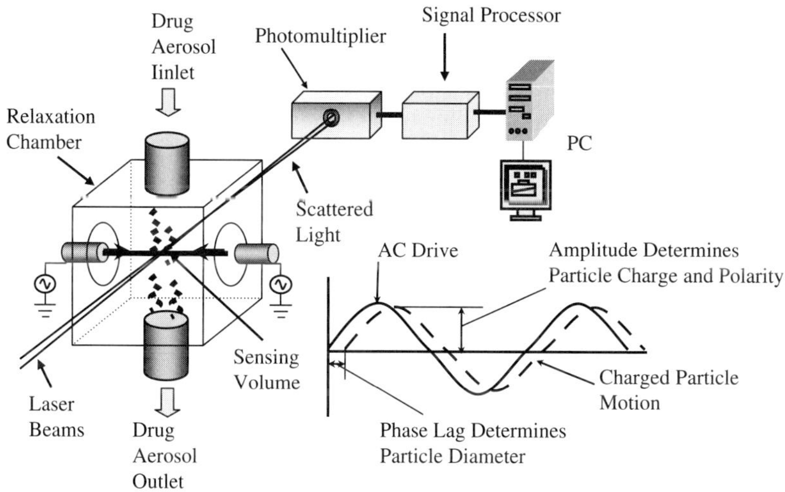

Figure 9.12 Schematic representation of the principle of operation of the electronic single-particle aerodynamic relaxation time (ESPART) analyzer. From Ali et al. (2009).

9.5 CHALLENGES AND ADVANTAGES

The goal of minimally invasive drug delivery targeted for PDD has become one of the most important areas of pharmaceutical research. In the early 1990s, expectations regarding pulmonary drug therapy were that the optimal formulations would be delivered via modified aerosol delivery devices to achieve reproducible and maximal doses to the lung that would eventually serve as alternatives to parenteral delivery (Byron, 1990). In the last few years, techniques and new drug delivery devices intended to deliver drugs into the lungs have been widely investigated, and many of those initiatives have been very successful. After more than 50 years of development, MDIs that deliver drugs to treat asthma and related airway diseases are an established multibillion dollar market serving some hundreds of million patients worldwide.

The critical success factors of maximum bioavailability of any delivered drug depends on (a) deposition sites, (b) receptors linked with sufficient blood and distribution of other fluid levels, (c) sufficiency in absorption rates, (d) biotransformation processes, (e) protein binding and (f) the rate of appearance in the blood. The large blood supply of the lung (i.e. virtually the entire cardiac output passes through the alveolar capillaries) makes it a potential route for the delivery of aerosolized systemic drugs with active agents including new-generation drugs that prefer non-invasive delivery methods. Drugs delivered in this fashion enter the arterial circuit directly and would not need to pass through hepatic beds, as do orally administered drugs. Intratracheal administration is commonly used as a first approach of *in vivo* drug delivery. Nasal drug delivery in aerosolized form can also be utilized for delivering proteins and peptides to treat osteoporosis, pain management, obesity, sexual dysfunction, Alzheimer's and Parkinson's diseases.

Drug delivery via the non-invasive pulmonary route still remains a challenge due to poor absorption and enzymatic instability. At present no drug delivery system achieves site-specific delivery while controlling the release kinetics of drug in a predictable manner. Furthermore, dosage forms containing the same amount of active pharmaceutical ingredients (chemically equivalent) may not necessarily provide the same therapeutic response. Rather, the rate at which the drug is liberated from the dosage form and the subsequent absorption, sufficiency in circulation, metabolism, and excretion kinetics determine the bioavailability at the receptor site. While systemic administration may achieve delivery of drugs to peripheral lung via diffusion from the pulmonary circulation, and also treat systemic aspects of COPD, there is a high risk of side-effects with many of the novel anti-inflammatory approaches currently in development, which may necessitate the development of novel inhalation approaches to achieve efficient peripheral drug deposition in the airways (Barnes, 2008).

Technological milestones of inhalation therapy in recent years have included several new developments in the field, including auxiliary systems to improve drug delivery from established devices and new categories of devices, notably single-breath aqueous systems (Dalby and Suman, 2003). Along with improvements in device technology, improved understanding of pulmonary diseases has led to new drugs, but the only barrier to therapy is the patient. Patient training and compliance are still the most important factors in the success or failure of inhalation therapy. Also, to transport a desired dose of drug to the lungs, suitable drug carriers are required, which can be either solid, liquid, or gaseous excipients. Liposomes, nano- and microparticles, cyclodextrins, microemulsions, micelles, suspensions, or solutions are all examples of this type of pharmaceutical carrier that have been successfully used to target drugs into the lungs (Courrier et al., 2002). Current particle and device engineering technologies have improved on delivering drugs to the target region of lungs, but still fail to recover the "wasted" fraction of the drug and deliver it deeper into the lungs, which is generally desirable (Ganguly et al., 2008).

Liposomal encapsulation of the API has been shown to be promising in sustaining the drug residence time within lung, improving therapeutic index, delaying systemic dilution, and thereby reducing side-effects (Chougule et al., 2007). Pulmonary drugs, those adopting liposomal technology, and received US patents are: delivery of corticostaroid for asthma (US5049389); ribonucleotides for respiratory influenza (US20050220720A1); amphiphatic drugs and their salts for tumor (US2006153217); aminoglycosides and other antibiotics for local pulmonary infections and cystic fibrosis (US5958449). Large porous particles (LPPs) are a drug delivery platform that manipulates particle properties by increasing geometric diameter and lowering particle mass density while maintaining an aerodynamic diameter that allows for entry and deposition in the lungs (Chougule et al., 2007). LPPs disperse more readily than classical small dense particles of inhaled pharmaceutical formulations, allowing efficient delivery of large drug masses from a relatively simple inhaler. At the other end of the size spectrum, nanoparticles have been pursued in drug delivery because of their ability to enhance drug absorption, traffic through tissues, and target cells. However, scale-up and handling of nanoparticles has been an issue since they tend to agglomerate due to their high surface-to-volume ratio (Chougule et al., 2007). Incorporating nanoparticles into the LPP forms porous nanoparticle-aggregate particles containing varying percentages of nanoparticles. It may be possible, by controlling the aerodynamic size of the nanoparticles, that they can be targeted to the lung, alveolar macrophages, or systemic circulation. Industry experts predict that all aspects of pharmaceutical development, analytical skills, medical device engineering, quality control, and delivery systems are multidisciplinary tasks that can be achieved through strategic partnerships.

9.6 CONCLUSIONS

During the past decade pulmonary drug delivery technology advanced substantially due to several inventions in propellants, excipients, and device technologies. The large surface area of the human lung, along with its rich blood supply, rapid onset of drug action with high bioavailability, and other physiological advantages, make it a potential route for treating asthma, COPD, and other pulmonary disorders. Minimally invasive drug delivery through the lung can be achieved using environment friendly propellants, non-aqueous inhalers, user-friendly dry powder inhalers, and jet or ultrasonic nebulizers. In order for these devices to effectively deliver micronized drug particles in aerosolized form, drug particles of the appropriate respirable size range and innate electrostatic charge are required. Characterization of pulmonary drug aerosols may be performed using several different methods. Those properties indicate stability and performance of the drug and the device. Assuming the quality of the drug aerosol is appropriate for lung deposition it is also necessary to demonstrate that sufficient drug will be delivered to accomplish the desired therapeutic effect in a reproducible manner.

It is clear that PDD has much potential to be successful for systemic inhalation delivery because (i) the basic science of pulmonary delivery of small molecules is well understood, (ii) good progress is being made to elucidate the pathways of absorption of macromolecules including assisted transport, (iii) systemic effects can be achieved even with modest bioavailability, (iv) a variety of mature dosage forms and device technologies are available, (v) considerations are being made for long-term safety of new excipients, and (vi) significant socioeconomic value propositions exist in systemic inhaled delivery of drugs and biologics.

Current understanding of *in vitro* and *in vivo* relationships for PDD products are incomplete and still evolving, but the correlations showed that *in vitro* data can be used more widely as a means of comparing different inhaler devices during product development (Newman and Chan, 2008). Pulmonary drug formulation and delivery device technologies are advancing very rapidly as the pharmaceutical industry moves to adopt the potential advantages of the pulmonary route as an alternative to oral and parenteral delivery methods. These advances are attracting new players and new investment, therefore accelerating the pace and number of new pulmonary drug candidates entering the drug development pipeline. The demand profile of inhaled drugs is producing new combination products such as drugs for treatment of endometriosis, several forms of cancer, hemophilia, neurological disorders, and pain management through development and clinical trials (Dubin, 2006). Several of these drugs will have significant impact on the PDD market. In response, regulatory bodies such as the US FDA, USP, and expert groups have devoted much attention to developing and introducing appropriate test methods. The recent inhalation product guidance from the European Medicines Agency, the draft pharmacopeia monographs of the US FDA, and the soon-to-be-completed International Standards Organization standards for drug delivery devices are examples of the ongoing nature of this work (Copley, 2008).

REFERENCES

Aggarwal, A.N., Gupta, D., Jindal, S.K., 2006. The relationship between FEV1 and peak expiratory flow in patients with airway obstruction is poor. Chest. 130 (5), 1454–1461.

Ali, M., Reddy, R.N., Mazumder, M.K., 2008a. Electrostatic charging effect on workplace aerosol particle deposition in a hollow throat cast. J. Electrostatics 66, 401–406.

Ali, M., Reddy, R.N., Mazumder, M.K., 2008b. Simultaneous characterization of aerodynamic size and electrostatic charge distributions of inhaled dry powder inhaler aerosols. Curr. Respir. Med. Rev. 4 (1), 2–5.

Ali, M., Mazumder, M.K., Martonen, T.B., 2009. Measurements of electrodynamic effects on the deposition of MDI and DPI aerosols in a replica cast of human oral–pharyngeal–laryngeal airways. J. Aerosol Med. Pul. Drug Deliv. 22 (1), 35–44.

Al-Showair, R.A.M., Pearson, S.B., Chrystyn, H., 2007. The potential of a 2Tone trainer to help patients use their metered dose inhalers. Chest. 131, 1776–1782.

Ashurst, I., Malton, A., Prime, D., Sumby, B., 2000. Latest advances in the development of dry powder inhalers. Pharm. Sci. Technol. 3, 246–256.

Asmus, M.J., Coowanitwong, I., Kwon, H.S., Khorsand, N., Hochhaus, G., 2003. In vitro performance of two common valve holding chambers with a chlorofluorocarbon free beclomethasone metered dose inhaler. Pharmacotherapy 23 (2), 1538–1544.

Barat, R.B., Callahan, V.L., Cedio-Fengyn, D.J., Stevens, J.G., 2002. The effect of particle size distribution on the deposition of charged particles in tubes. Adv. Env. Res. 7, 105–115.

Barnes, P.J., 2008. New treatments for COPD and the issue of drug delivery. In: Dalby, R.N., Byron, P.R., Peart, J., Suman, J.D., Farr, S.J., Young, P.M. (Eds), Respiratory Drug Delivery XIII. Virginia Commonwealth University, Richmond, pp. 1–10.

Barry, P.W., O'Callaghan, C., 1995. The effect of delay, multiple actuations spacer static charge on the in vitro delivery of budesonide from nebuhaler. Br. J. Clin. Pharmacol. 40, 76–78.

Boccuzzi, S.J., Wogen, J., Roehm, J.B., 2000. Use of hydrofluoroalkane propellant delivery system for inhaled albuterol in patients receiving asthma medications. Clin. Ther. 22 (2), 237–247.

Byron, P., 1990. Determinants of drug and polypeptide bioavailability from aerosols delivered to the lung. Adv. Drug Deliv. 5, 107–132.

Byron, P.R., Kelly, E.L., Kontny, M.J., et al., 1994. Recommendations of the USP advisory panel on aerosols on the USP general chapters on aerosols <601> and uniformity doses units <905>. Pharm. Forum 20, 7477–7505.

Cheng, Y.S., 2003. Modeling aerosol drug delivery. In: Gradon, L., Marijnissen, J. (Eds), Optimization of Aerosol Drug Delivery. Kluwer Academic, Boston, MA, pp. 165–188.

Chougule, M.B., Padhi, B.K., Jintukar, K.A., Misra, A., 2007. Developments of dry powder inhalers. Recent Patents Drug Deliv. Formul. 1, 11–21.

Clay, M.M., Clarke, S.W., 1987. Effect of nebulized aerosol size in lung deposition in patients with asthma. Thorax 42, 190–194.

Clifford, R.H., Ishii, I., Montaser, A., 1990. Dual beam light scattering interferometry for simultaneous measurements of droplet size and velocity distributions of aerosols from commonly used nebulizers. Chemistry 62, 309–394.

Colthorpe, P., 2003. Industry experience of the HFA transition. Drug Deliv. Syst. Sci. 3 (2), 41.

Copley, M., 2008. Pulmonary drug delivery: technology, regulation and testing. Drug Deliv. Technol. 8 (9), 60–67.

Courrier, H.M., Butz, N., Vandamme, T.F., 2002. Pulmonary drug delivery systems: recent developments and prospects. Crit. Rev. Ther. Drug Carrier Syst. 19 (228) 425–498.

Cross, J.A., 1987. Electrostatics: Principles, Problems and Applications. Institute of Physics, Bristol.

Dalby, R.N., Suman, J., 2003. Inhalation therapy: technological milestones in asthma treatment. Adv. Drug Deliv. Rev. 55 (7), 779–791.

Dalby, R.N., Tiano, S.L., Hickey, A.J., 1996. Medical devices for the delivery of therapeutic aerosol to the lungs. In: Hickey, A.J. (Ed.), Inhalation Aerosols. Marcel Dekker, New York, pp. 441–473.

de Boer, A.H., Molema, G., Frijlink, H.W., 2001. Pulmonary drug delivery: delivery to and through the lung. In: Molema, G., Meijer, D.K.F. (Eds), Drug Targeting. Wiley-VCH, Weinheim, pp. 53–87.

de Boer, A.H., Hagedoom, P., Gjaltema, D., Goede, J., Frijlink, H.W., 2006. Air classifier technology (ACT) in dry powder inhalation. Part 3: Design and development of an air classifier family for the novolyzer multi-dose dry powder inhaler. Int. J. Pharm. 310 (1–2), 72–80.

Dubin, C.H., 2006. Systemic pulmonary delivery finds its first stepping stone. Drug Deliv. Technol. 6 (4), 46–53.

Epstein, S., Maidenberg, A., Hallett, D., Khan, K., Chapman, K.R., 2001. Patient handling of a dry powder inhaler in clinical practice. Chest 120 (5), 1480–1484.
Escribano, A., Tutuncu, A., Lohr, I., Carlholm, M., Polanowski, T., 2006. Clinical comparability between the CFC and HFA budesonide pressurized metered-dose inhalers in paediatric patients with asthma: a randomized controlled trial. Curr. Med. Res. Opin. 22 (6), 1085–1092.
Fink, J.B., 2000. Metered dose inhalers, dry powder inhalers and transitions. Respir. Care 45 (6), 623–635.
Finlay, W.H., 2001. The Mechanics of Inhaled Pharmaceutical Aerosols: An Introduction. Academic Press, San Diego.
Fjeld, R.A., Wu, D., McFarland, A.R., 1990. Evaluation of continuum regime theories for bipolar charging of particles in the 0.3–13 μm diameter size range. IEEE Trans. Ind. Appl. 26 (3), 523–528.
Foss, S.A., Sladek, D.T., English, R.E., McGowen, P.D., 1997. Particle size comparison of MediSpacer and AeroChamber in the delivery of Proventil HFA and Proventil MDI. Respir. Care 42 (8), 987–995.
Frijlink, H.W., de Boer, A.H., 2004. Introduction to pulmonary drug delivery. Expert Opin. Drug Deliv. 1, 67–86.
Ganguly, S., Moolchandani, V., Roche, J.A., Shapiro, P.S., Somaraju, S., Eddington, N.D., Dalby, R.N., 2008. Phospholipid-induced *in vivo* particle migration to enhance pulmonary deposition. J. Aerosol Med. Pulm. Drug Deliv. 21 (4), 343–350.
Grossman, J., 1994. The evaluation of inhalation technology. J. Asthma 31, 55–64.
Heyder, J., 1975. Gravitational deposition of aerosol particles within a system of randomly oriented tubes. J. Aerosol Sci. 6, 133–137.
Hickey, A.J., 1990. An investigation of size deposition upon individual stages of cascade impactor. Drug Dev. Ind. Pharm. 16, 1911–1929.
Hickey, A.J. (Ed.), 1992. Methods of aerosol particle size characterization. In: Pharmaceutical Inhalation Aerosol Technology. Marcel Dekker, New York, pp. 219–254.
Hinds, W.C., 1998. Aerosol Technology: Properties, Behavior and Measurement of Airborne Particles, 2nd ed. John Wiley, New York.
Hofmann, W., Strum, R., Winkler-Heil, R., Pawlak, E., 2003. Stochastic model of ultrafine particle deposition and clearance in the human respiratory tract. Rad. Prot. Dosi. 105 (1–4), 77–80.
Ingham, D.B., 1975. Diffusion of aerosols from a stream flowing through a cylindrical tube. J. Aerosol Sci. 6, 125–132.
Janssens, H.M., Heijnen, E.M.E.W., de Jong, V.M., Hop, W.C.J., Holland, W.P.J., de Jonste, J.C., Tiddens, H.A.W.M., 2000. Aerosol delivery from spacers in wheezy infants: a daily life study. Eur. Respir. J. 16, 850–856.
Kellaway, I.W., 1988. The design of drug delivery systems. In: Smith, H.J. (Ed.), Introduction to the Principles of Drug Design and Action. Harwood Academic, Amsterdam, pp. 29–50.
Keskinen, J., Pietarinen, K., Lehtimaki, M., 1992. Electrical low pressure impactor. J. Aerosol Sci. 23 (4), 353–360.
Kim, C.S., 2000. Methods of calculating lung delivery and deposition of aerosol particles. Respir. Care J. 45 (6), 695–711.
Kim, C.S., Eldridge, M.A., Sackner, M.A., 1987. Oropharyngeal deposition and delivery aspects of metered dose inhaler aerosols. Am. Rev. Respir. Dis. 135, 157–164.
Kwok, P.C.L., Collins, R., Chan, H.K., 2006. Effect of spacers on the electrostatic charge properties of metered dose inhaler aerosols. J. Aerosol Sci. 37 (12), 1671–1682.
Lalor, C.B., Hickey, A.J., 1997. Generation and characterization of aerosols for drug delivery to the lungs. In: Adejei, A.L., Gupta, P.K. (Eds), Inhalation Delivery of Therapeutic Peptides and Proteins. Marcel Dekker, New York, pp. 235–277.
Landahl, H.D., 1963. Particle removal by the respiratory system. Notes on the removal of airborne particulates by the human respiratory tract with reference to the role of diffusion. Bull. Math. Biophys. 25, 29–39.

Larsen, J.S., Hahn, M., Ekholm, B., Wick, K.A., 1994. Evaluation of conventional press and breathe metered dose inhaler technique in 501 patients. J. Asthma 31 (3), 193–199.
Lawless, P.A., 1996. Particle charging bounds, symmetry relations, and an analytic charging rate model for the continuum regime. J. Aerosol Sci. 27 (2), 191–215.
Lippmann, M., Lee, D.H.K., Falk, H.L., Murphy, S.O., Geiger, S.R., 1977. Regional deposition of particles in the human respiratory tract. In: Handbook of Physiology, Reaction to Environmental Agents. American Physiological Society, Bethesda, MD.
Liu, Z.G., Wang, P.K., 1997. Pressure drop and interception efficiency of multifiber filters. Aerosol Sci. Technol. 26 (4), 313–325.
Mazumder, M.K., Wilson, J.D., Wankum, D.L., Cole, R., Northop, G.M., Neidhardt, L.T., Martonen, T.B., 1989. Dual laser Doppler system for real-time simultaneous characterization of aerosols by size and concentration. In: Crapo, J.D. (Ed.), Lung Dosimetry. Academic Press, San Diego, pp. 211–234.
Minter, B., 2003. Opportunities in Drug Delivery: Identifying Hot Technologies, Companies and Markets. Urch, London.
Newman, S.P., Chan, H.K., 2008. In vitro/in vivo comparisons in pulmonary drug delivery. J. Aerosol Med. Pul. Drug Deliv. 21 (1), 77–84.
Newman, S.P., Clarke, A.R., Talaee, N., Clarke, S.W., 1991. Lung deposition of 5 mg Intal from a pressurized metered dose inhaler assessed by radiotracer technique. Int. J. Pharm. 74, 203–208.
Newman, S.P., Steed, K., Hooper, G., et al., 1995. Comparison of gamma scintigraphy and a pharmacokinetic technique for assessing pulmonary deposition of terbutaline sulphate delivered by pressurized metered dose inhaler. Pharm. Res. 12 (2), 231–236.
Oberdorster, G., 2001. Pulmonary effects of inhaled ultrafine particles. Int. Arch. Occup. Environ. Health 74 (1), 237–247.
O'Callaghan, C., Nebrink, O., Vidgren, M.T., 2002. The history of inhaled drug therapy. In: Bisgaard, H., O'Callaghan, C., Smaldone, G.C. (Eds), Drug Delivery to the Lung. Marcel Dekker, New York, pp. 1–18.
Orban, J.C., Peart, J., 2004. Simultaneous electrostatic charge characterization and particle size analysis of metered dose inhalers (pMDIs) using the electrical low pressure impactor. In: Dalby, R.N., Byron, P.R., Peart, J., Suman, J.D., Farr, S.J. (Eds), Respiratory Drug Delivery IX. Virginia Commonwealth University, Richmond, pp. 877–880.
O'Riordan, T.J., 2002. Formulations and nebulizer performance. Respir. Care 47 (11), 1305–1312.
Lasers, Oxford, 2008. VisiSizer S100. Operating Manual. Oxford Laser Ltd.
Pauwels, R., Newman, S.P., Borgstrom, L., 1997. Airway deposition and airway effects of antiasthma drugs delivered from metered-dose inhalers. Eur. Respir. J. 10 (9), 2127–2138.
Peart, J., Kulphaisal, P., Orban, J.C., 2003. Relevance of electrostatics in respiratory drug delivery. Business Briefing: Pharmagenerics 84–87.
Phalen, R.F., Oldham, M.J., Wolff, R.K., 2008. The relevance of animal models for aerosol studies. J. Aerosol Med. Pul. Drug Deliv. 21 (1), 113–124.
Plaza, V., Sanchis, J., group, C.E.S.A., 1998. Medical personnel and patient skill in the use of metered dose inhalers: a multicentric study. Respiration. 65, 195–198.
ProAir HFA, 2008. What is different about HFA. <www.proairhfa.com>, accessed 18 October 2008.
Rau, J.L., 2002. Design principles of liquid nebulization devices currently in use. Respir. Care 47 (11), 1257–1275.
Rees, P.J., Clark, T.J.H., Moren, F., 1982. The importance of particle size in response to inhaled bronchodilators. Eur. J. Respir. Dis. Suppl. 119, 73–78.
Selroos, O., et al., 1996. Delivery devices for inhaled asthma medication: clinical implications of differences in effectiveness. Clin. Immunother. 6, 273–299.
Smith, H.J., Williams, H., 1998. Introduction to the Principles of Drug Design and Action, 3rd ed. Harwood Academic, Amsterdam.

Smyth, H.D.C., 2003. The influence of the formulation variables on the performance of alternative propellant driven metered dose inhalers. Adv. Drug Deliv. Rev. 55, 807–828.

Staniforth, J.N., 1994. The importance of electrostatic measurements in aerosol formulation and preformulation. In: Byron, P.R., Dalby, R.N., Farr, S.J. (Eds), Respiratory Drug Delivery IV. Interperm Press, Deerfield, pp. 303–311.

Su, W.C., Wu, J., Marijnissen, J.C.M., Cheng, Y.S., 2008. Deposition of man-made fibers in a human nasal airway. Aerosol Sci. Technol. 42 (3), 173–181.

Suarez, S., Hickey, A.J., 2000. Drug properties affecting aerosol behavior. Respir. Care J. 45 (6), 652–666.

Swift, D.L., 1996. Use of mathematical aerosol deposition models in predicting the distribution of inhaled therapeutic aerosols. In: Hickey, A.J. (Ed.), Inhalation Aerosols. Marcel Dekker, New York, pp. 51–81.

Thorsson, L., et al., 1994. Lung deposition of budesonide from Turbuhaler is twice that from a pressurized metered dose inhaler. Eur. Respir. J. 7, 1839–1844.

Thorsson, L., et al., 2001. Pharmacokinetics and systemic activity of fluticasone propionate and budesonide. Br. J. Clin. Pharmacol. 52, 1–15.

Tønnesen, F., et al., 1994. Bronchodilating effect of terbutaline powder in acute severe bronchial obstruction. Chest. 105, 697–700.

US EPA, 2006. Basic Concepts in Environmental Sciences. United States Environmental Protection Agency, Research Triangle Park, NC. <www.epa.gov/cgi-bin>, accessed March 2006.

US FDA, 2006a. Metered-Dose Inhalers. Food and Drug Administration. <http://www.fda.gov/cder/mdi/>, accessed 2 June 2006.

US FDA, 2006b. Questions and Answers on Final Rule of Albuterol MDIs. Food and Drug Administration. <http://www.fda.gov/cder/mdi/mdifaqs.htm>, accessed 2 June 2006.

US FDA, 2006c. Your Metered-Dose Inhaler is Changing to Help Improve the Environment. Food and Drug Administration. <http://www.fda.gov/cder/consumerinfo/metered-dose-inhaler-2pge.htm>, accessed 2 June 2006.

USP 22, 1992. United States Pharmacopeial Convention, Rockville, MD, pp. 1556–1557, 3122–3129, 3146–3148.

U.S.P. 24 Aerosols (601), 2000. Physical tests and determinations: aerosols, metered dose inhalers, and dry powder inhalers. United States Pharmacopeial Convention, Rockville, MD, pp. 1895–1912.

Vanbever, R., 2003. Optimization of drug powder aerosols for systemic drug delivery. In: Gradon, L., Marijnissen, J. (Eds), Optimization of Aerosol Drug Delivery. Kluwer Academic, Boston, MA, pp. 91–103.

Weibel, E.R., 1963. Morphometry of the Human Lung. Springer-Verlag, Berlin.

Wildhaber, J.H., Devadason, S.G., Hayden, M.J., et al., 1996. Electrostatic charge on a plastic spacer device influences the delivery of salbutamol. Eur. Respir. J. 9, 1943–1946.

Wildhaber, J.H., Devadason, S.G., Hayden, M.J., Eber, E., Summers, Q.A., 1997. Aerosol delivery to wheezy infants: comparison between a nebulizer and two small volume spacers. Pediatr. Pulmonol. 23, 212–216.

Yu, C.P., 1977. Precipitation of unipolarly charged particles in cylindrical and spherical vessels. J. Aerosol Sci. 8, 237–241.

Yu, C.P., Diu, K., 1982. A comparative study of aerosol deposition in different lung models. Am. Ind. Hyg. Assoc. J. 43, 54–65.

Yu, C.P., Pich, J., 1972. Theory of gravitational deposition of particles from laminar flows in channels. J. Aerosol Sci. 3, 351–361.

Zhao, J., 2006. Characterization of inhalation aerosols. Proc. ESA/IEJ/IEEE-IAS/SFE Joint Conference on Electrostatics, pp. 211–221.

CHAPTER 10

Regulatory Aspects of Nasal and Pulmonary Spray Drug Products

Dino J. Farina
Proveris Scientific Corporation, 290 Donald Lynch Boulevard, Suite 100, Marlborough, MA 01752, USA

Contents

10.1 Introduction 247
10.2 Setting specifications and performing product characterization studies 253
  10.2.1 Tests related to product labeling 253
  10.2.2 Tests related to patient use/misuse 259
  10.2.3 Tests related to general properties 267
10.3 Conclusion 287
References 287

10.1 INTRODUCTION

Drug safety and efficacy are of paramount importance when considering regulatory aspects for any drug, especially those intended to be delivered to the nose or lung. In the USA, the Food and Drug Administration (henceforth "FDA" or "Agency") has established a set of industry guidelines and regulations with which applicants must comply to market nasal and pulmonary spray drug products (otherwise known as orally inhaled and nasal drug products, or OINDPs) (Center for Drug Evaluation and Research (CDER), 1998, 2002, 2003). Similarly, the European Medicines Agency (EMEA) and Health Canada have established similar guidelines and regulations for OINDPs marketed in Europe and Canada respectively (Committee for Medicinal Products for Human Use (CHMP), 2006). The FDA has had more experience with and published more regulatory documents related to OINDPs than any other country. The guidelines adopted by the EMEA and Health Canada for OINDPs closely follow the guidelines published and enforced by the FDA, but there are some important differences between them, mainly regarding stability considerations, terminology (Table 10.1), and testing specifications (Table 10.2).

Handbook of Non-Invasive Drug Delivery Systems
ISBN 9780815520252

Table 10.1 Classification of OINDPs in regulatory guidelines

FDA terminology	**EMEA terminology**
Nasal spray	Non-pressurized metered-dose nasal spray Nasal single-use spray
Inhalation spray	Non-pressurized metered-dose inhaler
Inhalation solutions and suspensions	Product for nebulization (single and multiple use)
Metered-dose inhaler (MDI)	Pressurized metered-dose nasal spray Pressurized metered-dose inhaler
Dry powder inhaler (DPI), device metered	Dry powder inhaler, device metered Nasal powder, device metered
Dry powder inhaler (DPI), pre-metered	Dry powder inhaler, pre-metered

Adapted from Kim (2008).

Not to be overlooked are the regulatory enforcement capabilities of the FDA, EMEA, and HealthCanada for OINDPs. The most powerful example in recent times occurred in May 2002, when the FDA imposed a record $500 million "disgorgement" from Schering-Plough as part of a consent decree for repeated regulatory violations found at two of the company's manufacturing facilities (United States of America v. Schering-Plough Corporation, 2002). Note that this consent decree was officially dissolved in August 2007 after the company completed over 200 significant steps and 30 validation actions.

Though the above-mentioned consent decree may be an extreme case, as described by Parker (2005), there are 10 critical administrative enforcement areas that the FDA can leverage to protect the public's health, including recalls, application integrity, injunctions, and seizures, all of which can be considered for OINDPs, especially those that are life- or patient safety-critical such as rescue inhalers and those with controlled substances used for pain management (e.g. fentanyl). The Agency can take a number of actions to protect the public's health when a problem arises with an OINDP it regulates, but the process almost always starts with the Agency and the manufacturer working out a voluntary corrective action plan. If the corrective action plan fails to remedy the problem, then the Agency can take more drastic measures, including asking the manufacturer to recall the affected product(s), having federal marshals seize product if a voluntary recall is not done, or detaining

Table 10.2 Tests to be considered for product performance characterization studies and/or specifications

		Pulmonary				Nasal			
			DPI					Spray	
Test class	Test or specification name	Oral pMDI	Device-metered	Pre-metered	Inhalation spray	Nasal pMDI	Nasal powder	Single dose	Multi dose
Labeling	1. Appearance and color/	F	F	F	F	F	F	F	F
	description	E	E	E	E	E	E	E	E
	2. Cleaning instructions	F	F	F	F	F	F		F
		E	E	E	E	E	E		E
	3. Drug content (assay)	F	F	F	F	F	F	F	F
		E	E	E	E	E	E	E	E
	4. Identification	F	F	F	F	F	F	F	F
	5. Net content (fill)	F	F		F	F	F	F	F
	weight/number of	E	E	E	E	E	E	E	E
	actuations per container								
	6. Priming/repriming	F	F		F	F	F		F
		E			E	E		E	E
	7. Profiling of actuations	F	F		F	F	F		F
	near exhaustion (tailoff)	E	E	E	E	E	E		E
	8. Pump/valve delivery	F			F	F		F	F
	(shot weight)								
	9. Shaking (suspensions)	F			F	F		F	F
		E			E	E		E	E
	10. Stability of primary	F	F	F	F	F	F	F	F
	(unprotected) package								

(Continued)

Table 10.2 Tests to be considered for product performance characterization studies and/or specifications — Cont'd

		Pulmonary				Nasal			
			DPI					Spray	
Test class	**Test or specification name**	**Oral pMDI**	**Device-metered**	**Pre-metered**	**Inhalation spray**	**Nasal pMDI**	**Nasal powder**	**Single dose**	**Multi dose**
Patient use/misuse	1. Determination of appropriate storage conditions	F	F	F	F	F	F		
	2. Device robustness/ruggedness	F E	F E	F E	F E	F E	F E	F E	F E
	3. Dose buildup and flow resistance		F	F			F		
	4. Drug deposition on mouthpiece and/or accessories	F E	E	E	F E	F E	E	E	E
	5. Effect of moisture	E	F E	F E	E		F E		
	6. Effect of orientation/weight loss (stability)	F E	F E	F E	F E	F E	F E	F E	F E
	7. Effect of patient use	F	F	F	F	F	F	F	F
	8. Effect of resting time	F			F	F			
	9. Effect of storage on the particle size distribution	F	F		F	F	F	F	F
	10. Effect of varying flow rates	F	F E	F E	F	F	F		
	11. Low-temperature performance	E			E	E			
	12. Temperature cycling	F E			F E	F E		F E	F E

General properties	1. Chemical compatibility	F	F	F	F	F	F	F	F
	2. Dehydrated alcohol content	F				F			
	3. Delivery device	F	F	F		F	F	F	F
	development	E	E	E		E	E	E	E
	4. Dose content uniformity	F	F	F		F	F	F	F
		E	E	E		E	E	E	E
	5. Dose content uniformity	F	F			F	F		F
	through container life	E	E	E		E	E		E
	6. Droplet size distribution							F	F
	by laser diffraction							E	E
	7. Drug particle size	F	F	F		F	F	F	F
	distribution/fine particle mass	E	E	E		E	E	E	E
	8. Fine particle mass through container life	E	E	E					
	9. Fine particle mass with spacer use	E							
	10. Impurities and degradation products	F	F	F		F	F	F	F
	11. *In vitro* dose proportionality	F	F	F	F	F	F	F	F
	12. Leachables and	F	F	F		F	F	F	F
	extractables	E				E		E	E
	13. Leak rate	F				F			
		E				E			
	14. Microbial challenge	F				F			
		E				E			
	15. Microbial limits	F	F	F		F	F	F	F
		E	E	E		E	E	E	E

(Continued)

Table 10.2 Tests to be considered for product performance characterization studies and/or specifications — Cont'd

Test class	Test or specification name	Pulmonary				Nasal			
			DPI					Spray	
		Oral pMDI	Device-metered	Pre-metered	Inhalation spray	Nasal pMDI	Nasal powder	Single dose	Multi dose
	16. Microscopic evaluation/ foreign particulate matter	F	F	F		F	F	F	F
	17. Osmolality							F	F
	18. pH							F	F
	19. Photostability		F	F	F		F	F	F
	20. Physical characterization	E	E	E		E	E	E	E
	21. Plume geometry	F				F		F	F
	22. Preservative effectiveness and sterility maintenance				F	F		F E	F E
	23. Preservatives and stabilizing excipients assay							F E	F E
	24. Pressure testing	F				F			
	25. Single-dose fine particle mass	E	E	E					
	26. Spray pattern	F				F		F	F
	27. Sterility							E	E
	28. Viscosity							F	F
	29. Water or moisture content	F E	F E	F E		F E	F E		

This compilation is based on CDER (1998, 2002, 2003) and CHMP (2006).

Reader's guide: "F" and/or "E" denote that guidelines exist for the study/specification in an FDA (F) and/or EMEA (E) guidance document. If blank, no specific guidance exists. Specific section references in the guidance documents are given in text for each study/specification, as appropriate.

imports at the port of entry until problems are corrected. If warranted, the Agency can ask the courts to issue injunctions (as was done in the Schering-Plough case described above), or prosecute deliberate violations of the law, including seeking criminal penalties (even prison sentences) for flagrant or repeated offenses. All of which brings a high level of relevance to understanding and complying with the regulations for OINDPs.

10.2 SETTING SPECIFICATIONS AND PERFORMING PRODUCT CHARACTERIZATION STUDIES

Table 10.2 lists all of the specifications and product characterization studies that should be considered for OINDP development and production with respect to the FDA and EMEA/Health Canada guidelines. Detailed descriptions of the listed specifications and characterization studies along with stability considerations and regulatory references for each are given below.

10.2.1 Tests related to product labeling

10.2.1.1 Appearance and color/description

Background According to the FDA, the appearance of the content of the container (i.e. formulation) and the container closure system (e.g. *nasal sprays*: pump, container components; *metered-dose inhaler (MDI)/dry powder inhaler (DPI)*: the valve and its components, and the inside of the container) should conform to their respective descriptions (e.g. color and clarity of formulation, size and shape of pump components, texture of inside of the container) as an indication of the drug product integrity (CDER, 1998). If any color is associated with the formulation, either present initially or from degradative processes occurring during shelf-life, then a quantitative test with appropriate acceptance criteria should be established for the drug product (CDER, 1998, 2002).

Stability considerations Appearance and color/description should be assessed during stability studies at all stages of development, other than during preclinical studies. Studies conducted to support clinical trials should use methods that can be validated, while validated methods should be used during production and post-market approval stability studies (Pilewski, 2007; Kim, 2008).

References

Pulmonary – CDER (1998): III.F.1.a, III.F.2.a; CHMP (2006): 4.5.1.1
Nasal – CDER (1998): III.F.1.a, III.F.2.a; CDER (2002): III.F.1.a; CHMP (2006): 4.5.1.1.

10.2.1.2 Cleaning instructions

In-use studies should be performed to determine instructions to be included in the labeling for OINDPs regarding how to clean the particular product (see section 16.4 of Kim (2008) for more details). For NDAs, the FDA recommends that MDI devices used in clinical studies be sent after use for testing dose content uniformity and particle size distribution and, if feasible, be returned for continued patient use (CDER, 1998, 2002), though carrying out such a program is likely to be complex and cumbersome.

References

Pulmonary – CDER (1998): IV.A.8, IV.B.15; CDER (2002): IV.E; CHMP (2006): 4.2.1.14

Nasal – CDER (1998): IV.A.8, IV.B.15; CDER (2002): IV.E; CHMP (2006): 4.2.1.14.

10.2.1.3 Drug content (assay)

Background According to Kim (2008), the drug content assay specification should be considered as follows.

For multi-dose products, the EMEA guideline states that the amount of drug substance should be determined per weight unit or per volume unit as applicable (CHMP, 2006). For single-dose products, assay results should be expressed as mass per dosage unit – in other words, concentration.

According to the FDA (CDER, 1998), drug content assay for MDIs may be performed indirectly by determining concentration and actual net content (i.e. fill weight/volume), whereas for DPIs the amount of drug substance in each individual dosage unit should be determined for pre-metered devices and in the reservoir for device-metered inhalers (CHMP, 2006). Therefore, for MDIs and device metered DPIs, the FDA describes assay as total content rather than concentration.

Stability considerations There are many considerations for drug product assay testing during stability and the reader is encouraged to refer to the specific regulations for details. However, drug content (assay) tests should be performed during stability studies at all stages of development. Studies conducted to support clinical trials should use methods that can be validated, while validated methods should be used during production and post-market approval stability studies (Pilewski, 2007; Kim, 2008).

References

Pulmonary – CDER (1998): III.F.1.g, III.F.2.f; CHMP (2006): 4.5.1.2
Nasal – CDER (1998): III.F.1.g, III.F.2.f; CDER (2002): III.F.1.c; CHMP (2006): 4.5.1.2.

10.2.1.4 Identification

Background The FDA has the following recommendations for performing identification test(s) to verify the identity of the drug substance in the drug product:

> **"Identification using a single chromatographic procedure is not considered to be specific. A second, independent and complementary procedure (e.g. UV spectroscopy, IR), two chromatographic procedures where the separation is based on different principles, or a combination of tests into a single procedure (e.g. high-performance liquid chromatography/mass spectrometry) should be used. If the drug substance is a salt, an identification test should be included for the counterion. If the drug substance is chiral, then at least one of the methods used for identification should be specific for this property."**
>
> **(CDER, 1998)**

The EMEA guidance does not directly address this specification.

Stability considerations Identification tests should be performed during stability studies at all stages of development, other than during preclinical studies. Studies conducted to support clinical trials should use methods that can be validated, while validated methods should be used during production and post-market approval stability studies (Kim, 2008).

References

Pulmonary – CDER (1998): III.F.1.b, III.F.2.b
Nasal – CDER (2002): III.F.1.b.

10.2.1.5 Net content (fill) weight/number of actuations per container

Background This test is intended to confirm that the containers are sufficiently filled to allow the device to meet the label claim number of actuations. This test can be conducted in a variety of ways, including dose content uniformity, shot weight, or by specifically weighing the formulation during filling. Shot weight is likely to be the simplest and most conclusive method for high fill volume products. The net content (fill) weight test can be done in combination with tailoff profiling (see "Profiling of actuations near exhaustion (tailoff)"). This test does not apply for unit-dose nasal spray formats or DPI capsules.

Stability considerations The FDA and EMEA guidelines differ on whether or not to include the number of actuations per container test during stability testing. However, this test should be considered during stability studies at all stages of development because of its importance to reducing wastage. Studies conducted to support clinical trials should use methods that can be validated, while validated methods should be used during production and post-market approval stability studies (Pilewski, 2007; Kim, 2008).

References

Pulmonary – CDER (1998): III.F.1.f, III.F.2.e, IV.B.13; CHMP (2006): 4.2.1.2, 4.5.1.13

Nasal – CDER (1998): III.F.1.f, III.F.2.e, IV.B.13; CDER (2002): III.F.1.m; CHMP (2006): 4.2.1.2, 4.5.1.13.

10.2.1.6 Priming/repriming

Background As appropriate, studies should be performed to determine both initial priming requirements for the drug product and repriming requirements after various periods of non-use. The repriming time may also be derived from studies similar to the study described in "Effect of resting time" later in this chapter. Priming and repriming information will be used to support proposed labeling statements, and hence the information will be under close scrutiny by regulators.

Stability considerations Priming/repriming tests should be performed during stability studies at all stages of development. Studies conducted to support clinical trials should use methods that can be validated, while validated methods should be used during production and post-market approval stability studies (Pilewski, 2007; Kim, 2008).

References

Pulmonary – CDER (1998): IV.A, IV.A.5, IV.B.12; CHMP (2006): 4.2.1.12, 4.2.1.13

Nasal – CDER (1998): IV.A, IV.A.5, IV.B.12; CDER (2002): IV.A; CHMP (2006): 4.2.1.12, 4.2.1.13.

10.2.1.7 Profiling of actuations near exhaustion (tailoff)

Background According to the FDA (CDER, 1998), a study should be conducted to determine the profiles of the delivered amount of drug product and, if appropriate, the aerodynamic particle size distribution of the drug substance in each actuation after the point at which the labeled number of actuations have been dispensed and until no more actuations are available (i.e. the canister is empty). Alternatives such as content uniformity can be used in certain circumstances in place of aerodynamic particle size distribution data. For nasal sprays, laser diffraction droplet sizing techniques can also be used to determine tailoff profiles (Malvern Instruments, 2007). In principle, these studies help to determine if a proposed overfill of the containers is justified and give a profile of the dose delivery after the labeled number of actuations. Finally, the overfill specification needs to be carefully considered because it is directly related to excess product cost, and is connected to any dose-counting mechanism that might be used on the device.

Stability considerations Tailoff tests are not required during stability studies. However, these tests should be conducted early in development to justify the proposed overfill and to avoid costly wastage of formulation and for incorporation into dose-counting mechanisms used in the device.

References

Pulmonary – CDER (1998): IV.A.9, IV.B.11; CDER (2002): IV.I; CHMP (2006): 4.2.1.4

Nasal – CDER (1998): IV.A.9, IV.B.11; CDER (2002): IV.I; CHMP (2006): 4.2.1.4.

10.2.1.8 Pump/valve delivery (shot weight)

Background This test is directly related to the metering ability of the pump/valve, and it evaluates pump-to-pump/valve-to-valve reproducibility of the delivery performance of the product. This test is commonly done by pump/valve manufacturers, because they typically are responsible for assembling the pump/valve. The FDA recommends verifying the pump/valve delivery with the drug product because the drug product properties may affect performance of the pump/valve (CDER, 1998, 2002). The FDA further recommends that metered-dose pumps and valves should have dose delivery acceptance criteria of not more than ±15% for individual actuations, and not more than ±10% for the mean of the actuations relative to the target delivered dose (CDER, 1998). Note that this test is an FDA-only requirement.

Stability considerations Although for stability studies the measurement of dose during the determination of dose content uniformity may be more appropriate than pump/valve delivery (shot weight), the FDA recommends shot weight as a simple and effective way of identifying potential causes of dose variability, particularly in nasal sprays. It is therefore wise to discuss the stability testing strategy with regulatory agencies prior to registration of stability studies (Pilewski, 2007).

References

Pulmonary – CDER (1998): III.F.1.p
Nasal – CDER (2002): III.F.1.f.

10.2.1.9 Shaking (suspensions)

Background For products requiring shaking before use according to their instructions, such as suspensions, a study should be conducted to demonstrate that the shaking instructions provided to the consumer are adequate. This study should be carefully constructed because of the subjective nature of "shaking," and should be factored into the overall testing regime for the product, particularly when drug content measurements are made. Examining consumer instruction interpretation followed by product shaking after reading, together with examining delivered dose uniformity after shaking, may be an example of such a study, though quantifying the shaking act should also be considered for testing consistency.

Stability considerations Shaking is not a particular requirement of the FDA or EMEA during stability testing. However, the possibility of excessive shaking, which can lead to foaming and inaccurate dosing among other problems, should be examined by testing delivered dose uniformity after excessive shaking, preferably by some consistent shaking means.

References

Pulmonary – CDER (1998): V.A.4; CHMP (2006): 4.2.1.11
Nasal – CDER (2002): V.A.2; CDER (2002): V.A.5; CHMP (2006): 4.2.1.11.

10.2.1.10 Stability of primary (unprotected) package

Background For a drug product labeled for storage at room temperature, if additional packaging is used to protect the drug product from degradation and/or evaporative effects, such as foil overwrap for low-density polyethylene (LDPE)-containing product, adequate stability data conducted at a minimum of 25°C and a maximum of 40% relative humidity (RH) for nasal sprays or 75% RH for MDIs and DPIs should be generated for these units without the protective packaging. It is especially important for stability data to include the parameters needed to establish the maximum length of time the drug product may be used by patients after the protective packaging is removed. This information will be provided with the product labeling instructions, and hence will be under close scrutiny by regulators and patient advocacy groups. The FDA recommends that both newly manufactured drug products and drug products near the end of their proposed expiration date should be evaluated, preferably as part of an overall stability testing regime. It is particularly important for MDIs and DPIs to be periodically reassessed immediately after product manufacture and near the proposed expiry date to ensure continued integrity of the primary packaging (CDER, 1998).

Stability considerations See section 16.4 of Kim (2008) for information on designing an appropriate stability program for evaluating primary (unprotected) packaging.

References

Pulmonary – CDER (1998): IV.A.2, IV.B.2; CDER (2002): IV.O
Nasal – CDER (1998): IV.A.2, IV.B.2; CDER (2002): IV.O.

10.2.2 Tests related to patient use/misuse

10.2.2.1 Determination of appropriate storage conditions

Background The FDA (CDER, 1998) provides detailed information on how to determine the appropriate storage conditions for MDIs and DPIs. The basic concepts involve: (1) conducting studies to assess (a) the formulation and container and closure system, and (b) the necessity for secondary or additional protective packaging; and (2) determining whether a significant change occurs during stability testing under specific conditions. The following changes would generally be considered significant:

- A 5% change from the initial drug content assay value of a batch
- A failure to meet established stability acceptance criteria except for dose content uniformity and particle size distribution criteria
- For dose content uniformity, a 10% change in the mass of the mean dose (beginning, middle, and end means determined separately) at any test interval relative to the initial time-point value, or failure to meet the established acceptance criteria for the first tier of testing
- For particle size distribution, generally a greater than 10% change in the total mass of relevant fine particles (e.g. particles less than 5 μm) within the particle size distribution, or a shift in the profile for these particles.

Stability considerations Storage conditions are described in detail in Kim (2008), and depend on the region in which the product is to be registered. The one difference in requirements for MDIs and DPIs is described in the draft FDA guideline and is for products needing to be packaged in moisture protective packaging (CDER, 1998). In that case, the FDA recommends storage in conditions of 25°C/75% RH for one-third of the shelf-life to ensure that the packaging is adequate to protect the product. However, storage at the long-term conditions of 30°C/75% RH would essentially meet the strictest scenario and therefore should cover registration in all zones (WHO, 2008). Thus, stability testing at 25°C/75% RH might be satisfactorily addressed by testing the product at the more severe conditions, though this concept should be confirmed with the regulatory authority in the particular region for applicability. If implemented, this approach would eliminate additional stability storage condition testing needed later if the product is redeployed for a market in another, stricter region (Kim, 2008; see section 16.4 for additional information on stability considerations for OINDP storage conditions).

References

Pulmonary – CDER (1998): IV.A.1, IV.B.1
Nasal – CDER (1998): IV.A.1, IV.B.1.

10.2.2.2 Device robustness/ruggedness

Background Based on the current FDA and EMEA guidelines, it is generally understood that studies for nasal and inhalation spray drug device robustness should address the following:

- For devices that can be reused repeatedly with replaceable reservoirs, or for pre-metered DPIs, a study should be conducted to establish the product performance characteristics for dose content uniformity or emitted dose, and particle/droplet size distribution, or other appropriate metric, throughout the device's claimed number of sprays and likely beyond for a truer ruggedness assessment.
- Use limits related to failure of critical device mechanisms should be studied to determine the appropriate replacement intervals for the device. Note that these limits will have a direct impact on product cost and should be considered carefully for product success in the market.
- Performance characteristics of the device should be studied after different handling situations such as dropping, shaking, and/or vibrating, though these characteristics should be quantifiable to make the study repeatable, objective, and data-driven.

Stability considerations Device robustness/ruggedness tests should be performed during stability studies at all stages of development. Studies conducted to support clinical trials should use methods that can be validated, while validated methods should be used during production and post-market approval stability studies (Pilewski, 2007; Kim, 2008).

References

Pulmonary – CDER (1998): III.F, IV.F, IV.B.14; CDER (2002): IV.F; CHMP (2006): 4.2.1.18

Nasal – CDER (1998): III.F, IV.F, IV.B.14; CDER (2002): IV.F; CHMP (2006): 4.2.1.18.

10.2.2.3 Dose buildup and flow resistance (DPIs only)

Background The FDA recommends conducting a study to determine any applicable dose buildup and flow resistance characteristics of DPI products because of the strong linkage that exists between these characteristics and the performance of DPIs. At a minimum, the total flow resistance of the device should be characterized and, if possible, the flow resistance of each of the individual components involved in the flow paths within the inhaler should also be characterized. The FDA expects supportive information to be included in the regulatory filing for the product. Note that this test is recommended only in FDA guidance documents (CDER, 1998).

Stability considerations Dose buildup and flow resistance tests should be performed during stability studies at all stages of development. Studies conducted to support clinical trials should use methods that can be validated, while validated methods should be used during production and post-market approval stability studies (Kim, 2008).

References

Pulmonary – CDER (1998): IV.B.5, III.G.2.d
Nasal – CDER (1998): IV.B.5, III.G.2.d.

10.2.2.4 Drug deposition on mouthpiece and/or accessories

Background

> **"The amount of drug deposited per actuation on the mouthpiece and any other drug product accessory [e.g. spacer for MDIs] should be established and documented in the application."**
>
> **CDER (1998)**

> **"The amount of drug deposited on the actuator or mouthpiece should be determined and, where applicable, demonstrated to be consistent with any correction factor used to support ex-valve (or ex-delivery device) label claims."**
>
> **CHMP (2006)**

Stability considerations Drug deposition on mouthpiece and/or accessories tests are not specifically required for stability by either the FDA or EMEA.

References

Pulmonary – CDER (1998): IV.A.7; CDER (2002): IV.A.7; CHMP (2006): 4.2.1.9
Nasal – CDER (1998): IV.A.7; CHMP (2006): 4.2.1.9.

10.2.2.5 Effect of moisture

Background Moisture effects are particularly important for DPI products and a study should be conducted to determine these effects at various high- and low-humidity conditions on critical performance parameters (e.g. emitted dose content uniformity, drug particle size distribution, microscopic evaluation, and water content). The main purpose of such a study is to assess the overall effects of different environmental conditions on device and/or powder, which together are responsible for aerosol behavior and performance of the product (CDER, 1998). These environmental moisture effects on product performance should be investigated early and often during development because of the potential negative impact they can have on product performance. For pre-metered products that incorporate powder-filled capsules, special attention should be paid to brittleness of the capsules under various humidity conditions because premature breakage of the capsule would negatively affect the performance of the product (CHMP, 2006).

Stability considerations Moisture effects should be factored into a complete stability testing regime for all the reasons described above and the high potential liability that adverse moisture effects can have on OINDPs, particularly DPIs. As described in Kim (2008), moisture effects can be effectively managed during stability testing but it requires careful attention, particularly in the early development stages, for a successful product. According to the FDA, if moisture effects are determined to be relevant for the product, they should be performed during stability studies at all stages of development. Studies conducted to support clinical trials should use methods that can be validated, while validated methods should be used during production and post-market approval stability studies (Kim, 2008).

References

Pulmonary – CDER (1998): IV.B.9; CHMP (2006): 4.2.1.17
Nasal – CDER (1998): IV.B.9; CHMP (2006): 4.2.1.17.

10.2.2.6 Effect of orientation/weight loss (stability)

Background Orientation and weight loss effects during stability testing should be included as part of the overall robustness assessment for the product. This assessment should be based on evaluating the critical performance metrics for the product (e.g. metered and emitted dose content uniformity, particle size distribution, and priming/repriming) at various dosing orientations and under different handling situations (e.g. dropping, shaking) (CDER, 1998; CHMP, 2006). Nasal spray drug products should include acceptance criteria for weight loss on stability. Since storage orientation plays a role in assessment of the sealing characteristics of the container closure system, weight loss for the drug product stored in upright and inverted, or upright and horizontal positions should be part of the overall evaluation protocol (CDER, 2002).

Stability considerations Effect of orientation/weight loss tests should be performed during stability studies at all stages of development. Studies conducted to support clinical trials should use methods that can be validated, while validated methods should be used during production and post-market approval stability studies (Kim, 2008).

References

Pulmonary – CDER (1998): IV.A.4, IV.B.6; CDER (2002): IV.G; CHMP (2006): 4.2.1.4

Nasal – CDER (1998): IV.A.4, IV.B.6; CDER (2002): III.F.1.n, IV.G; CHMP (2006):4.2.1.4.

10.2.2.7 Effect of patient use

Background The FDA strongly encourages carrying out studies for OINDPs to identify the effects of patient use on the characteristics of the drug product. For NDAs, the FDA recommends that devices used in clinical studies be tested for pertinent performance parameters and physical attributes after use (e.g. emitted dose, particle size distribution, moisture content, microbial limits) and, if feasible, then returned for continued patient use (CDER, 1998), though this may be logistically impractical in some cases. Additionally, the FDA recommends factoring in patient usage in the selection process for automated actuation settings used for a variety of *in vitro* tests (e.g. shot weight/dose content uniformity, spray pattern, droplet/particle size distribution, plume geometry) (CDER, 2003).

Stability considerations Patient usage tests should be factored into a stability program at the very earliest stages, particularly to establish automated actuation settings that simulate patient usage. Quality by design principles can be useful to establish the design, control, and operating spaces of the devices in terms of patient usage (Farina, 2007, 2008). Studies conducted to support clinical trials should use methods that can be validated, while validated methods should be used during production and post-market approval stability studies (Kim, 2008).

References

Pulmonary – CDER (1998): IV.B.8
Nasal – CDER (1998): IV.B.8; CDER (2003): V.A.

10.2.2.8 Effect of resting time

Background Resting time effects are normally evaluated as part of a complete priming/repriming evaluation. These evaluations should include the following steps: (a) a thorough examination of the first actuation of unprimed MDI or nasal spray devices followed immediately by the second and the third actuations where the devices are only primed prior to initiation of the study; (b) after resting for increasing periods of time (e.g. 6, 12, 24, 48 hours), content uniformity of the first, second, and third actuations should be measured to define the drug delivery profile per actuation. For a complete evaluation, unprimed MDI or nasal spray containers that have been stored in different orientations (i.e. upright, inverted, and/or horizontal) should be included. The study time may be shortened by concurrently testing separate samples with progressively longer resting periods (CDER, 1998).

Stability considerations Effect of resting time tests should be performed during stability studies at all stages of development, and should likely be coupled with priming/repriming tests. Studies conducted to support clinical trials should use methods that can be validated, while validated methods should be used during production and post-market approval stability studies (Kim, 2008).

References

Pulmonary – CDER (1998): IV.A.4; CDER (2002): IV.B
Nasal – CDER (1998): IV.A.4; CDER (2002): IV.B.

10.2.2.9 Effect of storage on the particle size distribution

Background During primary stability studies for suspension-based OINDPs, the effect of storage on particle size distribution from the initial actuation to the labeled number of actuations should be evaluated to determine any trends (CDER, 1998). Refer to section "Drug particle size distribution/fine particle mass" for details.

Stability considerations For suspension-based OINDP products, the stability studies on the primary stability batches should determine the effect of storage time and conditions on drug particle size distribution through unit life (beginning to end for device-metered products). If stability studies demonstrate an orientation effect on the drug particle size distribution within unit life, then the routine stability protocol should include orientation-based drug particle size distribution testing through unit life (CDER, 2002).

References

Pulmonary – CDER (1998): IV.A.6, IV.B.4; CDER (2002): IV.J
Nasal – CDER (1998): IV.A.6, IV.B.4; CDER (2002): IV.J.

10.2.2.10 Effect of varying flow rates

Background for MDIs According to the FDA (CDER, 1998), if an MDI is intended to be marketed with a spacer or similar accessory, a study should be performed to characterize the MDI's emitted dose and drug particle size distribution as a function of different flow rates (true patient experience) at constant volume (e.g. 2 liters). This intent of this study is to assess the sensitivity of the drug product to varying flow rates that are generated by patients of different age, gender, and disease severity, though faithfully reproducing the patient experience during laboratory testing is a significant challenge. Furthermore, a study to assess the effect of increasing waiting periods (e.g. 0, 5, 10 seconds) between actuation and initiation of in-flow on the emitted dose and the particle size distribution is encouraged for understanding any adverse formulation settling effects.

Background for DPIs (CDER, 1998) DPIs should be evaluated in a similar fashion to MDIs as described above, particularly with regard to replicating the patient experience (e.g. age, gender, and disease severity) under different flow rates at constant volume (e.g. 2 liters). For NDAs, the FDA recommends relating these *in vitro* tests to *in vivo* performance for DPIs (which are dependent on patient effort for de-aggregation and dose delivery). These patient experience studies should also be conducted to determine what flow characteristics are obtained through the device by adult and pediatric subjects with normal lung function and by adult and pediatric patients with varying degrees of obstructed lung function, using a severely affected chronic obstructive pulmonary disease (COPD) patient's forced expiratory volume in 1 second (FEV_1) as the basis for the comparison.

Stability considerations Effect of varying flow rate tests should be performed during stability studies at all stages of development. Studies conducted to support clinical trials should use methods that can be validated, while validated methods should be used during production and post-market approval stability studies (Kim, 2008).

References

Pulmonary – CDER (1998): IV.A.13, IV.B.3; CDER (2002): IV.H; CHMP (2006): 4.2.1.5
Nasal – CDER (1998): IV.A.13, IV.B.3; CHMP (2006): 4.2.1.5.

10.2.2.11 Low-temperature performance

Background According to the FDA (CHMP, 2006), a study should be conducted to determine the effect of low-temperature storage on the performance of OINDPs. Containers should be stored in various orientations for at least 3 hours at a temperature below freezing (0°C), and then immediately tested to determine the number of actuations required until the subsequent doses meet the drug product specification limits for delivered dose uniformity and fine particle mass. If the product does not perform satisfactorily (e.g. repriming actuations required exceed the number required according to the instructions for use), an additional study should be conducted to determine the method and length of time needed to adequately warm the containers so that satisfactory performance is achieved. The product labeling instructions should carry cold-temperature performance information even if the study is not conducted. Keep in mind that the OINDP should tolerate low temperatures because patients are likely to store them in such conditions during the product's lifetime (e.g. in a car glove-box overnight in winter), and full justification will need to be provided to regulatory authorities if the OINDP cannot meet this condition.

Stability considerations See section 16.4 of Kim (2008) for more information regarding implementing a successful low-temperature stability program.

References

Pulmonary – CHMP (2006): 4.2.1.15
Nasal – CHMP (2006): 4.2.1.15.

10.2.2.12 Temperature cycling

Background (CDER, 1998, 2002; CHMP, 2006) The FDA and EMEA recommend temperature cycling tests be carried out for all OINDPs. The intent of the tests is to determine what effect, if any, extreme temperature cycling has on the performance of the product (e.g. particle size distribution, microscopic evaluation, physical appearance of the content, valve component integrity, dose content uniformity, water content, and leak rate). The FDA suggests that the temperature cycling study may consist of three or four 6-hour cycles per day, between subfreezing temperature and 40°C for a period of up to 6 weeks. At the end of predetermined cycles, the samples should be analyzed for appropriate parameters (as described above) and compared with products that were not temperature cycled (e.g. control units). The study should examine the following conditions, as appropriate: discoloration of the contents, microscopic evaluation, distortion or elongation of valve components, valve clogging, canister corrosion, and adherence of the drug to the walls of the container or valve components.

Stability considerations See "Background."

References

Pulmonary – CDER (1998): IV.A.3; CDER (2002): IV.C; CHMP (2006): 4.2.1.16

Nasal – CDER (1998): IV.A.3; CDER (2002): IV.C; CHMP (2006): 4.2.1.16.

10.2.3 Tests related to general properties

10.2.3.1 Chemical compatibility

Background The FDA and EMEA recommend performing thorough chemical compatibility studies to ensure that no incompatibilities exist between the drug and any container closure components that come into contact with it (e.g. MDI metering valve or nasal spray metering pump).

Stability considerations According to the FDA (CDER, 1998), the performance of all of the OINDP's drug contacting container closure components should be thoroughly investigated for chemical compatibility before initiating critical clinical and/or bioequivalence studies.

References

Pulmonary – CDER (1998): III.G.1.b, III.G.2.a

Nasal – CDER (1998): III.G.1.b, III.G.2.a; CDER (2002): III.G.

10.2.3.2 Dehydrated alcohol content

Background According to the FDA (CDER, 1998), a specific assay with acceptance criteria should be used when dehydrated alcohol (e.g.

dehydrated alcohol, United States Pharmacopeia, USP) is used as a co-solvent in the OINDP's formulation. The FDA further recommends that additional discriminatory specifications for water content (e.g. Karl Fischer) and impurities (e.g. excipients) should be included in the assay.

Stability considerations Dehydrated alcohol content tests should be performed during stability studies at all stages of development, if applicable. Studies conducted to support clinical trials should use methods that can be validated, while validated methods should be used during production and post-market approval stability studies (Kim, 2008).

References

Pulmonary – CDER (1998): III.F.1.e
Nasal – CDER (1998): III.F.1.e.

10.2.3.3 Delivery device development

Background (CDER, 1998) One significant difference between OINDPs and other, more conventional drug products is that the clinical efficacy of these drug products may be directly dependent on the design, reproducibility, and performance characteristics of the container and closure system (e.g. container/bottle, the nozzle tip/mouthpiece, the valve/pump and its components, and any additional accessories (e.g. spacer), as well as protective packaging if applicable). Because of the performance-critical interaction between the device and the formulation, delivery device development should be thoroughly evaluated, documented, and understood for the OINDP with particular emphasis on patient experience and efficacy.

Stability considerations The delivery device used in OINDPs should be factored into stability studies with particular attention focused on the interaction of the drug with the container closure system components.

References

Pulmonary – CDER (1998): III.G.1, III.G.2
Nasal – CDER (1998): III.G.1, III.G.2; CDER (2002): III.G; CHMP (2006): 4.2.1.19.

10.2.3.4 Dose content uniformity

Background The dose content uniformity test is arguably the most important test in characterizing OINDPs. Various terms and standards are used for this test, including dose content uniformity, delivered-dose uniformity, and emitted dose uniformity, as well as spray content uniformity for nasal sprays. The EMEA guideline refers to the relevant pharmacopoeia for guidance on requirements, but adds the "Mean delivered dose" and "Content uniformity/uniformity of dosage units" specifications to the test, which in turn add to the complexity of carrying out the test and confusion to the novice in this field (Kim, 2008).

Recently, much discussion and collaboration between industry and regulatory authorities has been carried out to try to simplify the FDA requirements using parametric tolerance intervals. See Olsson (2003) and Novick et al. (2009) for the latest research on this topic.

However, the latest editions of the national pharmacopoeias have not adopted the parametric tolerance interval requirements, and the current harmonized United States Pharmacopoeia (USP, 2008) and European Pharmacopoeia (EP, 2007) limits are wider than those outlined in the FDA guidelines (CDER, 1998, 2002). Additionally, for metered-dose nasal sprays the pharmacopoeias differ: the EP contains the same requirements as those for MDIs and DPIs while the USP requirements are more challenging, as per those provided in CDER (2002).

Stability considerations The delivered dose uniformity requirements contained in the FDA guideline for MDIs and DPIs are challenging and a consistent approach should be taken for how to perform the testing and set acceptance limits. Note also that, according to Kim (2008), stability testing can become complicated for OINDPs because using increased numbers of test samples during stability could result in an increase in the number of out-of-specification results even when the product is stable. See Kim (2008) and Novick et al. (2009) for a description of the various interpretations of this test and the impact they have on stability sample preparation.

References

Pulmonary – CDER (1998): III.F.1.i, III.F.2.h; CHMP (2006): 4.5.1.4, 4.5.1.5

Nasal – CDER (1998): III.F.1.i, III.F.2.h; CDER (2002): III.F.1.g; CHMP (2006): 4.5.1.4, 4.5.1.5.

10.2.3.5 Dose content uniformity through container life

Background See "Dose content uniformity."

Stability considerations See "Dose content uniformity."

References

Pulmonary – CDER (1998): III.F.1.j, III.F.2.i; CHMP (2006): 4.5.1.4
Nasal – CDER (1998): III.F.1.j, III.F.2.i; CDER (2002): III.F.1.g; CHMP (2006): 4.5.1.4.

10.2.3.6 Droplet size distribution by laser diffraction (nasal sprays)

Background Droplet size distribution (DSD) is an important property influencing the deposition of suspension and solution nasal sprays, and laser diffraction is the recommended technique for measuring DSD in these products (CDER, 2002, 2003; CHMP, 2006). Note, however, that the major hurdle for using laser diffraction in aerosols has to do with laser beam steering caused by refractive index effects due to evaporation of propellant. DSD would be characterized at distances from the nozzle tip that eliminate or minimize beam steering, if possible. If a correction algorithm is used, then it must be fully described to regulatory authorities and validated prior to use (CDER, 2002). Much research has been done to address beam steering complications (e.g. Haynes et al., 2004; Zhao et al., 2008; Malvern Instruments, Westborough, MA 2007).

Laser diffraction is a non-aerodynamic optical method of droplet sizing that measures the geometric size of droplets in flight using well-known light scattering principles (e.g. Mie scattering theory indicates that the scattering angle is inversely proportional to the size of the droplet when the droplet is illuminated). Modern laser diffraction instrumentation can provide time history profiles of obscuration (optical concentration) or percentage transmission (%T, which is the opposite of obscuration), and automatically compute the DSD profiles based on measurement statistics (e.g. D_{10}, D_{50}, D_{90}) over the entire life of a single spray while accounting for multiple scattering from the droplets (though multiple scattering or high droplet concentration situations are preferred to be avoided if possible – CDER (2002)). Other measurements such as the "span" (defined as $[(D_{90} - D_{10})/D_{50}]$) are typically reported to quantify the spray in consistent terms.

In general, nasal spray DSD time history profiles can be characterized by three relatively distinct phases: the formation phase (rapid decrease in %T), followed by the fully developed phase (steady %T), and finally the dissipation phase (gradual increase in %T back to pre-spray levels).

Changes in DSD are directly correlated with the changes in obscuration or %T levels, with DSD attaining plateau values within the same approximate time period as the plateau in obscuration or %T. The FDA recommends determination of the complete time history profiles of the DSD statistics (e.g. D_{10}, D_{50}, D_{90}) and obscuration or %T at specific distances from the nozzle tip, but that DSD data be reported only during the fully developed phase. For bioequivalence purposes, two distances should be used between 2 and 7 cm from the nozzle tip, with 3 cm between each selected distance (CDER, 2003). For NDAs, only one distance between 2 and 7 cm is typically needed to establish DSD specifications. The FDA also recommends that appropriate and validated and/or calibrated droplet size analytical procedures should be described in sufficient detail to allow accurate assessment by Agency laboratories (e.g. apparatus and accessories, calculation theory, correction principles, software version, sample placement, laser trigger condition, measurement range, beam width, automated actuation parameters) (CDER, 2002).

Stability considerations Droplet size distribution by laser diffraction tests should be performed during stability studies at all stages of development for suspension and solution nasal sprays. Studies conducted to support clinical trials should use methods that can be validated, while validated methods should be used during production and post-market approval stability studies (Kim, 2008).

References

Pulmonary – none

Nasal – CDER (1998): III.F.1.i; CDER (2003): V.B.2; CHMP (2006): 4.5.2.1.

10.2.3.7 Drug particle size distribution/fine particle mass

Background Particle size distribution (PSD) is a complex and multivariate parameter with regard to OINDPs. A significant amount of work has been done on PSD for these products over the past 50 years and PSD still remains the most important property of an aerosol, affecting lung deposition and drug efficacy (see Nagao et al. (2005) for an overview of regulatory aspects of pharmaceutical particle science). In early development, PSD was often described by a single point control known as fine particle mass, typically being the mass of particles less than or equal to 5 μm (optimal range for pulmonary inhalation is generally known to be 1–5 μm in terms of aerodynamic particle size) and as development proceeds, more complex specifications are developed whereby the particle size distribution is represented by a number of particle size fractions between 1 and 10 μm, with requirements linked to clinical batches (Kim, 2008). Suspension formulations have the potential to produce particle agglomeration or to undergo particle size changes (Berry et al., 2003), potentially altering the PSD profile significantly. From a pharmaceutical viewpoint, the most important parameter for an inhalation product is usually the aerodynamic particle size distribution (APSD) of the emitted aerosol. For suspension-based OINDPs, it is particularly important to understand that the APSD is influenced by many factors (including but not limited to the formulation itself, the metering valve/pump, and mouthpiece/nozzle tip) and is not solely determined by the size of the individual drug substance particles initially suspended in the formulation (CDER, 1998). In DPI products humidity during storage can affect powder properties including fine particle mass and electrostatic charge, the latter of which may pose significant challenges for cascade impaction-based APSD measurements (see below), particularly with hygroscopic growth and aggregation of particles (Borgström et al., 2005; Chi Lip Kwok and Chan, 2008). To minimize variability from these sources, the FDA recommends collecting APSD data in temperature- and relative humidity-controlled environments with equilibrated air (CDER, 1998).

The APSD is practically measured using a multistage cascade impactor (see Mitchell and Nagel (2003) for a thorough overview of the technology related to cascade impaction-based measurements of APSD for OINDPs). A multistage cascade impactor fractionates and collects particles of one or more drug components by aerodynamic diameter through serial multistage impactions. A properly operating multistage cascade impactor (equipped with all associated accessories) allows determination of a size distribution throughout the whole dose including, in particular, the small particle size fraction of the dose (e.g. fine particle mass). A multistage cascade impactor also provides information that allows for the complete mass balance of the total labeled dose to be determined, though this topic has been openly discussed in recent journal articles and industry presentations (see PQRI (2003) for an overview of the mass balance issue and how it has been approached by industry and regulatory authorities). However, to minimize distortions and to ensure reproducibility, it is important to specify and control the following aspects during testing: information on the calibration of the impactor and its accessories, flow rate, duration, the size and shape of the expansion chamber or inlet stem, the selection of impaction surfaces, and the method, accessories, and adapters by which the aerosol is introduced into the impactor. These important parameters should be selected to obtain a complete profile of the dose (CDER, 1998). Recently, industry presentations have shown alternatives to simplifying and consolidating APSD measurements from cascade impactors (Tougas et al., 2008; Mitchell et al., 2009).

From a regulatory perspective, criteria should be provided in the application for the qualification of each cascade impactor and the FDA recommends that all cascade impactors used in support of the drug product in the application be of the same design (e.g. Andersen or next generation impactor style). Other critical variables that should be specified and controlled in such a test procedure are relative humidity and temperature, because particles may undergo changes during their passage into or through the cascade impactor depending on humidity and temperature conditions, as described earlier (CDER, 1998).

In practice, the FDA recommends that the number of actuations needed to determine APSD using a multistage cascade impactor should be kept to the minimum justified by the sensitivity of the analytical method used to quantify the deposited drug substance. The amount of drug substance deposited on the critical stages of the cascade impactor should be sufficient for reliable assay, but not so excessive as to bias the results by masking individual actuation variation (CDER, 1998). The APSD analysis and the mass balance obtained (drug substance deposited on surfaces from the valve to the cascade impactor filter) should be calculated (refer to the preceding paragraph for more information on cascade impactor mass balance approaches). The FDA has set the following acceptance criteria for APSD measurements: the total mass of drug collected on all stages and accessories should be between 85% and 115% of label claim on a per-actuation basis, with some accountability being paid to proposed groupings of individual stage and/or accessories results (a minimum of three to four groupings should ensure future batch-to-batch consistency). Furthermore, the FDA considers acceptance criteria expressed in terms of mass median aerodynamic diameter (so-called MMAD) and geometric standard deviation (GSD) alone, as well as in terms of respirable fraction, respirable dose, or fine particle mass inadequate to characterize the particle size distribution of the whole dose (CDER, 1998).

Stability considerations Drug particle size distribution/fine particle mass tests should be performed during stability studies at all stages of development for all MDI and DPI products. Studies conducted to support clinical trials should use methods that can be validated, while validated methods should be used during production and post-market approval stability studies (Kim, 2008).

References

Pulmonary – CDER (1998): III.F.1.k, III.F.2.j; CHMP (2006): 4.2.1.8, 4.5.1.7

Nasal – CDER (1998): III.F.1.k, III.F.2.j; CDER (2002): III.F.1.j; CHMP (2006): 4.2.1.8, 4.5.2.1.

10.2.3.8 Fine particle mass through container life

Background See "Drug particle size distribution/fine particle mass" above. Note, however, that this specification is only described in the EMEA guidance (CHMP, 2006).

Stability considerations See "Drug particle size distribution/fine particle mass" above.

References

Pulmonary – CHMP (2006): 4.2.1.4
Nasal – none.

10.2.3.9 Fine particle mass with spacer use

Background See "Drug particle size distribution/fine particle mass" above. Note, however, that this specification is only described in the EMEA guidance (CHMP, 2006).

Stability considerations See "Drug particle size distribution/fine particle mass" above.

References

Pulmonary – CHMP (2006): 4.2.1.6
Nasal – none.

10.2.3.10 Impurities and degradation products

Background According to the FDA (CDER, 1998, 2002), the levels of impurities and degradation products should be determined by a validated analytical procedure or procedures. Acceptance criteria should be set for individual and total impurities and degradation products with specification of all related impurities appearing at the level specified in the ICH guidelines based upon the maximum daily dose (CDER/CBER, 2006). Note that specified impurities and degradation products are those, either identified or unidentified, that are individually listed and limited in the drug product specification. Refer to CDER (2005) and CDER/CBER (2006) for identification and qualification thresholds and other relevant information for impurities in drug products. Note that the EMEA directly references the ICH guidelines (CDER/CBER, 2006) for impurities testing.

Stability considerations Impurities and degradation products tests should be performed during stability studies at all stages of development, as indicated in Table 10.2. Studies conducted to support clinical trials should use methods that can be validated, while validated methods should be used during production and post-market approval stability studies (Kim, 2008).

References

Pulmonary – CDER (1998): III.F.1.h, III.F.2.g
Nasal – CDER (1998): III.F.1.h, III.F.2.g; CDER (2002): III.F.1.d.

10.2.3.11 *In vitro* dose proportionality

Background For OINDPs with multiple-strength doses, the FDA recommends performing studies to characterize the *in vitro* dose proportionality in terms of the emitted dose content uniformity and the particle size distribution (CDER, 1998, 2002).

Stability considerations *In vitro* dose proportionality tests should be performed during stability studies at all stages of development, as indicated in Table 10.2. Studies conducted to support clinical trials should use methods that can be validated, while validated methods should be used during production and post-market approval stability studies (Kim, 2008).

References

Pulmonary – CDER (1998): IV.A.12, IV.B.7; CDER (2002): IV.D
Nasal – CDER (1998): IV.A.12, IV.B.7; CDER (2002): IV.D.

10.2.3.12 Leachables and extractables

Background Device components, as part of a drug–device combination product, may contain polymers, elastomers, and other components from which minute quantities of material may migrate (leach) into the medicinal product over time and thus may affect the quality and safety of the product (Kim, 2008). There is much information available from regulatory authorities and industry groups related to leachables and extractables in general (CDER/CBER, 1999; CHMP, 2005), and with particular emphasis on OINDPs (ITFG/IPAC-RS Collaboration, 2001; PQRI, 2006).

Kim (2008) also identifies two key points from PQRI (2006) with regard to leachable studies: (a) a well-thought-out leachables program should be conducted on the drug product packaging configuration employed for long-term stability studies (e.g. capsule with blister, low-density polyethylene vial with overwrap); and (b) an in-use study should also be conducted in order to determine the leachables derived from components that are in contact with either the formulation or the patient's mouth or nasal mucosa only during administration (such as mouthpieces and actuators).

Stability considerations Leachables studies can be conducted as part of the stability studies to support product registration as described in PQRI (2006), Kim (2008), and Norwood et al. (2008) in addition to the FDA and EMEA guidelines listed below. Leachables and extractables tests should be performed during stability studies at all stages of development, as indicated in Table 10.2. Studies conducted to support clinical trials should use methods that can be validated, while validated methods should be used during production and post-market approval stability studies (Kim, 2008).

References

Pulmonary – CDER (1998): III.F.1.q, III.G.1, III.G.2; CDER (2002): IV.D; CHMP (2006): 4.5.1.11, 4.2.1.3
Nasal – CDER (1998): III.F.1.q, III.G.1, III.G.2; CDER (2002): III.F.o; CHMP (2006): 4.2.1.3.

10.2.3.13 Leak rate (MDIs)

Background According to the FDA, to maintain optimal performance characteristics for the drug product, acceptance criteria for the leak rate should be based on historical data, including primary stability data using the test and sampling plan described in USP <601>. Leak rate testing should be performed in addition to both the online leak test, which identifies and eliminates the occasional gross leakers, and the testing that follows the lag or equilibration time instituted before product release (CDER, 1998).

Stability considerations According to the FDA (CDER, 1998), leak rate testing is important in stability studies because it may provide information on pressure loss and may predict, at subsequent test stations, failures in testing for dose content uniformity throughout container life. It should be noted, however, that leak rates are not necessarily constant over time. Leak rates for propellants within the same drug product line are usually independent of the formulation fill volume, since the containers and closures (i.e. seals) used are usually the same. As a result, selective leakage of the propellants may concentrate the content of a smaller container faster than that of a larger container, to a point where, for example, dose content uniformity or drug particle size distribution, or both, would be outside of the acceptance criteria. An important result of this behavior is that smaller containers may have shorter expiration dating periods than larger containers of the same drug product when the same seals are used.

References

Pulmonary – CDER (1998): III.F.1.n; CHMP (2006): 4.5.1.8
Nasal – CDER (1998): III.F.1.n; CHMP (2006): 4.5.1.8.

10.2.3.14 Microbial challenge

Background The FDA recommends performing a study to determine the viability of microorganisms in OINDP formulations that have been inoculated intentionally (CDER, 1998). Note that microbial challenge testing is an FDA-only requirement.

Stability considerations Microbial challenge tests should be performed during stability studies at all stages of development, as indicated in Table 10.2. Studies conducted to support clinical trials should use methods that can be validated, while validated methods should be used during production and post-market approval stability studies (Kim, 2008).

References

Pulmonary – CDER (1998): IV.A.11
Nasal – CDER (1998): IV.A.11.

10.2.3.15 Microbial limits

Background According to the FDA (CDER, 1998), the microbial quality should be controlled by appropriate tests and acceptance criteria for total aerobic count, total yeast and mold count, and freedom from designated indicator pathogens. Acceptance criteria should reflect the data for the tested batches and be consistent with benchmarks listed in international pharmacopoeias. Furthermore, appropriate testing should be done to show that the OINDP does not support the growth of microorganisms and that microbial quality is maintained throughout the OINDP's expiration period. The minimum sample size should be 10 grams or the full content of 10 containers, as recommended in USP <61>.

Stability considerations Microbial limit tests should be performed during stability studies at all stages of development, as indicated in Table 10.2. Studies conducted to support clinical trials should use methods that can be validated, while validated methods should be used during production and post-market approval stability studies (Kim, 2008).

References

Pulmonary – CDER (1998): III.F.1.c, III.F.2.c; CHMP (2006): 4.5.1.9
Nasal – CDER (1998): III.F.1.c, III.F.2.c; CDER (2002): III.F.1.l; CHMP (2006): 4.5.1.9.

10.2.3.16 Microscopic evaluation/foreign particulate matter

Background According to the FDA (CDER, 1998), microscopic examination of the formulation has certain merits and therefore should be used during release and stability as appropriate. For example, microscopic examination can provide information on the presence of large particles, changes in morphology of the drug substance particles, extent of agglomerates, crystal growth, and foreign particulate matter. Validated methods that emphasize user training and experience should be employed for microscopic evaluation and foreign particulate matter to reduce unwanted analyst subjectivity.

Stability considerations Microscopic evaluation methods are recommended to control and monitor the morphic form if changes are observed on stability (e.g. where the crystalline form of the drug substance can affect the bioavailability, performance, stability, or other properties of the drug product) (CDER, 1998). Additionally, microscopic evaluation can be useful as a qualitative investigative tool to understand root causes and sources of foreign particulate formation, particularly if problems arise during stability (Kim, 2008).

References

Pulmonary – CDER (1998): III.F.1.l, III.F.2.k
Nasal – CDER (1998): III.F.1.c, III.F.2.c; CDER (2002): III.F.1.k.

10.2.3.17 Osmolality (nasal products only)

Background The FDA (CDER, 2002) recommends that the osmolality of nasal formulations should be tested and controlled at release with an appropriate procedure and acceptance criterion for formulations that contain tonicity-controlling agents or for products having a label claim regarding tonicity. Studies have shown that lower osmolality levels can provide better drug permeability for nasal suspension products and eliminate the need for enhancers or other additives (Yoshihisa et al., 2002). Other studies have investigated the joint effects of tonicity and viscosity on nasal spray performance (Dua et al., 1997).

Stability considerations No specific guidance exists for osmolality tests during stability. However, since osmolality is considered a physical property of the formulation, it can likely be tested during development to demonstrate that no adverse stability issues exist, which can then justify its inclusion only in the product specification.

References

Pulmonary – none
Nasal – CDER (2002): III.F.1.q.

10.2.3.18 pH

Background pH or apparent pH, as appropriate, of nasal suspension and solution formulations should be tested and an appropriate acceptance criterion should be established because the pH level can significantly influence efficacy of nasal formulations (Ahmed et al., 2000; CDER, 2002).

Stability considerations No specific guidance exists for pH tests during stability. However, since pH is considered a physical property of the formulation, it can likely be tested during development to demonstrate that no adverse stability issues exist, which can then justify its inclusion only in the product specification.

References

Pulmonary – none
Nasal – CDER (2002): III.F.1.p.

10.2.3.19 Photostability

Background The intrinsic photostability characteristics of drug substances and products should be evaluated to demonstrate that, as appropriate, light exposure does not result in unacceptable change to the performance of the drug substance or product (CDER/CBER, 1996). Photostability studies for OINDPs should be performed using appropriate test conditions, if warranted (e.g. if capsules or clear blisters are used for premetered DPIs or if the reservoir containing the formulation in a device-metered DPI can receive light exposure by the immediate container) and these studies should be conducted in the absence of any additional packaging (e.g. foil overwrap) (CDER, 1998).

Stability considerations Photostability tests should be performed during stability studies at all stages of development, as indicated in Table 10.2. Studies conducted to support clinical trials should use methods that can be validated, while validated methods should be used during production and post-market approval stability studies (Kim, 2008). However, consultation with the latest ICH guidelines related to photostability studies should be done prior to establishing the stability protocols.

References

Pulmonary – CDER (1998): IV.B.10; CDER (2002): IV.N
Nasal – CDER (1998): IV.B.10; CDER (2002): IV.N.

10.2.3.20 Physical characterization

Background According to the EMEA (CHMP, 2006), physical characteristics such as solubility, size, shape, density, rugosity, charge, and crystallinity of the drug substance and/or excipients may influence the homogeneity and reproducibility of the finished product. Development studies should include physical characterization of drug substance and excipients, relevant to their effect on the functionality of the product. If applicable, the effect of pre-processing the material on the physical characteristics should be evaluated. Note that physical characterization in these terms is an EMEA-only requirement, but should be considered part of normal product development for other target countries.

Stability considerations Physical characterization tests should be performed during stability studies at all stages of development, as indicated in Table 10.2, and are normally considered part of in-use testing (see section 16.4 of Kim (2008) for more details).

References

Pulmonary – CHMP (2006): 4.2.1.1
Nasal – CHMP (2006): 4.2.1.1.

10.2.3.21 Plume geometry

Background Plume geometry testing involves visualizing and/or measuring the shape of the emitted plume from a nasal or pulmonary spray drug product from a sideward looking view (e.g. parallel to the plume axis). Two plume geometry visualization techniques are currently described in the regulatory guidelines: (1) high-speed photography, and (2) laser light sheet and high-speed digital camera. The high-speed photography technique involves more art than science to capture a few frames of the fast-moving plume typically generated by nasal spray pumps and MDIs (Purewal and Grant, 1997). Laser light sheet and high-speed digital imaging provides a high-resolution time history of the emitted plume that can be used for much better

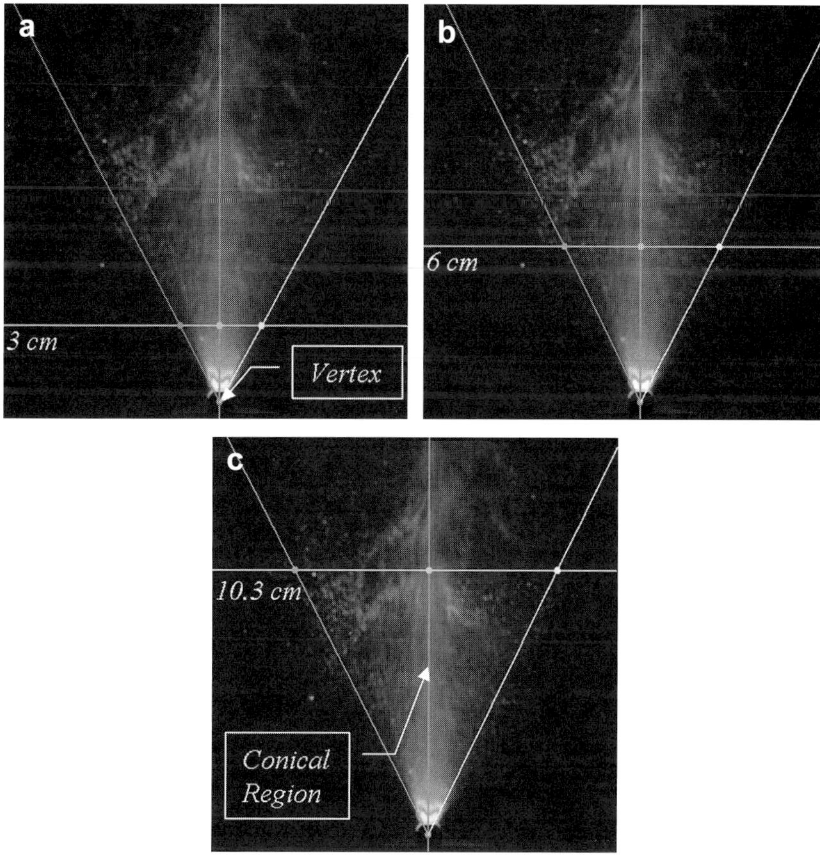

Figure 10.1 Plume geometry measurements for an aqueous nasal spray taken with Proveris Scientific's SprayVIEW® instrument at single delay time after automated actuation with consistent plume angle = 54°. (a) The vertex of the plume at the pump orifice, and plume width = 32.2 mm at 3 cm from vertex. (b) Plume width = 61.3 mm at 6 cm from vertex. (c) Conical region length = 10.3 cm.

analysis of the conical region of the spray and for creating a scientific basis for selecting spray pattern and droplet size distribution distances (CDER, 2003; Farina, 2004) (see Figure 10.1a–c for an example of plume geometry analysis of a nasal spray). The time history provided by laser light sheet and high-speed digital imaging plume geometry analysis can also be used to easily quantify the fully developed region of the spray as required by droplet size distribution by laser diffraction (Farina et al., 2000), and for assessing the effects of automated actuation on product performance (Guo and Doub, 2006).

Metrics commonly used for plume geometry include the plume angle (measured with a vertex that occurs at or near the nozzle tip), plume width (the width of the plume at a specified height from the mouthpiece edge (MDIs) or nozzle tip (nasal sprays)), and the delay time from the start of actuation used for the analyzed image. The FDA recommends selecting a delay time for image analysis corresponding to the fully developed phase of the plume while the plume is still in contact with the mouthpiece edge or nozzle tip (CDER, 2003). Note, particularly, that the use of the plume width metric can correlate with and complement spray pattern (e.g. D_{min} or D_{max} as defined below in "Spray pattern") data collected at the same distance, adding to the robustness and value of the collected data during product development or troubleshooting.

Stability considerations Plume geometry testing is normally performed only during product development and not during stability studies.

References

Pulmonary – CDER (1998): III.F.1.m, IV.A.10
Nasal – CDER (1998): III.F.1.m, IV.A.10; CDER (2002): III.F.1.h, IV.K; CDER (2003): V.B.6.

10.2.3.22 Preservative effectiveness and sterility maintenance

Background Preservative effectiveness testing is required to demonstrate that the minimum content limit of the preservative is microbiologically effective for the drug formulation (generally, the use of preservatives is discouraged for inhalation formulations). The test is normally performed with a microbial challenge assay of the drug formulated with an amount of preservative equal to or less than the minimum specified amount (CDER, 2002).

For device-metered, aqueous-based inhalation spray drug products, studies should be performed to demonstrate the appropriate microbiological quality through the life of the reservoir and during the period of reservoir use, with the intent of assessing the ability of the container closure system to prevent microbial ingress into the formulation and/or the growth-inhibiting properties of the formulation (CDER, 2002).

Stability considerations As described above, stability testing for this specification would be covered by a microbial challenge (see "Microbial challenge" for details).

References

Pulmonary – CDER (2002): IV.L
Nasal – CDER (1998): IV.A.11; CDER (2002): IV.L; CHMP (2006): 4.2.1.20.

10.2.3.23 Preservatives and stabilizing excipients assay

Background The FDA recommends the use of a specific assay for preservatives (e.g. antioxidants, chelating agents) or other stabilizing excipients (e.g. benzalkonium chloride, phenylethyl alcohol, edetate) if they are used in the OINDP formulation (e.g. nasal applications). Acceptance criteria for the chemical content of preservatives at the time of product release and through the product shelf-life should be included in the drug product specification (CDER, 2002).

Stability considerations Preservatives and stabilizing excipients assay tests should be performed during stability studies at all stages of development, as indicated in Table 10.2. Studies conducted to support clinical trials should use methods that can be validated, while validated methods should be used during production and post-market approval stability studies (Pilewski, 2007; Kim, 2008).

References

Pulmonary – none
Nasal – CDER (2002): III.F.1.e; CHMP (2006): 4.5.1.12.

10.2.3.24 Pressure testing

Background The FDA recommends performing pressure testing for MDI products that are formulated using a co-solvent and/or more than one propellant. The purpose of pressure testing is to verify the internal pressure of the canister containing the formulation/propellant mixture and to ensure the use of proper propellants or propellant mixture ratio (CDER, 1998).

Stability considerations Pressure testing should be performed during stability studies at all stages of development, as indicated in Table 10.2. Studies conducted to support clinical trials should use methods that can be validated, while validated methods should be used during production and post-market approval stability studies (Kim, 2008).

References

Pulmonary – CDER (1998): III.F.1.o
Nasal – CDER (1998): III.F.1.o.

10.2.3.25 Single-dose fine particle mass

Background See "Drug particle size distribution/fine particle mass" for detailed background on this test. Note that this specification is only required under the EMEA guideline.

Stability considerations See "Drug particle size distribution/fine particle mass" for details on stability considerations for this test.

References

Pulmonary – CHMP (2006): 4.2.1.7
Nasal – none.

10.2.3.26 Spray pattern

Background Spray pattern testing involves measuring the axial cross-sectional uniformity of the spray at a fixed distance from the mouthpiece edge (MDIs) or nozzle tip (nasal sprays). Similar to droplet size distribution testing, the distance(s) used for spray pattern analysis are normally between 2 and 7 cm from the mouthpiece edge or nozzle tip with a 3-cm gap if two distances are used (CDER, 2002, 2003). Two spray pattern measurement methods are currently described in regulatory guidelines: impaction and non-impaction (CDER, 2003). The impaction technique, the older of the two methods, involves firing the spray or aerosol at a thin-layer chromatography plate and analyzing the resultant splatter pattern. The impaction technique also requires the addition of a reagent to the formulation to allow the splatter pattern to be visualized on the plate, making the technique much easier to use reliably. Additionally, impaction techniques have relied on human interpretation of the splatter pattern and this subjectivity has made the technique very difficult to validate in pharmaceutical environments (Purewal and Grant, 1997).

Most modern spray pattern analysis is done with the non-impaction technique that involves using a system based on a laser light sheet and high-speed digital camera to enable visualization of a pattern perpendicular to the axis of the nasal spray or aerosol (CDER, 2003). The non-impaction technique is far simpler to use and validate, and can produce high-quality images in seconds (Farina, 2004; Marx, 2009) (see Figure 10.2 for a spray pattern example from an aqueous nasal spray). Non-impaction-based spray pattern techniques use patented, validated image-processing algorithms to objectively analyze the spray pattern images and produce metrics such as D_{min}, D_{max}, ovality (D_{max}/D_{min}), and spray pattern area, as defined by the FDA (CDER, 2003). Spray pattern testing is most useful as a quality-control metric and for helping to determine the bioequivalence of different spray devices/formulations (CDER, 2003; Pennington et al., 2008; Guo and Doub, 2006; Guo et al., 2008).

Stability considerations Spray pattern testing should be performed during stability studies at all stages of development, as indicated in Table

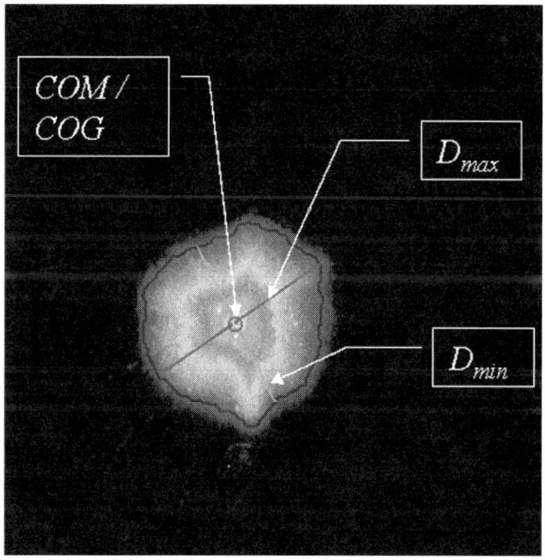

Figure 10.2 Spray pattern measurement taken at 3 cm from an aqueous nasal spray pump orifice with Proveris Scientific's SprayVIEW® instrument. The contour (solid outline), center of gravity (COG), center of mass (COM), D_{max}, and D_{min} are measurement parameters recommended in the 2003 FDA draft guidance for nasal spray and nasal aerosols (CDER, 2003). Image courtesy of Proveris Scientific Corporation.

10.2. Studies conducted to support clinical trials should use methods that can be validated, while validated methods should be used during production and post-market approval stability studies (Kim, 2008).

References

Pulmonary – CDER (1998): III.F.1.m

Nasal – CDER (1998): III.F.1.m; CDER (2002): III.F.1.h; CDER (2003): V.B.5.

10.2.3.27 Sterility

Background According to CHMP (2006), as appropriate, sterility testing should be conducted according to a test listed in accepted pharmacopoeia (e.g. Test 3.2.1 of the International Pharmacopoeia (WHO, 2007)).

Stability considerations As appropriate, sterility testing should be performed during stability studies at all stages of development, as indicated in Table 10.2. Studies conducted to support clinical trials should use methods that can be validated, while validated methods should be used during production and post-market approval stability studies (Pilewski, 2007).

References

Pulmonary – none
Nasal – CHMP (2006): 4.5.1.10.

10.2.3.28 Viscosity

Background Numerous investigators have studied the effects of viscosity on the performance of nasal spray formulations (e.g. Harris et al., 1988; Dua et al., 1997; Kippax et al., 2008). Viscosity plays a particularly strong role in the spray dynamics (e.g. spray pattern, plume geometry, and droplet size distribution) and can significantly affect the patient experience with a mechanical pump-based nasal spray product (Guo et al., 2008). For these reasons, the FDA recommends testing and controlling the formulation viscosity at release and on stability with an appropriate procedure and acceptance criterion (CDER, 2002).

Stability considerations Viscosity testing should be performed during stability studies at all stages of development, as indicated in Table 10.2. Studies conducted to support clinical trials should use methods that can be validated, while validated methods should be used during production and post-market approval stability studies (Pilewski, 2007).

References

Pulmonary – none
Nasal – CDER (2002): III.F.1.r.

10.2.3.29 Water or moisture content

Background According to the FDA (CDER, 1998), testing for the presence of water in the container should be performed, particularly for suspension formulations. Furthermore, water or moisture should be strictly limited to prevent changes in particle size distribution, morphic form, and other changes such as crystal growth or aggregation, all of which can significantly affect the stability of the product.

Stability considerations It is often difficult, if not impossible, to set appropriate acceptance criteria for water or moisture content during early development stages because the effects of these parameters on critical product attributes such as particle size, degradation, or microbial growth are not known until later in the development process. However, moisture should still be measured at key stability checkpoints because its effects can be linked to critical product attributes (as mentioned above). If a negative correlation is found between moisture and a critical parameter, then moisture itself must be controlled during the manufacturing process (e.g. through raw material controls or through appropriate packaging/storage). If moisture is not an issue or is less indicative of an issue than measurement of a critical attribute itself, then moisture control may be omitted, pending regulatory approval, from the specification and future post-approval stability studies (Pilewski, 2007).

References

Pulmonary – CDER (1998): III.F.1.d, III.F.2.d; CHMP (2006): 4.5.1.3

Nasal – CDER (1998): III.F.1.d, III.F.2.d; CHMP (2006): 4.5.1.3.

10.3 CONCLUSION

This chapter highlighted all of the tests currently required by the FDA and EMEA/Health Canada for marketing and sale of OINDPs in the USA, Europe, and Canada respectively. The list of tests is quite long and complex, so a thorough understanding of them, including how and when to conduct the tests, is crucial to acceptance by these regulatory agencies. Using a risk-based approach based on product understanding and targeted indication/patient group is highly recommended to streamline the development process, produce high-quality test data and, ultimately, to help patients.

This chapter also showed that numerous discrepancies exist between the FDA and EMEA/Health Canada guidelines with regard to OINDP development and testing. These discrepancies can have a significant impact on how the product is developed, tested, documented, and presented to the appropriate regulatory agency for approval.

Lastly, the FDA, EMEA, Health Canada and industry groups such as IPAC-RS, PQRI, and the AAPS/INTFG have been very active in the area of OINDP development, producing numerous documents to help clarify current thinking on many of the tests and approaches that should be used. Hence, it is highly recommended to maintain awareness of emerging developments in this area. All of the above-mentioned regulatory and industry groups maintain effective websites and conduct regular symposia on issues related to OINDP development and testing.

REFERENCES

Ahmed, S., Sileno, A.P., deMeireles, J.C., Dua, R., Pimplaskar, H.K., Xia, W.J., Marinaro, J., Langenback, E., Matos,., F.J., Putcha, L., Romeo, V.D., Behl, C.R., 2000. Effects of pH and dose on nasal absorption of scopolamine hydrobromide in human subjects. Pharm. Res. 17 (8), 974–977.

Berry, J., Kline, L.C., Hart, J.L., Sequeira, J., 2003. Influence of the storage orientation on the aerodynamic particle size of a suspension metered dose inhaler containing propellant HFA-227. Drug Dev. Ind. Pharm. 29 (6), 631–639.

Borgström, L., Asking, L., Lipniunas, P., 2005. An in vivo and in vitro comparison of two powder inhalers following storage at hot/humid conditions. J. Aerosol Med. 18 (3), 304–310.

Center for Drug Evaluation and Research (CDER), 1998. Guidance for Industry: Metered Dose Inhaler (MDI) and Dry Powder Inhaler (DPI) Drug Products. Chemistry, Manufacturing, and Controls Documentation. US Department of Health and Human Services, Food and Drug Administration.

Center for Drug Evaluation and Research (CDER), 2002. Guidance for Industry: Nasal Spray and Inhalation Solution, Suspension, and Spray Drug Products. Chemistry, Manufacturing, and Controls Documentation. US Department of Health and Human Services, Food and Drug Administration.

Center for Drug Evaluation and Research (CDER), 2003. Guidance for Industry: Bioavailability and Bioequivalence Studies for Nasal Aerosols and Nasal Sprays for Local Action. US Department of Health and Human Services, Food and Drug Administration.

Center for Drug Evaluation and Research (CDER), 2005. Guidance for Industry: ANDAs: Impurities in Drug Products. US Department of Health and Human Services, Food and Drug Administration.

Center for Drug Evaluation and Research (CDER)/Center for Biologics Research and Evaluation (CBER), 1996. Guidance for Industry: Q1B Photostability Testing of New Drug Substances and Products. US Department of Health and Human Services, Food and Drug Administration.

Center for Drug Evaluation and Research (CDER)/Center for Biologics Research and Evaluation (CBER), 1999. Guidance for Industry: Container Closure Systems for Packaging Human Drugs and Biologics, CMC. US Department of Health and Human Services, Food and Drug Administration.

Center for Drug Evaluation and Research (CDER)/Center for Biologics Research and Evaluation (CBER), 2006. Guidance for Industry: Q3B(R2) Impurities in New Drug Products. US Department of Health and Human Services, Food and Drug Administration.

Chi Lip Kwok, P., Chan, H.K., 2008. Effect of relative humidity on the electrostatic charge properties of dry powder inhaler aerosols. Pharm. Res. 25 (2), 277–288.

Committee for Medicinal Products for Human Use (CHMP), 2005. Guideline on Plastic Immediate Packaging Materials. European Medicines Agency, 1 December.

Committee for Medicinal Products for Human Use (CHMP), 2006. Guideline on the Pharmaceutical Quality of Inhalation and Nasal Products. European Medicines Agency, 1 October.

Dua, R., Zia, H., Needham, T., 1997. The influence of tonicity and viscosity on the intranasal absorption of salmon calcitonin in rabbits. Int. J. Pharm. 147 (2), 233–242.

European Pharmacopoeia (EP), 2007. Preparations for Inhalation, 5th ed. EP, pp. 2843–2847.

Farina, D., 2004. Advancing the science of in vitro testing and laboratory data management for nasal sprays. Drug Deliv. Tech. 4 (4), 46–51.

Farina, D., 2007. Characterization of an aqueous nasal spray using a novel, quality by design approach. Presented at AAPS 2007, San Diego, CA.

Farina, D., 2008. Characterization of an HFA pressurized metered dose inhaler using a novel, quality by design approach. Presented at AAPS 2008, Atlanta, GA.

Farina, D.J., Fallon, T.M., Kalogrianitis, S., Brouet, G., Carlotti, P., 2000. Evaluation of the spray characteristics of a novel unit dose nasal dry powder device using the SprayVIEW® technique. Presented at the 2000 AAPS Annual Meeting, Indianapolis, IN.

Guo, C., Doub, W.H., 2006. The influence of actuation parameters on in vitro testing of nasal spray products. J. Pharm. Sci. 95 (9), 2029–2040.

Guo, C., Stine, K.J., Kauffman, J.F., Doub, W.H., 2008. Assessment of the influence factors on in vitro testing of nasal sprays using Box-Behnken experimental design. Eur. J. Pharm. Sci. 3 (5), 417–426.

Harris, A.S., Svensson, E., Wagner, Z.G., Lethagen, S., Nilsson, I.M., 1988. Effect of viscosity on particle size, deposition, and clearance of nasal delivery systems containing desmopressin. J. Pharm. Sci. 77 (5), 405–408.

Haynes, A., Shaik, M.S., Krarup, H., Singh, M., 2004. Evaluation of the Malvern Spraytec® with inhalation cell for the measurement of particle size distribution from metered dose inhalers. J. Pharm. Sci. 93 (2), 349–363.

ITFG/IPAC-RS Collaboration, 2001. Leachables and extractables testing: points to consider. Available from IPAC-RS at: <http://www.ipacrs.com/PDFs/Points_to_Consider_FINAL.PDF>.

Kim, H.-B. (Ed.), 2008. Handbook of Stability Testing in Pharmaceutical Development. Regulations, Methodologies, and Best Practices. Springer Science & Business Media, LLC, pp. 333–340.

Kippax, P., Suman, J., Williams, G., 2008. Enhancing the in vitro assessment of nasal sprays: can different polymers improve stability and prolong retention in model nasal spray formulations? Pharm. Tech. Europe.

Malvern Instruments Ltd, 2007. Monitoring Nasal Spray Tail-Off Using the Malvern Spraytec. Spraytec Application Note MRK844–01.

Malvern Instruments Ltd, 2007. Measuring sprays containing propellants or volatile components: detection and elimination of beam steering. Spraytec Application Note MRK881–01.

Marx, D., 2009. Selecting a nasal spray pump. Inhalation. 3 (3), 18–21.

Mitchell, J.P., Nagel, M.W., 2003. Cascade impactors for the size characterization of aerosols from medical inhalers: their uses and limitations. J. Aerosol Med. 16 (4), 341–377.

Mitchell, J.P., Nagel, M.W., Copley, M., 2009. The abbreviated impactor measurement concept. Inhalation. 3 (3), 26–30.

Nagao, L.M., Lyapustina, S., Munos, M.K., Capizzi, M.D., 2005. Aspects of particle science and regulation in pharmaceutical inhalation drug products. Crystal Growth and Design (Am. Chem. Soc.). 5 (6), 2261–2267.

Norwood, D.L., Paskiet, D., Ruberto, M., Feinberg, T., Schroeder, A., Poochikian, G., et al., 2008. Best practices for extractables and leachables in orally inhaled and nasal drug products: an overview of the PQRI recommendations. Pharm. Res. 25 (4), 727–739.

Novick, S., Christopher, D., et al., 2009. A two one-sided parametric tolerance interval test for control of delivered dose uniformity – Part 2. Effect of changing parameters. AAPS Pharm. Sci. Tech.10.1208/s12249-009-9269-3.

Olsson, B., 2003. Parametric tolerance interval (PTI) test for improved control of delivered dose uniformity (DDU) in orally inhaled and nasal drug products (OINDP). Presentation to IPAC-RS, 13 March, Rockville, MD.

Parker, F.P., 2005. FDA Administrative Enforcement Manual. CRC Press, Boca Raton, FL.

Pennington, J., Pandey, P., Tat, H., Willson, J., Donovan, B., 2008. Spray pattern and droplet size analyses for high-shear viscosity determination of aqueous suspension corticosteroid nasal sprays. Drug Dev. Ind. Pharm. 34 (9), 923–929.

Pilewski, S., 2007. Product Dossier Preparation: Navigating the Regulatory CMCs for Pulmonary Products. Presented at RDD Europe 2007.

PQRI, 2003. Particle Size Distribution Mass Balance Working Group. Considerations for the development and practice of cascade impaction testing, including a mass balance failure investigation tree. J. Aerosol Med. 16 (3), 235–247.

PQRI, 2006. Leachables and Extractables Working Group. Safety thresholds and best practices for extractables and leachables in orally inhaled and nasal drug products. Available from PQRI at: <http://www.pqri.org/pdfs/LE_Recommendations_to_FDA_09-29-06.pdf>.

Purewal, T.S., Grant, D.J.W. (Eds), 1997. Metered Dose Inhaler Technology. CRC/Interpharm, pp. 176–177.
Tougas, T., Christopher, D., Lyapustina, S., Mitchell, J., Patel, R., 2008. Capabilities of aerodynamic particle size distribution (APSD) measurements based on analysis of a blinded database. Presented at RDD 2008, Scottsdale, AZ, May. <http://www.ipacrs.com/PDFs/RDD2008%20TT.pdf>.
United States of America v. Schering-Plough Corporation and Schering-Plough Products, LLC, corporations, and Richard J. Kogan, and Steven C. Chellevold, Individuals. Filed 20 May 2002 in United States District Court, District of New Jersey, William T. Walsh, Clerk.
United States Pharmacopoeia (USP), 2008. Aerosols, nasal sprays, metered dose inhalers, and dry powder inhalers. USP 30-NF 25 <601>.
World Health Organization (WHO), 2007. The International Pharmacopoeia. WHO, Geneva.
World Health Organization (WHO), 2008. Stability testing of active pharmaceutical ingredients and pharmaceutical products. Working document QAS/06.179/Rev. 3.
Yoshihisa, N., Atsuhiro, N., Yasuhide, U., 2002. Nasal suspensions with low osmolality and drug permeability. Oto-Rhino-Laryngology Tokyo 45 (Suppl. 1), 46–49.
Zhao, P., Bakrenev, V., Calderone, D., Dilone, E., Tesler, S., 2008. Elimination of beam steering and determination of droplet size distribution for nitroglycerin oral spray (NitroMist) drug product. Presented at AAPS Annual Meeting 2008, Atlanta, GA. <http://www.aapsj.org/abstracts/AM_2008/AAPS2008-001667.PDF>.

INDEX

Absorption enhancers for nasal delivery, 205–6
N-Acetyl-L-cysteine (NAC), 205
ACI *see* Andersen cascade impactor
Active pharmaceutical ingredient (API):
  dry powder inhalers, 225
  liposomal encapsulation, 240
  micellar nanoparticles, 39–42, 44–5, 48–9
  thermodynamic activity, 81–3
Actuations (net content) regulations, 255–6
Acyclovir (Zovirax®), 42, 54
AeroChamber Plus® Z STAT® valve-holding chamber, 220–1
Aerodynamic diameter (d_a), 213
Aerodynamic particle size distribution (APSD), 272–3
Aerodynamic particle sizer (APS) for aerosols analysis, 237–8
AeroEclipse nebulizer, 228
Aerosols:
  analysis, 231–9
  drug delivery devices, 210, 218–31, 231–9
  oropharyngeal region, 218–28
  regulation, 247–90
  see also DPIs; MDIs
AET *see* antimicrobial effectiveness test
Age-related macular degeneration (AMD), 182
Akten, 171
Albuterol, 223–4
Alcohols, 22–4
Aldara®, 84, 86
Alprostadil, 49
AMD *see* Age-related macular degeneration
Anatomy
  eye, 167–8
  skin, 1–2
Andersen cascade impactor (ACI), 233–5
Androsorb™, 47
Anti-microbial properties of micellar nanoparticles, 44–5
Antimicrobial effectiveness test (AET), 44
API *see* active pharmaceutical ingredient
APS *see* aerodynamic particle sizer (APS)
APSD *see* aerodynamic particle size distribution
Aqueous humor (eye), 68
Arterial blood vessels and microneedles, 157
Asthma, 212, 218, 226, 239
AzaSite®, 171
Azone, 20, 84, 107

Bacillus anthracis protective antigen (Anthrax) vaccine, 160
Bio-Cor®, 176
Bioadhesive polymers (Chitosan), 174, 175, 203–5
Bioavailability and eye medications, 169
Biodegradable microneedles, 140
Biodegradable polymers for nasal delivery, 205–6
BioMEMS, 159
Blood glucose monitoring and diabetes, 114, 128–9
Blood–retinal barrier (BRB), 168
Bolus injections and retinal treatment, 182
Bordetella bronchiseptica, 204
Brownian diffusion, 216
Brownian motion, 213
Budesonide, 223

Carbopol, 86
Carboxymethylcellulose (CMC), 144, 170
Cascade impactor for aerosols analysis, 233–5
Cavitation and ultrasound, 121–2, 125–7
Central cavity (eye), 168
Central nervous system (CNS) and nasal delivery, 203
Cetirizine, 49
CFCs *see* chlorofluorocarbons

Characterization of emulsions/ microemulsions, 63
Chemical enhancers:
 iontophoresis, 106–7
 permeation, 16–26
 ultrasound, 130
Chitosan, 174, 175, 203–5
Chlorofluorocarbons (CFCs) propellants for MDIs, 220, 223–4
Chondroitin sulfate, 174
Choroid (eye), 168
Chronic obstructive pulmonary disease (COPD), 211, 212, 218, 240, 241
Ciliary neurotrophic factor (CNTF), 181
Clonidine, 5, 48
CMC *see* carboxymethylcellulose; critical micelle concentration
CMV *see* cytomegalovirus
CNTF *see* ciliary neurotrophic factor
Coat and poke technique for microneedles, 141
Coating strategies in transdermal permeation, 143–8
Collagen shields, 176
Combination strategies fro iontophoresis, 105–10, 114
Commercialization of microneedle technology, 157–60
Complexation, 8–9
Contact lenses, 176
Controlled release and encapsulation in nasal delivery, 201
COPD *see* chronic obstructive pulmonary disease
Cornea, 167–8, 169
Creams, 4, 80
Critical micelle concentration (CMC), 107
Cutaneous drug absorption, 6–26
Cyclobenzaprine, 49
Cyclodextrins, 8–9, 203
Cystic fibrosis, 212
Cytomegalovirus 180 (CMV)

d_a *see* aerodynamic diameter
DDDs *see* drug delivery devices
Decylmethylsulfoxide, 19
Dendrimers and ophthalmic drug delivery, 179
Depot effects, 54
Depth of permeation with microneedles, 143
Dermis and skin, 3
Diabetes and blood glucose monitoring, 128–9
Differin®, 84, 86
Dimethyl sulfoxide (DMSO), 19
Dimethylformamide (DMF), 19
Dip and scrape technique for microneedles, 141–2
DMF *see* dimethylformamide
DMSO *see* dimethly sulfoxide
Dose applied and drug transport, 77–8
Dose content uniformity regulations, 268–9, 269–70
Dosing frequency and drug transport, 77–8
DPIs *see* dry powder inhalers
Droplet size distribution (DSD) regulations, 270–1
Drug aerosol particles, 212–14
Drug delivery devices (DDDs) for pulmonary airways, 210, 218–31, 231–9
Drug transport:
 diffusion across stratum corneum, 76–7
 dose applied, 77–8
 dosing frequency, 77–8
 emulsion to skin, 75–6
 iontophoresis, 100–2
 skin, 71–8, 100–2
 solubility, 72–4
 thermodynamic activity, 72–4
Drug/ion properties and iontophoresis, 103
Dry powder inhalers (DPIs):
 active pharmaceutical ingredient, 225
 ease of use, 219
 electrostatic charge, 218
 metered-dose inhalers/nebulizers comparison, 229–31, 232
 optical microscopy, 237
 pulmonary drug delivery, 225–7

regulation, 248, 249–52, 254, 258–62, 265, 269
Spinhaler, 210
DSD *see* droplet size distribution

E-TRANS®, 113–14
ECT *see* encapsulated cell technology
Electrical low pressure impactor (ELPI) for aerosols analysis, 235–7
Electronic single-particle aerodynamic relaxation time analyser (ESPART), 238–9
Electroplating (Galvanik) in fabrication of microneedles, 137
Electroporation and iontophoresis, 26, 106
Elocon®, 86
ELPI *see* electrical low pressure impactor
EMEA *see* European Medicines Agency
Emulsifiers for emulsions, 85, 86
Emulsions:
characterization, 63
dose, 77–8, 82
drug release, 74–5
drug transport (emulsion to skin), 75–6
emulsifiers, 85, 86
estradiol, 79
formulation strategies, 80–6
interfacial tension measurements, 68–9
microemulsions comparison, 86–7
particle size/dispersion, 65–6
phase behavior, 63–4
properties, 59–61
pulsed-field-gradient spin echo NMR, 67–8
rheology, 64–5
stability evaluation, 69–71
stabilization, 85–6
topical/transdermal drug delivery, 71–8, 78–80
zeta potential, 66–7
Encapsulated cell technology (ECT), 181
EPO *see* erythropoetin
Erodible implants for retinal treatment, 181–3
Erythropoietin (EPO), 154
ESPART *see* electronic single-particle aerodynamic relaxation time analyser
Essential oils, 25–6
Estradiol, 5, 44, 46, 79
Estrasorb™, 38–9, 45–7, 56
Estrogen cream, 80
Ethanol and penetration enhancers, 22–3
Ethosomes, 13
Ethylcellulose, 205
European Medicines Agency (EMEA), 247, 253–5, 261
Eutectic systems, 7
Excipients:
coating strategies, 144
formulation strategies, 83–5
Exhaustion (tailoff) regulations, 256–7
Extractables regulations, 275–6
Eye:
anatomy, 167–8
bioavailability of medications, 169
drug delivery, 165–6

Fabrication of microneedles, 137–8, 141
Fatty acids, 21–2
Fatty alcohols, 22–4
FDA *see* Food and Drug Administration
Fentanyl, 48, 114
FEV_1 *see* forced expiratory volume in one minute
FFEM *see* freeze-fracture electron microscopic
Finacea®, 84
FITC *see* fluorescein isothiocyanate
Fluorescein isothiocyanate (FITC), 205
Food and Drug Administration (FDA) *see* US FDA
Force of penetration of microneedles, 148–51
Forced expiratory volume in one minute (FEV_1), 211–12
Formulations:
emulsions
active pharmaceutical ingredient, 81–3
excipients, 83–5
project timeline, 81
stabilization, 85–6
thermodynamic activity, 81–3

Formulations (*continued*)
nasal delivery, 194–7
topical comparison with microemulsions, 89–91
Franz cell, 46, 51–3, 127
Freeze-fracture electron microscopic (FFEM) images, 43

Gastrointestinal tract (GI), 96, 210, 214
Gatifloxacin, 173, 175, 176
Gelrite®, 171
Gels, 171–3, 205
Gene delivery by microneedles, 155–6
General properties regulations, 251–2, 267–87
Geometry of microneedles, 136–7, 138, 148–51
GI *see* gastrointestinal tract
GlucoWatch®, 96, 113–14
Glucose, 114, 128–9
Glycol, 22–4
Gravitational settling in distal airways, 216
Growth hormone, 5

Heating and ultrasound, 122
Helicobacter pylori, 204
Heparin, 108, 128
HFAs *see* hydrofluoroalkanes
hGH *see* human growth hormone
HLB *see* hydrophilic–lipophilic balance
Hollow microneedles, 139
Hormone replacement therapy (HRT), 78–9
HPMC *see* hydroxypropylmethylcellulose
HTCC *see* *N*-[(2-hydroxy-3-trimethylammonium)propyl]chitosan chloride
Human growth hormone (hGH), 154, 206
Human respiratory system and pulmonary drug delivery, 211–14
Human subjects and topical/transdermal drug delivery, 78–80
Hybresis™, 96, 113–14
Hydrofluoroalkanes (HFAs) propellants for aerosols, 223–4
Hydrophilic–lipophilic balance (HLB), 40
Hydroxypropylmethylcellulose (HPMC), 170, 173
Hypodermic needles:
design and microneedles, 156
injection and hollow microneedles, 139
Hypodermis and skin, 3

Inertial impaction technique for aerosols analysis, 233, 236
Inserts for ophthalmic drug delivery, 176–7
Insulin:
dry powder inhalers, 226
microneedles, 151–4
sonophoresis, 123–4
transdermal systems, 5
ultrasound, 128
Interfacial tension measurements for emulsions/microemulsions, 68–9
International Pharmaceutical Excipients Council (IPEC), 84
Ion pair formation, 8
Ion-exchange and iontophoresis, 109–10
IONSYS™, 96, 114
Iontophoresis:
advantages/limitations, 97–8
combination strategies for transdermal drug delivery, 105–10, 114
cortical bone, 112–13
development, 113–14
drug transport, 100–2
drug/ion properties, 103
history/development, 97
macromolecules, 112
ophthalmic drug delivery, 178–9
pH and current, 103–5
principles/components, 98–9
reverse, 110–11, 128–9
transdermal drug delivery, 95–105, 105–10, 110–14
transport pathways, 102–3
transungal transport, 113
IPEC *see* International Pharmaceutical Excipients Council
IPM surfactant, 88–91

Jet nebulizers, 227–8

Klaron®, 86

Lacrisert®, 176
Langerhans cells, 130
Large porous particles (LPPs), 241
Large unilamellar vesicles (LUV), 9–13
Large-molecular-weight compounds and microneedles, 154–5
Laser diffraction and droplet size distribution, 270–1
Laser-induced tools for aerosols analysis, 237
Leaches regulations, 275–6
Lidocaine, 5, 113–14
LidoSite®, 113–4
LIGA (Lithographie, Galvanik and Abformung) technique for microneedles, 137–8, 139
"Like charges repel" principle, 98
Lipid-polycation-pDNA (LPD) nonviral gene therapy, 156
Liposomes:
  cutaneous delivery, 26
  description, 9–13
  microemulsions, 91
  ophthalmic drug delivery, 177–8
  plasmid DNAs, 178
Liquid crystalline systems, 16
Lithography (Lithographie) in fabrication of microneedles, 137–8, 139
Low-frequency ultrasound, 124–5, 127–8, 130–1
LPD *see* Lipid-polycation-pDNA
LPPs *see* large porous particles
Lucentis®, 180, 182
Lutrol® F-68 NF surfactant, 143
LUV *see* large unilamellar vesicles

Macroflux® technology, 108, 158
Macromolecules and ultrasound, 127–8
Macugen®, 180, 182
MDIs *see* metered-dose inhalers
MDP *see* muramyl dipeptide
Medidur, 181
Medihaler, 210
Medilens®, 176
MEMS *see* microelectromechanical systems (MEMS)
L-Menthol, 25
Metered-dose inhalers (MDIs):
  asthma and related diseases, 239
  dry powder inhalers/nebulizers comparison, 229–32
  Medihaler, 210
  optical microscopy, 237
  pulmonary drug delivery, 218–19, 219–25, 239
  regulation, 249–52, 255, 258–9, 264–5, 269, 276–7, 280
Methylparaben (MP), 67
Micellar nanoparticles (MNP):
  active pharmaceutical ingredient, 49–50
  anti-microbial properties, 44–5
  composition/structure, 39–41
  introduction, 38–9
  physicochemical characterization, 42–4
  topical drug delivery, 53–4
  transdermal drug delivery, 45–53, 54–6
Micro/nano encapsulated drugs for nasal delivery, 201–6
Microbial challenge regulations, 277
Microbial limits regulations, 277–8
Microelectromechanical systems (MEMS), 137–8, 142
Microemulsions:
  characterization, 63
  description, 13–14
  emulsions comparison, 86–7
  interfacial tension measurements, 68–9
  particle size/dispersion, 65–6
  phase behavior, 63–4
  properties, 61–3
  pulsed-field-gradient spin echo nuclear magnetic resonance, 67–8
  rheology, 63–5
  stability evaluation, 69–71
  topical drug delivery systems, 86–91
  topical formulations comparison, 89–91
  transdermal drug delivery systems, 86–91
  zeta potential, 66–7

Microneedles:
arterial blood vessels, 157
Bacillus anthracis protective antigen (Anthrax) vaccine, 160
biodegradable, 140
coat and poke technique, 141
coating strategies, 143–8
commercialization, 157–60
conclusions, 160–1
depth of permeation, 143
dip and scrape technique, 141–2
drug delivery, 151–7
electroplating (Galvanik), 137
erythropoietin, 154
fabrication, 137–8, 141
force of penetration, 148–51
gene delivery, 155–6
geometry, 136–7, 138, 148–51
hollow, 139
hypodermic needle design, 156
insulin, 151–4
iontophoresis, 107–8
large-molecular-weight compounds, 154–5
lithography (Lithographie), 137
molding (Abformung), 137
pain, 139
poke with patch technique, 141
polymer, 140–2, 155
riboflavin, 146–7
solid, 137–9
stainless steel, 138–9, 145
targeted drug delivery, 156–7
thermal inkjet technology, 160
transdermal permeation, 143–51, 146
vaccination, 160, 161
MicronJet® (microneedle device), 159
Microspheres and ophthalmic drug delivery, 178–85
Microstreaming and ultrasound, 122
Minimally invasive transdermal drug delivery, 135–64
MNP *see* micellar nanoparticles
Molding (Abformung) in fabrication of microneedles, 137
Moxifloxacin, 176
MP *see* methylparaben
Multiphasic MNP, 42
Muramyl dipeptide (MDP), 204

N-methyl-2-pyrrolidone (NMP), 20–1
NAC *see* *N*-acetyl-L-cysteine
Naltrexone, 49
Nanoemulsions, 38
Nanosuspensions and ophthalmic drug delivery, 175–6
Nanotechnology, 37–8
Nasal delivery:
absorption enhancers, 205–6
advantages, 199
central nervous system, 203
Chitosan, 203–5
controlled release, 201
drug delivery formulation, 199–201
formulations, 194–7
gels, 205
micro/nano encapsulated drugs, 201–6
nose physiology, 198–9
plasmid DNA, 200–1
PLGA, 201–3
polyethylene glycol, 202
vaccines, 200–1, 205
Nasal spray drug products:
patient use/misuse, 259–67
product labeling, 253–9
regulation, 247–90
tests, 249–52
NDA *see* New Drug Application
Nebulizers, 227–9, 229–31
NEEP *see* New Excipient Evaluation Procedure
New Drug Application (NDA), 55
New Excipient Evaluation Procedure (NEEP), 84
Nicoderm®, 52
Nicotine, 5, 48, 51–3
NMP *see* N-methyl-pyrrolidone
Non-erodible implants for retinal treatment, 180–1
Non-invasive drug delivery for pulmonary airways, 209–10

Non-invasive scintigraphic techniques, 232–3
Nose, physiology, 198–9

O/W *see* oil in water
Ocusert®, 176
Oil in water (O/W):
  emulsions, 60, 70, 84, 86
  microemulsions, 62
Oils, essential, 25–6
OINDPs *see* orally inhaled and nasal drug products
Ophthalmic drug delivery:
  anterior segment delivery systems, 166–79
  bioadhesive polymers, 173–5
  Chitosan, 174, 175
  conclusions, 186
  dendrimers, 179
  gels, 171–3
  inserts, 176–7
  iontophoresis, 178–9
  liposomes, 177–8
  microspheres, 178, 185
  nanosuspensions, 175–6
  new techniques, 186
  penetration enhancers, 178
  periocular route, 165–6
  punctal plugs, 179
  retinal systems, 179–86
  trans-scleral, 166
  viscous solutions, 170–1
OPL *see* oropharyngeal
"Opposite charges attract" principle, 98
Optical microscopy for aerosols analysis, 237
Orally inhaled and nasal drug products (OINDPs), 247–54, 260, 263, 268, 272
Organogels, 14–15
Orientation/weight loss (stability) regulations, 262–3
Oropharyngeal (OPL) region and aerosols, 218, 228
Osmolality regulations, 278–9
Oswald ripening, 44
Oxybutynin, 5, 48
Ozone layer, 223
Ozurdex®, 181

Pain:
  management, 113–4
  microneedles, 139
Pari LC Plus nebulizer, 228
Particle size distribution regulations, 271–2
Particle size/dispersion, emulsions/ microemulsions, 65–6
Passive cutaneous drug absorption, 6–26
Passive transdermal drug delivery, 45–53
Patches *see* transdermal drug delivery systems
Patchless-patch, 46
Patient use/misuse of nasal/pulmonary products, 259–65
Patient use/misuse tests, 250, 259–65
PCP-CYS *see* polycarbopil-cysteine
PDD *see* pulmonary drug delivery
PDMS *see* polydimethylsiloxane
Peak expiratory flow (PEF), 211–12
PEF *see* peak expiratory flow
PEG *see* polyethylene glycol
Pemulen™, 70–1, 86
Penetration:
  enhancers, 174
  microneedles, 148–51
Percutaneous absorption pathways, 5–6
Periocular drug delivery, 165–6, 182–6
Permeation:
  depth, 143
  enhancement, 22–3
  microneedles, 143–51
  skin, 83
  stratum corneum, 5, 26, 131
  transdermal and microneedles, 146
PEs *see* penetration enhancers
pH and iontophoresis, 103–5
Phase behavior of emulsions/ microemulsions, 63–4
Phonophoresis *see* sonophoreisis
Physical properties of drug aerosol particles, 212–14

Physicochemical characterization of MNP, 42–4
Physiology of nose, 198–9
Piezoelectric effect, 120
Pilocarpine, 174, 176
Pilopine HS®, 11
Plasmid DNA and nasal delivery, 200–1
PLGA *see* poly(lactic-co-glycolic acid)
Plume geometry regulations, 280–2
Pluronic, 172, 185
PMMA *see* polymethylmethacrylate
Poke with patch technique for microneedles, 141
Polycarbopil-cysteine (PCP-Cys), 206
Polydimethylsiloxane (PDMS), 140
Polyethylene glycol (PEG), 14, 145, 202, 205
Poly(lactic-co-glycolic acid) (PLGA):
  ganciclovir/forscarnet, 182
  gatifloxacin, 175
  microneedles, 145–6
  nasal delivery, 201–3, 205
  scleral plugs, 185
Polymers:
  microneedles, 140–2, 155
  nasal delivery, 205–6
Polymethylmethacrylate (PMMA), 138
Polyvinylpyrrolidone (PVP), 145
Priming/repriming regulations, 256
Prodrug approach, 7–8
Product labeling tests for nasal/pulmonary spray drug products, 249, 253–9
Project timeline and formulation strategies, 81
Propylene glycol, 14, 23–4, 25, 40
Proshield®, 176
Pseudo-patches, 46
Pulmonary airways and non-invasive drug delivery, 209–10
Pulmonary drug delivery (PDD):
  analytical methods, 231–9
  challenges/advantages, 239–41
  conclusions, 241–2
  devices, 210, 225–7, 231
  dry powder inhalers, 225–7
  human respiratory system, 211–14
  mechanisms, 214–18
  metered-dose inhalers, 219–25
  methods comparison, 229–31
  nebulizers, 227–9
  non-invasive, 209–11
Pulmonary spray drug products:
  patient use/misuse, 259–65
  product labeling, 253–9
  regulation, 247–90
  tests, 249–52
Pulsed-field-gradient spin echo NMR for emulsions/microemulsions, 67–8
Pump/valve delivery (shot weight) regulations, 257–8
Punctal plugs and ophthalmic drug delivery, 179
PVP *see* polyvinylpyrrolidone
PyraDerm™, 158
2-Pyrolidone (2-P), 20–1
Pyrrolidones, 20–1

Raloxifene, 47–51, 49
Regulation:
  actuations, 255–6
  conclusions, 287
  dose content uniformity, 268–9, 269–70
  droplet size distribution, 270–1
  exhaustion (tailoff), 256–7
  extractables, 275–6
  general properties, 251–2, 267–87
  leaches, 275–6
  microbial challenge, 277
  microbial limits, 277–8
  nasal/pulmonary product tests, 249–52
  orientation/weight loss (stability), 262–3
  osmolality, 278–9
  particle size distribution, 271–4
  patient use/misuse, 251–2, 259–65
  plume geometry, 280–2
  priming/repriming regulations, 256
  product labeling, 249, 253–9
  pump/valve delivery (shot weight), 257–8
  shaking (suspensions), 258

spray pattern, 284–6
stability of primary (unprotected) packages, 258–9
storage conditions, 259–60
viscosity, 286
Retina:
description, 168
drug delivery systems, 179–86
Retinal pigment epithelium (RPE), 166, 168, 180, 183
Retisert, 180–1
Reverse iontophoresis, 110–11, 128–9
Rheology of emulsions/microemulsions, 64–5
Riboflavin, 146–7
Rotigotine, 5
RPE *see* retinal pigment epithelium

Salmon calcitonin (SCT), 205
Scintigraphic techniques (non-invasive), 232–3
Sclera (eye), 168, 183
Scopolamine, 5
SCT *see* salmon calcitonin
Shaking (suspensions) regulations, 258
Shot weight (pump/valve delivery) regulations, 255, 257–8
Skin:
anatomy, 1–2
barrier permeabilization (ultrasound), 119–30, 130–1
depot effects, 54
dermis, 3
drug transport, 71–8, 102–3
hypodermis, 3
permeation, 83
structure, 1–3
ultrasound, 119–30, 130–1
SLNs *see* solid lipid nanoparticles
Small unilamellar vesicles (SUV), 9–13
Sodium alginate, 173
Sodium nonivamide acetate, 108
Solid lipid nanoparticles (SLNs), 15–16
Solid microneedles, 137–9
Solubility and drug transport, 72–4
Sonophoresis:
definition, 119
enhanced drug delivery, 123–7
iontophoresis, 108–9
low-frequency, 130–1
topical genomic therapy, 131
transducers, 129–30
see also ultrasound
SonoPrep®, 129
Spinhaler, 210
Spray pattern regulations, 284–6
Stability evaluation for emulsions/ microemulsions, 69–71
Stability of primary (unprotected) package regulations, 258–9
Stabilization of emulsions, 85–6
Stainless steel microneedles, 138–9, 145
Storage conditions regulations, 259–60
Stratum corneum:
cavitation, 126
emulsions/microemulsions comparison, 91
drug diffusion, 76–7
permeation, 5, 26, 131
transdermal drug delivery, 135–6
Sulfoxides, 19
Supersaturation of drugs, 6–7
Surfactants, 24
Suspension shaking regulations, 258
SUV *see* small unilamellar vesicles

Tailoff (exhaustion) regulations, 256–7
Targeted drug delivery by microneedles, 156–7
TDDS *see* transdermal drug delivery systems
TEM *see* transmission electron microscopy
Terpenes, 25–6
Terpenoids, 25–6
Testosterone, 48
Tests for nasal/pulmonary products:
general, 251–2, 267–87
labeling, 249, 253–9
patient use/misuse, 250, 259–65
Tetronic, 172

Thermal inkjet technology and
microneedles, 160
Thermodynamic activity:
drug transport, 72–4
formulation strategies, 81–3
Timoptic XE®, 171–2, 174
TMC *see* *N*-trimethylchitosan (TMC)
Topical drug delivery:
conclusions, 26–7
cutaneous drug absorption, 6–16, 16–26
emulsions, 71–8, 71–85, 85–6
formulations/microemulsions comparison,
89–90
genomic therapy and sonophoresis, 131
human subjects, 78–80
microemulsions, 86–91
MNP technology, 53–4
percutaneous absorption airways, 5–6
skin structure, 1–3
transdermal comparison, 3–5
Topical formulations comparison with
microemulsions, 89–91
Trans-scleral:
diffusion, 183–4
drug delivery, 166
permeabilities, 184
Transderm-Scop®, 38
Transdermal drug delivery:
conclusions, 26–7
cutaneous drug absorption, 6–16, 16–26
emulsions, 71–8, 78–85, 85–6
human subjects, 78–80
iontophoresis, 95–118
micellar nanoparticles, 37–58
microemulsions, 86–91
microneedles, 135–64
minimally invasive (microneedles),
135–64
patch systems, 4–5
percutaneous absorption airways, 5–6
skin structure, 1–3
topical comparison, 3–5
Transderm-Scop, 38
Transdermal drug delivery systems (TDDS),
4–5, 38, 96
Transducer fabrication for sonophoresis,
129–30
Transmission electron microscopy (TEM), 42
Tri-Luma®, 84
N-Trimethylchitosan (TMC), 204–5
Triptolide, 14

Ultrasonic nebulizers, 228–9
Ultrasound:
bio-effects, 121–3, 131
cavitation, 125–7
challenges/trends, 130–1
definition, 120
developments, 127–30
history, 122–3
low-frequency, 124–5, 127–8, 130–1
propagation, 120–3
reverse sonophoresis for transdermal
diagnostics, 128–9
skin, 124–5
skin barrier permeabilization, 119–30, 130–3
sonophoresis and enhanced drug delivery,
123–7
transducers, 129–30
United States Pharmacopeia (USP), 44
United States (US), FDA:
APSD, 273
AzaSite, 171
chlorofluorocarbons, 223
Lidosite®, 113
MDI and DPI drug products, 231, 242
nasal/pulmonary products, 247–8, 253–7,
258–9, 261–3
New Excipient Evaluation Procedure, 84
Ozurdex, 181
SonoPrep®, 129
transdermal products, 38
Urea, 24–5
US *see* United States
USP *see* United States Pharmacopeia

Vaccines, nasal delivery, 200–1, 205
Valve-holding chamber (VHC) for MDIs, 220–3
Vasoactive intestinal peptide (VIP), 203
"Vesosomes", 12

VHC *see* valve-holding chamber
VIP *see* vasoactive intestinal peptide (VIP)
Virgan®, 171
Viscosity regulations, 286
Viscous solutions for ophthalmic drug
delivery, 170–1
Vitrasert, 180–1

W/O *see* water in oil
Water as drug permeability enhancer, 17–18
Water in oil (W/O):
emulsions, 60, 84
microemulsions, 62
WGA *see* wheatgerm agglutinin (WGA)
Wheatgerm agglutinin (WGA), 203

Zeta potential of emulsions/microemulsions,
66–7
Zovirax® (acyclovir), 42, 54